Lecture Notes in Computer Science   11456

Commenced Publication in 1973
Founding and Former Series Editors:
Gerhard Goos, Juris Hartmanis, and Jan van Leeuwen

FoLLI Publications on Logic, Language and Information
Subline of Lectures Notes in Computer Science

More information about this series at http://www.springer.com/series/7407

Alexandra Silva · Sam Staton ·
Peter Sutton · Carla Umbach (Eds.)

Language, Logic, and Computation

12th International Tbilisi Symposium, TbiLLC 2017
Lagodekhi, Georgia, September 18–22, 2017
Revised Selected Papers

Springer

Editors
Alexandra Silva
Department of Computer Science
University College London
London, UK

Sam Staton
Department of Computer Science
University of Oxford
Oxford, UK

Peter Sutton
Department of Linguistics
Heinrich Heine University Düsseldorf
Düsseldorf, Germany

Carla Umbach
Leibniz-Zentrum Allgemeine
Sprachwissenschaft
Berlin, Germany

ISSN 0302-9743   ISSN 1611-3349   (electronic)
Lecture Notes in Computer Science
ISBN 978-3-662-59564-0   ISBN 978-3-662-59565-7   (eBook)
https://doi.org/10.1007/978-3-662-59565-7

LNCS Sublibrary: SL1 – Theoretical Computer Science and General Issues

This Springer imprint is published by the registered company Springer-Verlag GmbH, DE
part of Springer Nature
The registered company address is: Heidelberger Platz 3, 14197 Berlin, Germany

Preface

The 12th International Tbilisi Symposium on Language, Logic, and Computation was held during September 18–22, 2017, at Lagodekhi, Georgia. The symposium was organized by the Centre for Language, Logic, and Speech at the Tbilisi State University, the Georgian Academy of Sciences, the Institute for Logic, Language and Computation (ILLC) of the University of Amsterdam, and the Collaborative Research Center 991 of Heinrich Heine University, Düsseldorf. The biennial conference series and the proceedings are representative of the aims of the organizing institutes: to promote the integrated study of logic, information, and language. While the conference is open to contributions from any of the three fields, it aims to foster interaction among them by achieving stronger awareness of developments in the other fields, and of work that embraces more than one field or belongs to the interface between fields. The scientific program consisted of tutorials, invited lectures, contributed talks, and two workshops.

The symposium offered three tutorials, given on each of the three major disciplines of the conference and aimed at students as well as researchers working in the other areas:

Language: "Generalized Quantifiers. Logical, Computational, and Cognitive Approaches," by Jakub Szymanik (University of Amsterdam)

Logic: "Machines, Models, Monoids, and Modal Logic," by Sam van Gool (City College of New York and University of Amsterdam)

Computation: "Semantics for Probabilistic Systems," by Ana Sokolova (University of Salzburg)

Six invited lectures were delivered at the symposium: two on logic, by Alexander Kurz (University of Leicester) and Eric Pacuit (University of Maryland), two on language, by Gemma Boleda (Universitat Pompeu Fabra) and Ruth Kempson (King's College, London), and two on computation by Dexter Kozen (Cornell University) and Alex Simpson (University of Ljubljana).

The workshop on Transmodal Perspectives on Secondary Meaning, organized by Fabian Bross, Daniel Hole, Daniel Gutzmann, and Katharina Turgay featured six contributed talks. The workshop on Logic, Algebra, Categories and Quantitative Models, organized by Alexander Kurz and Alex Simpson, featured four invited talks.

This volume contains a selection of papers that went through a rigorous, two-stage refereeing process during which each paper was reviewed by at least two anonymous referees. Here we give a brief overview of their contributions.

"Compounds or Phrases? Pattern Borrowing from English into Georgian," by Nino Amiridze, Rusudan Asatiani, and Zurab Baratashvili. In this paper, Amiridze et al. discuss a case of borrowing of English noun–noun constructs into Georgian. In contrast to English, there are no established tests for Georgian to distinguish

compounds from phrases in noun–noun constructs. Amiridze et al. suggest a number of tests based on phonological, morphological, syntactic, and semantic criteria.

"A Study of Subminimal Logics of Negation and Their Modal Companions," by Nick Bezhanishvili, Almudena Colacito, and Dick de Jongh. Bezhanishvili, Colacito, and de Jongh study propositional logical systems arising from the language of Johansson's minimal logic and obtained by weakening the requirements for the negation operator. The authors first prove that there are uncountably many such logical systems. They then provide model-theoretic and algebraic definitions of filtration for minimal logic and show that they are dual to each other. Finally, they investigate bi-modal companions with non-normal modal operators for some relevant subminimal systems, and give infinite axiomatizations for these bi-modal companions.

"Bare Nouns and the Hungarian Mass/Count Distinction," by Kurt Erbach, Peter R. Sutton, and, and Hana Filip. Erbach et al. argue that, in Hungarian, notionally count, singular nouns like *könyv* ("book"), *toll* ("pen"), and *ház* ("house") are semantically number-neutral. This departs from the view that such nouns are dual-life with respect to being count or mass, as has recently been argued. The paper provides a novel analysis in which Hungarian has many count nouns and many mass nouns, rather than many dual-life and mass nouns, but few count nouns.

"Why Aktionsart-Based Event Structure Templates Are Not Enough – A Frame Account of Leaking and Droning," by Jens Fleischhauer, Thomas Gamerschlag, and Wiebke Petersen. The focus of the paper by Fleischhauer et al. is the gradability of emission verbs (e.g., *leak, drone*). They motivate a dynamic frame-based analysis of such verbs based on the independence of gradability from a verb's event structure. By adopting more fine-grained representations (verb frames), Fleischhauer et al. can capture more data, for instance, the distinction between event-dependent and event-independent gradation as exemplified by German examples such as *Der Motor dröhnt sehr* ("The engine is droning a lot") and *Das Rohr hat sehr geleckt* ("The pipe has leaked a lot").

"The Athlete Tore a Muscle: English Locative Subjects in the Extra Argument Construction," by Katherine Fraser. Fraser addresses cases in which changes of state verbs (e.g., *tear, break*) are a part of constructions that convey that the change of state is unintentional (*the skier tore a muscle, the boat tore a sail*). Fraser argues that such constructions have an extra LOCATION argument that is part of the subject (*the skier tore a muscle in her calf, the boat tore a sail on its mast*).

"An Axiomatization of the d-Logic of Planar Polygons," by David Gabelaia, Kristina Gogoladze, Mamuka Jibladze, Evgeny Kuznetsov, and Levan Uridia. The paper introduces the modal logic of planar polygonal subsets of the plane, with the modality interpreted as the Cantor–Bendixson derivative operator. The authors prove the finite model property of this logic and provide a finite axiomatization for it.

"An Ehrenfeucht–Fraisse Game for Inquisitive First-Order Logic," by Gianluca Grilletti and Ivano Ciardelli. Grilletti and Ciardelli develop an Ehrenfeucht–Fraisse game for inquisitive first-order logic, an extension of classic first-order logic with questions. From a mathematical point of view, formulas in this logic express properties

of sets of relational structures. The authors show that the developed game characterizes the distinguishing power of the logic and use this result to show a number of unde-finability results.

"Two Neighborhood Semantics for Subintuitionistic Logics," by Dick de Jongh and Fatemeh Shirmohammadzadeh Maleki. De Jongh and Maleki study two types of neighborhood models for weak subintuitionistic logics, as introduced by the authors in 2016. The paper contains a comparison and detailed study of the relationship between the two types of models. Thereby modal companions for various logics are recognized. Many of the extensions of the basic logics are discussed and characterized.

"Finite Identification with Positive and with Complete Data," by Dick de Jongh and Ana Lucia Vargas Sandoval. De Jongh and Sandoval study the differences between finite identifiability of recursive languages with positive and with complete data. They show that in finite families the difference lies exactly in the fact that for positive identification the families need to be anti-chains, while in in the infinite case it is less simple, being an anti-chain is no longer a sufficient condition. The authors show that with complete data there are no maximal learnable families whereas with positive data there usually are, but there do exist positively identifiable families without a maximal positively identifiable extension.

"Language as Mechanisms for Interaction: Toward an Evolutionary Tale," by Ruth Kempson, Eleni Gregoromichelaki, and Christine Howes. Kempson et al. argue against a static code-based view of language as form-meaning mappings. Instead, they propose a view of grammar as a coordinating mechanism for dynamic interactivity among situated agents. They propose dynamic syntax as a domain-general theoretical archi-tecture that models syntax/meaning/processing in terms of predictive actions and they defend the view that, while the model is empirically motivated by the need to capture, inter alia, dialogue data, it is also necessary to get a handle on human evolution, specifically, on the group-forming properties of linguistic interactions.

"Computational Model of the Modern Georgian Language and Search Patterns for an Online Dictionary of Idioms," by Irina Lobzhanidze. This paper describes the use of finite state technology for the morphological analysis of the modern Georgian language and the application of a morphological transducer to address issues of lemmatization and alphabetization noticed in Georgian dictionaries. Information on lemmas and the morphological structures of words was used to solve lexicographic problems in an online dictionary of idioms.

"Bridging Inferences in a Dynamic Frame Theory," by Ralf Naumann and Wiebke Petersen. In their article, Naumann and Petersen develop a theory of bridging infer-ences in a dynamic frame theory that is an extension of incremental dynamics. In contrast to previous approaches, bridging is seen as based on predictions/expectations that are triggered by discourse referents in particular contexts. Predictions are modeled as extensions of a frame representing a discourse referent and are constrained by a probability distribution on the domain of frames.

"Misfits: On Unexpected German ob-Predicates," by Kerstin Schwabe. Schwabe investigates clause-embedding predicates in German focusing on ones that embed

interrogatives headed by *ob* ("whether"/"if"), although only in negative contexts. She explains the unexpected behavior of these predicates by the fact that negative contexts turn them into subjectively nonveridical predicates even though in positive contexts they are subjectively veridical. The analysis is spelled out in a propositional logical form.

"A Non-Factualist Semantics for Attributions of Comparative Value," by Andrés Soria Ruiz. Soria Ruiz proposes characterizing evaluative adjectives (e.g., *good, bad, courageous, elegant*) as a type of gradable adjective, the instances of which can be subject to faultless disagreement. Building on Gibbard's hyperplan approach, he analyzes adjectives in this class as conceptually related to action (and to motivations for action). For example, part of what is communicated by positive evaluative adjectives is defined in relation to a set of instructions for action (a hyperplan) that supports certain actions to a sufficient degree relative to the context.

"Spectra of Goedel Algebras," by Diego Valota. Valota exploits the duality between the variety of Goedel algebras and the category of forests and open maps in order to compute the duals of finite k-element Goedel algebras. From this construction the author obtains a recurrence formula to compute the fine spectrum of the variety of Goedel algebras and, as a corollary, derives that the set of cardinalities of finite Goedel algebras is the set of positive integers.

"From Semantic Memory to Semantic Content," Henk Zeevat. In this paper, frames are considered as representing semantic content built from concepts. Particular frames are seen as a collection of stochastic variables. Zeevat develops a simple but powerful notion of semantic memory on the basis of lexical knowledge under the frame hypothesis and discusses the question of whether the stochastic information in semantic memory contributes to conceptual content.

"Explaining Meaning: The Interplay of Syntax, Semantics, and Pragmatics," by Yulia Zinova. Zinova provides a rational speech act model paired with a frame-based lexical representation to predict the most probable meanings of certain Russian verbal prefixes, given a context of use and a set of alternatives. For example, Zinova can explain why *po-gret* ("PO-heat") receives a low-degree reading despite being semantically neutral between high- and low-degree readings.

We would like to thank all the authors for their contributions, and the anonymous reviewers for their high-quality reports. We would also like to express our gratitude to the organizers of the symposium, who made the event an unforgettable experience for all of its participants. The Tbilisi symposia are renowned not only for their high scientific standards, but also for their friendly atmosphere and heartwarming Georgian hospitality, and the 12th symposium was no exception. Finally, we thank the ILLC (University of Amsterdam), the Department of Computational Linguistics at

Düsseldorf University, and Johan van Benthem for their generous financial support for the symposium.

March 2019   Alexandra Silva
                                                                Sam Staton
                                                               Peter Sutton
                                                              Carla Umbach

Organization

Standing Committee

Rusiko Asatiani	Tbilisi State University, Georgia
Matthias Baaz	TU Vienna, Austria
Guram Bezhanishvili	New Mexico State University, USA
George Chikoidze	Georgian Technical University, Georgia
Dick de Jongh (Chair)	ILLC, University of Amsterdam, The Netherlands
Paul Dekker	ILLC, University of Amsterdam, The Netherlands
Hans Kamp	University of Stuttgart, Germany
Manfred Krifka	Zentrum für Allgemeine Sprachwissenschaft Berlin, Germany
Temur Kutsia	Johannes Kepler University Linz, Austria
Sebastian Loebner	Heinrich Heine University of Düsseldorf, Germany
Barbara Partee	University of Massachusetts Amherst, USA

Program Committee

Samson Abramsky	Oxford University, UK
Kata Balogh	University of Düsseldorf, Germany
Guram Bezhanishvili	New Mexico State University, USA
Nick Bezhanishvili	ILLC, University of Amsterdam, The Netherlands
Rajesh Bhatt	UMass Amherst, USA
Filippo Bonchi	University of Pisa, Italy
Valeria De Paiva	University of Birmingham, UK
David Gabelaia	TSU, Razmadze Mathematical Institute, Georgia
Brunella Gerla	University of Insubria, Italy
Nina Gierasimczuk	ILLC, University of Amsterdam, The Netherlands
Helle Hvid Hansen	Delft University of Technology, The Netherlands
Daniel Hole	University of Stuttgart, Germany
George Metcalfe	University of Bern, Switzerland
Alessandra Palmigiano	Delft University of Technology, The Netherlands
Wiebke Petersen (Chair)	University of Düsseldorf, Germany
Mehrnoosh Sadrzadeh	Queen Mary University of London, UK
Alexandra Silva (Chair)	University College London, UK
Sonja Smets	ILLC, University of Amsterdam, The Netherlands
Rui Soares Barbosa	University of Oxford, UK
Luca Spada	University of Salerno, Italy
Carla Umbach	Zentrum für Allgemeine Sprachwissenschaft, ZAS, Berlin, Germany
Galit W. Sassoon	Bar-Ilan University, Israel
Henk Zeevat	ILLC, University of Amsterdam, The Netherlands

Organizing Committee

Rusiko Asatiani	Tbilisi State University, Georgia
Almudena Colacito	Mathematical Institute, Universität Bern, Switzerland
Anna Chutkerashvili	Tbilisi State University, Georgia
David Gabelaia	TSU Razmadze Mathematical Institute, Georgia
Kristina Gogoladze	ILLC, University of Amsterdam, The Netherlands
Gianluca Grilletti	ILLC, University of Amsterdam, The Netherlands
Marina Ivanishvili	Tbilisi State University, Georgia
Nino Javashvili	Tbilisi State University, Georgia
Mamuka Jibladze	TSU Razmadze Mathematical Institute, Georgia
Ramaz Liparteliani	Tbilisi State University, Georgia
Liana Lortkipanidze	Tbilisi State University, Georgia
Ana Kolkhidashvili	Tbilisi State University, Georgia
Evgeny Kuznetsov	Tbilisi State University, Georgia
Peter van Ormondt	ILLC, University of Amsterdam, The Netherlands
Khimuri Rukhaia	Tbilisi State University and Sokhumi State University, Georgia

Additional Reviewers

Aguzzoli, Stefano	Grossi, Davide	Soselia, Ether
Aloni, Maria	Kalocinski, Dariusz	Szymanik, Jakub
Balogh, Kata	Kirillovich, Alexander	Tomaszewicz, Barbara
Bastings, Joost	Kupke, Clemens	Turrini, Paolo
Blutner, Reinhard	Lauridsen, Frederik M.	van Ditmarsch, Hans
Buecking, Sebastian	Liang, Fey	Vanvalin, Robert
Canavotto, Ilaria	Meurer, Paul	Velazquez-Quesada,
Dekker, Paul	Osswald, Rainer	Fernando R.
Dell'Orletta, Felice	Pinosio, Riccardo	Verelst, Karin
Feys, Frank	Pourtskhvanidze,	Yang, Fan
Flaminio, Tommaso	Zakharia	Zhao, Zhiguang
Gardani, Francesco	Przepiorkowski, Adam	

Contents

Compounds or Phrases? Pattern Borrowing from English into Georgian

Nino Amiridze[✉], Rusudan Asatiani, and Zurab Baratashvili

Ivane Javakhishvili Tbilisi State University, Tbilisi, Georgia
nino.amiridze@gmail.com

Abstract. In this paper we investigate a case of borrowing of English noun-noun (NN) constructs into Georgian. The phenomenon has been observed lately in Georgian in sequences of two nouns, where the first noun, always marked by nominative, represents the dependent noun and the second is the head of the construct.

In English, NN constructs can potentially be analyzed as phrases or compounds. There have been no tests developed for Georgian so far that would help to decide the status of such sequences. We try to address this problem and propose several phonological, morphological, syntactic, and semantic criteria to distinguish compounds from phrases in NN constructs. The tests indicate that the borrowed pattern represents compounds in Georgian. This result raises some interesting research questions about category change in pattern borrowing.

Keywords: Language contact · Noun-noun construct · Compounds · Phrases · Georgian · English

1 Introduction

The goal of this contribution is to study a contact-induced syntactic phenomenon, a pattern borrowing (a replication of a structure from a donor language into a recipient language, [14,15]) from English into modern spoken Georgian, documented throughout social media. The sources are social networks, blogs, forums, and printed and online media.

In Georgian, if a possessor relation exists between the referents of two nouns, it is necessary to use a genitive (henceforth, GEN[1]) marking on the dependent noun, as illustrated below in (1a) vs. (1b).[2] According to numerous descriptions of the language, non-genitive possessors are not permitted (see [10,18] among many others):

---

[1] Other abbreviations: DAT = dative; ERG = ergative; M = masculine; NOM = nominative; PL = plural; REL = relative clause marker; SUP = superlative; VOC = vocative.

[2] In the examples we indicate all language names except Georgian and English.

© Springer-Verlag GmbH Germany, part of Springer Nature 2019
A. Silva et al. (Eds.): TbiLLC 2018, LNCS 11456, pp. 1–20, 2019.
https://doi.org/10.1007/978-3-662-59565-7_1

(1) a. lilo-s   bazroba-$\emptyset$[3]   b. *lilo-$\emptyset$ bazroba-$\emptyset$
         Lilo-GEN market-NOM   Lilo-NOM market-NOM
         'Lilo market' (proper name)   'Lilo market'

In the contact situation with English (a brief overview of which is in Sect. 2), however, Georgian started using previously non-existent NOM marking of possessors (2a), (2b):

(2) a. lilo-$\emptyset$ mol-i   b. tbilis-i   central-i
         Lilo-NOM (Eng.)mall-NOM   Tbilisi-NOM (Eng.)central-NOM
         'Lilo Mall' (proper name)   'Tbilisi Central Station' (proper name)

Such sequences of nouns with a NOM-marked dependent noun as in (2) have a possessive reading and are used interchangeably with corresponding possessive phrases in Georgian (cf. (2a) vs. (3a), (2b) vs. (3b)):

(3) a. lilo-s   mol-i[4]   b. tbilis-is   central-i[5]
         Lilo-GEN (Eng.)mall-NOM   Tbilisi-GEN (Eng.)central-NOM
         'Lilo Mall' (proper name)   'Tbilisi Central Station' (proper name)

It seems that Georgian has borrowed the English rule/pattern of combining two nouns to express a possessive relationship without marking the possessor by GEN, illustrated in (4).

(4) a. school property   b. *school's property

In English, such sequences are called *noun-noun (NN) constructs*[6] and express a variety of meanings, not just a possessive one.

Notably, in Georgian, many NN constructs have an English loan word as the head (2a), (2b). However, by analogy, the construct has been spreading in everyday language use with native head nouns as well (5a). Despite the absence of GEN marking of the dependent element, there is a possessive relation between the referents of the nouns in (5a), which makes the NN construct easily interchangeable with the corresponding possessive phrase in (5b), as one can observe on the property developer's web page:[7]

(5) a. krcanis-i   tqup-eb-i   b. krcanis-is   tqup-eb-i
         Krtsanisi-NOM twin-PL-NOM   Krtsanisi-GEN twin-PL-NOM
         DEPENDENT   HEAD   DEPENDENT   HEAD
         'the Krtsanisi Twins'   'the Krtsanisi Twins'
         (Twin buildings of Krtsanisi)   (Twin buildings of Krtsanisi)

---

[3] The -$\emptyset$ represents one of the allomorphs of the NOM case morpheme which is exclusively associated with vowel-final stems. Compare consonant-final stems, which get marked by -*i* in NOM (see *tbilis-i* or *central-i* in (2b)).

[4] https://ss.ge/ka/udzravi-qoneba/iyideba-3-otaxiani-bina-dampalos-dasaxlebashi-2081600.

[5] http://www.tabula.ge/ge/story/56928-shoping-festivali-211.

[6] Some authors refer to them as *noun+noun sequences/constructions* [4] or *noun-plus-noun constructions* [8].

[7] http://bestdevelopers.ge/properties/.

What is important is that NN constructs in Georgian always have their dependent element marked by NOM, whatever the case marker of the head element (see (6a), (6b)).

(6)   a. The DAT-marked head of the theme argument in Future Indicative

        kompania-∅   aašenebs   krçanis-i   tqup-eb-s.
        company-NOM it.will.build.it/them Krtsanisi-NOM twin-PL-DAT
                                                DEPENDENT   HEAD

        'The company will build the Krtsanisi Twins.'

    b. The ERG-marked head of the subject argument in Aorist Indicative

        krçanis-i   tqup-eb-ma ganacvipres   turist-eb-i
        Krtsanisi-NOM twin-PL-ERG they.amazed.them tourist-PL-NOM
        DEPENDENT   HEAD

        'The Krtsanisi Twins amazed the tourists.'

In the past two decades, it has become trendy in Georgia to give businesses English names. Many of them are registered under names that consist either partly (7a) or fully of elements of English origin (7b, 7c).

(7)   Examples of names of Georgian companies

    a. anaklia-∅   development-i[8]
       Anaklia-NOM (Eng.)development-NOM

      'Anaklia Development' (*Anaklia* is a toponym)

    b. jorjia-∅   palas-i[9]
       (Eng.)Georgia-NOM (Eng.)palace-NOM

      'Georgia Palace' (a hotel)

    c. muvi-∅   taim-i[10]
       (Eng.)movie-NOM (Eng.)time-NOM

      'Movie Time' (a movie theater)

There are two main ways recipient languages borrow from donor languages [14,17]: (i) by the replication of phonological shapes from donor languages, also referred to as *matter borrowing*, and (ii) by the re-shaping of language internal structures according to the structures of the donor languages, referred to as *pattern borrowing*.

The examples in (7) reflect the direct replication of phonological shapes of the corresponding English forms, and thus, represent matter borrowing. Namely, the Georgian *development-* (7a), *jorjia- palas-* (7b), and *muvi- taim-* (7c) are direct replications of the following corresponding English forms: *development, Georgia palace*, and *movie time*. As for examples such as *lilo-∅ mol-i* (2a), *tbilis-i central-i* (2b), and *krçanis-i tqup-eb-i* (5a), they are a replication of the English pattern/rule that combines two nouns to express a possessive relationship without marking the possessor by GEN (as in (4)).

---

[8] http://gov.ge/files/253_33479_569174_62_2_.pdf.

[9] http://gph.ge/georgian/home.

[10] https://www.city24.ge/ge/tbilisi/organizations/1058/movie-time/.

The NN constructs discussed in this paper could not evolve in Georgian out-side of contact settings. The following points could be used to argue that the use of NN constructs in Georgian represents pattern borrowing and is caused by the influence of English:

- possessive relations have always been expressed by GEN marking of the depen-dent elements in Georgian;
- the use of the rule of combining two nouns to express a possessive relationship without marking the possessor by GEN started with NN constructs that have English loan words as the head;
- the rule was extended to NN constructs with native Georgian nouns as the head.

Observe the following examples in (8a) and (8b), which illustrate a further exten-sion of the use of NN constructs to those situations when there is no possessive semantics involved:

(8)   Names of the cards used by two Georgian pharmaceutical companies

    a. mṭred-i   barat-i   b. zɣarb-i   barat-i
      pigeon-NOM card-NOM   hedgehog-NOM card-NOM
      'Pigeon card'   'Hedgehog card'

(9)   The stadium named after the late soccer player Mikheil Meskhi[11]

    mesx-i   arena-∅
    Meskhi-NOM arena-NOM

    'Meskhi Arena'

The referent of the head element *barat-i* 'card' in (8a) does not belong to a particular pigeon, nor does the one in (8b) belong to any particular hedgehog. They rather represent consumer cards (each with an image of a pigeon and a hedgehog) that are used at two different pharmacy chains in Georgia.[12] As for the NN construct *mesx-i arena-∅* in (9), it refers to a stadium that belongs neither to the late soccer player Mikheil Meskhi, nor to his family/associates, but rather is named after him.

As is known, some sequences of two nouns in English can be treated as compounds and others as noun phrases. Bauer in [4] reviews the criteria that are assumed in the literature to distinguish between the two construction types. He argues that these criteria do not draw a clear-cut distinction between these types.

---

[11] https://www.worldsport.ge/ge/page/26-tebervali-saqartvelos-erovnuli-safexburto-chempionati.

[12] One can recall other similar cards, also with a depicted image of an animal or a plant. For instance, the American Eagle credit card for customers of *American Eagle Outfitters* or the Rose gift card of the Japanese department store chain *Takashimaya*.

A more recent work by Haspelmath and Sims [9] discusses several phonological, morphological, syntactic, and semantic tests that have been used successfully on cross-linguistic data to distinguish phrases from compounds. We need to have similar decision criteria for Georgian to identify the status of the borrowed pattern – sequences of two nouns with a NOM dependent noun (see, e.g., (2) or (5a)). Currently, there are no tests elaborated for this language which would distinguish whether such a sequence represents a compound or a syntactic phrase. The only exception is the so-called traditional test, which is based on the rule of case marking (see Sect. 4.2).

To address this problem, we first analyze the existing tests for various languages reported in [9], and consider to what extent they apply to Georgian. We show that such an adaptation gives morphological, syntactic, and semantic criteria which can successfully distinguish compounds from phrases in Georgian.

The next step is the application of the obtained tests to our data of the borrowed pattern. We perform such an analysis, which reveals an interesting result: in Georgian, such a borrowed pattern, i.e., a sequence of two nouns with a NOM dependent noun, is always a compound.

This paper is organized as follows: In Sect. 2, we give a brief overview of Georgian-English language contact and characterize it from the point of view of the borrowing scale [20,21]. Section 3 discusses how syntactic phrases are distinguished from compounds in the linguistic literature. In Sect. 4, the criteria used to distinguish between phrases and compounds in Georgian are introduced. Section 5 examines the pattern borrowed from English into Georgian according to the criteria discussed in Sect. 4. Concluding remarks and further research questions are stated in Sect. 6.

2   Georgian-English Language Contact

In this section we give a brief analysis of the Georgian-English language contact situation. The contact is relatively new and little studied.[13] It is interesting to look at Georgian-English language contact from the perspective of the *borrowing scale* of Thomason and Kaufman [21], further elaborated in [20]. The scale consists of the following four stages:

Stage 1. *Casual contact.* Borrowers need not be fluent in the source language, and/or there are few bilinguals among borrowing-language speakers. Only lexical borrowing of content words (most often nouns, but also verbs, adjectives, and adverbs) takes place. No change to language structure.

Stage 2. *Slightly more intense contact.* Borrowers must be bilinguals, but they are probably a minority among borrowing-language speakers. Function words as well as content words are borrowed, still non-basic vocabulary. Only minor structural borrowing takes place.

---

[13] Among the few works on Georgian-English language contact, see [1] and [2].

Stage 3. *More intense contact.* More bilinguals, attitudes, and other social factors favoring borrowing. More function words as well as basic and non-basic vocabulary is borrowed. Moderate structural borrowing occurs (no major typological changes).

Stage 4. *Intense contact.* Very extensive bilingualism among borrowing-language speakers, social factors strongly favoring borrowing. Continuing heavy lexical borrowing in all sections of the lexicon, heavy structural borrowing.

English has high prestige as a donor language for Georgian. Over the past two decades, Georgian has borrowed English lexical items used in different domains, technology, business, entertainment, sport, etc., however all constituting non-basic vocabulary. There is no major influence on the language structure but some minor structural borrowing [2], namely, the introduction of the NN construct in parallel to the traditional genitive construction, to be studied in the present paper.

From the borrowing scale point of view [20, 21], we could argue that this is a transitional period from the first stage of borrowing to the slightly more intense contact, leading to the second stage of borrowing.

3   Existing Tests to Distinguish between Phrases and Compounds

To identify a NN construct as a phrase or a compound, there are established phonological, morphological, syntactic, and semantic tests. We briefly review them here, following [9].

3.1   Phonological Tests

Among the phonological tests are *stress assignment* and *vowel harmony.* In languages in which stress plays a role, main stress can apply only once per word. If both components of a NN construct receive main stress, it means that the construct consists of two separate words and thus, represents a syntactic phrase, as is the case in (10a). However, if such a NN construct gets only one main stress, then it represents a word and thus, a compound, as in (10b).

(10)   [9, p. 192]

    a. whíte kníght   b. Whíte Hòuse

In some languages with vowel harmony, like Chukchi [9, p. 193], an expression is a compound if both of its components are affected by harmony. Namely, in a phonological word, vowels have to belong to either of the two sets of vowels: {i, e, u} or {ə, a, o}. All the vowels in the word *kupre-n* 'net' (11a) belong to the first set but have to change according to the phonological rule when the word is compounded with another word with vowels from the other set. See, for instance, the compound in (11b):

(11) Chukchi, from [9, p. 193]

a. kupre-n	b. pəlvəntə + kopra-n /	*kupre-n
net	metal + net	net
'net'	'metal net'	

However, this criterion is not universal for categorizing a NN construct as a compound, as there are languages with vowel harmony like Turkish and Finnish that treat elements of compounds as separate words. In these languages only one component of a compound undergoes the phonological rule:

(12) a. Finnish [11, p. 241]   b. Turkish [13, p. 231]
      vesi+pullo[14] / *vesi+püllö   göz+kulak/ *gözkülak / *gözküläk
      water+bottle water+bottle   eye+ear

      'water-bottle'   'alert, interested'

Whichever rule works for a particular language, it seems compounds are phonologically more cohesive, while phrases are less so.

3.2 Morphological Tests

Morphological tests also follow cohesion criteria: syntactic phrases cannot be morphologically cohesive, while compounds are. This means that when there is a compound, affixation and cliticization would treat it as a whole word, while the same affixation and cliticization would be applied to the different components of a phrase rather than to the phrase as a whole.

For instance, in (13a), there is a Finnish noun phrase, *nuori mies* 'a young man'. Both the head noun *mies* 'man' and the modifier *nuori* 'new' can be encliticized by an enclitic particle *kin* 'also' (see (13b) and (13c)). However, the compound *nuori+mies* 'an unmarried man', consisting of the same adjective and the noun, can only get encliticized by the particle once, at the end of the compound (cf. (13d) vs. (13e)).

(13) Finnish, from [12, p. 192]

a. nuori mies	c. nuori=kin mies	e. *nuori=kin+mies
young man	young=also man	young=also+man
'a young man'	'also a young man'	'also an unmarried man'
b. nuori mies=kin	d. nuori+mies=kin	
young man=also	young+man=also	
'also a young man'	'also an unmarried man'	

---

[14] The form *vesi+pullo* is a compound according to the criteria used to distinguish between compounds and multilexical expressions in Finnish. Among these criteria are the rule of encliticizing particles and the prosodic rule of main stress application. Namely, enclitic particles can be attached to every word of multiword expressions but only to the end of compounds [12, p. 192].

3.3 Syntactic Tests

Apart from phonological and morphological tests, there are syntactic tests to make a clear distinction between phrases and compounds. For instance, there is the criterion of *separability*. Unlike compounds, phrases are separable. There can be new items inserted in between the components of phrases, but not between those of compounds.

To illustrate this, Haspelmath and Sims [9, p. 194] give an example from Hausa [16]. In this language, when modified by an adjective, compounds are inseparable, while phrases are separable. For instance, the compound *gida-n-sauroo* (14a) when modified by the adjective *bàbba* 'big', is inseparable (15a). As for the phrase *gida-n Muusaa* 'Musa's house' (14b), the components of the phrase can be separated by the modifying adjective (15b):

(14) a. Hausa compound   b. Hausa phrase

      gida-n-sauroo   gida-n   Muusaa
      house-REL.M-mosquito   house-REL.M Musa

      'mosquito net'   'Musa's house'

(15) Hausa compound (a) and phrase (b) modified by an adjective

    a. gida-n-sauroo   bàbba   b. gidaa bàbba na   Muusaa
       house-REL.M-mosquito big   house big   REL.M Musa
       'big mosquito net'   'Musa's big house'

The authors list *expandability* of the dependent element as one more syntactic criterion that makes it possible to distinguish between phrases and compounds [9, p. 194]. The dependent element of phrases can be expanded by a modifier, while that of compounds cannot be. One of the examples discussed there is the English compound *crispbread*, which cannot be modified by the adverb *very* (**very crispbread*). However, the phrase *crispy bread* can be modified by the same modifier: *very crispy bread*.

One more test discussed in [9, pp. 194–195] is to check whether *coordination ellipsis* is possible. In coordinated phrases one of the elements can be deleted (cf. (16a) vs. (16b)). With compounds, such a deletion is impossible (cf. (17a) vs. (17b)):

(16) [9, p. 195]

    a. Coordinated phrase
       Large fish and small fish were mistakenly placed in the same tank.

    b. Coordination ellipsis in the phrase
       Large ∅[15] and small fish were mistakenly placed in the same tank.

---

[15] Note that we distinguish two zeros: by ∅ we indicate deleted heads in phrases (see examples (16b), (33b), (40b)) or compounds (see examples (17b), (34b)). As for ∅, as we have already mentioned earlier, it is used to denote zero allomorphs (see, for instance, (1) among many examples in this paper).

(17) [9, p. 195]
    a. The compound *flying fish* in coordination
       Flying fish and small fish were mistakenly placed in the same tank.

    b. Ellipsis of the head of the compound *flying fish*
       *Flying ∅ and small fish were mistakenly placed in the same tank.

3.4 Semantic Tests

Between phrases and compounds there are important semantic differences, e.g., the dependent element of compounds is always generic, while that of phrases may be referential [9, p. 195]. Therefore, one of the semantic tests would be to check the referential status of the dependent element. For instance, the dependent element *church* in the NN construct *church-goer* cannot be referential. It has a generic meaning, as a part of a compound.

Note that the dependent element of phrases can be referential, which does not exclude those cases where it could be generic. As the authors note in [9, p. 192], the dependent element *Holz* 'wood' of the German syntactic phrase *Haus aus Holz* 'house from wood' is generic like the dependent element of the compound *Holzhaus* 'wood house'. Thus, genericity of the dependent element of a syntactic phrase cannot be a necessary criterion.

Another semantic test given in [9, p. 195], which from our point of view is rather a syntactic test (and will be discussed among other syntactic tests later in Sects. 4.2 and 5.3), is the replacement of the head element by an anaphoric pronoun. While the head of phrases may be replaced by such a pronoun (18a), the head of compounds may not be (18b):

(18) [9, p. 194]
    a. My aunt has one gold watch and three silver ones (i.e., three silver watches).
    b. *My aunt knows one goldsmith and three silver ones (i.e., three silversmiths).

In the next section, the criteria for distinguishing phrases and compounds for Georgian will be discussed. The tests considered in the present section will be checked with Georgian data, and alternative tests that are relevant for this language will be proposed.

4 Tests to Distinguish Phrases vs Compounds in Georgian

In this section we adapt the existing tests for Georgian. We start with a brief overview of phrases and compounds in Georgian, and then consider the tests themselves.

4.1   Brief Overview

This section gives a brief description of those nominal sequences that are recognized as (i) noun phrases and (ii) compounds in traditional grammatical literature on Georgian.

In Modern Georgian, the head of a phrase is preceded by a modifier. The latter can be represented by adjectives (19a), numerals (19b), pronouns (19c), participles (19d) [18, p. 81], and verbal nouns (20) [18, p. 82].

(19)   a. mċipe-∅ vašl-i
           ripe-NOM apple-NOM

           'a ripe apple'

       b. sam-i   avṭor-i
           three-NOM author-NOM

           'three authors'

       c. čem-i   saxl-i
           my-NOM house-NOM

           'my house'

       d. gamokveqnebul-i našrom-i
           published-NOM   work-NOM

           'published work'

(20)   sam-i   ċvela-∅   qvel-i
       three-NOM milking-NOM cheese-NOM

       'cheese made out of milk gathered over three milking sessions'

Georgian has endocentric, exocentric, copulative, and appositional compounds. Endocentric compounds consist of a head and a modifier, which can manifest itself as attributive (21a) or possessive (21b).

(21)   Endocentric compounds

       a. [18, p. 153]

           av+dar-i
           bad+weather-NOM

           'bad weather'

       b. [18, p. 155]

           jar+is+ḳac-i
           army+GEN+man-NOM

           'soldier'

There is a small group of non-productive exocentric compounds, the meaning of which is not predictable from the meaning of the components. For instance, the elements of the compound or+γobe in (22a) refer to the numeral or- 'two' and the noun γobe 'fence'. However, the whole form refers to a narrow path that runs between two parallel fences, each belonging to two neighboring houses.

Copulative compounds are more in number and their meaning derives from the sum of the meaning of their parts, as in (22b). As for appositional compounds, they consist of elements all of which refer to the same referent (22c):

(22)   a. Exocentric compound [18, p. 154]

           or+γobe-∅
           two+fence-NOM

           'path running between two fences'

       b. Copulative compound [18, p. 87]

           mepe-dedopal-i
           king-queen-NOM

           'royal couple'

       c. Appositional compound

           mecnier-tanamšromel-i
           scientist-co.worker-NOM

           'research fellow'

Traditionally, Georgian linguistic literature recognizes compounds as one word items, composed of two or more elements. When the head is recognizable, like it is in endocentric and exocentric compounds, the order is head-final.

Some compounds show a head-initial order though (as in *xorc+met-i* [meat+more-NOM] 'tumor' [18, p. 154]), reflecting the Old Georgian word order in phrases and compounds (i.e., modifiers preceded by heads).

4.2 Tests for Georgian

Tradition. The main test that is traditionally used in distinguishing between compounds and phrases is a morphological test, namely, the rule of case marking. While both components of phrases get marked by case markers in NOM (see (19b) vs. (23a), (19c) vs. (23b), (19d) vs. (23c)), ERG (24a), DAT (24b), and VOC (24c), only the second component of compounds gets case-marked (cf. (21a) vs. (25a), (22a) vs. (25b), (22c) vs. (25c)). Thus, compounds get treated as one word entities:

(23) a. *sam avṭor-i
      three author-NOM
      'three authors'

    b. *čem saxl-i
      my   house-NOM
      'my house'

    c. *gamokveqnebul našrom-i
      published   work-NOM
      'published work'

(24) a. sam-ma avṭor-ma / *sam avṭor-ma
      three-ERG author-ERG three author-ERG
      'three authors'

    b. čem-s   saxl-s   / *čem saxl-s[16]
      my-DAT house-DAT my house-DAT
      'to my house'

    c. saocar-o   adamian-o / *saocar   adamian-o
      fascinating-VOC person-VOC fascinating person-VOC
      '[Oh, you] fascinating person!'

(25) a. *av-i+dar-i
      bad-NOM+weather-NOM
      'bad weather'

    b. *or-i+γobe-∅
      two-NOM+fence-NOM
      'path running between fences'

    c. *mecnier-i+tanamšromel-i
      scientist-NOM+co.worker-NOM
      'research fellow'

Phonological Tests. If we try applying the general tests for distinguishing between phrases and compounds considered in Sect. 3, the phonological tests such as stress placement and vowel harmony (see Subsect. 3.1) are not very helpful for Georgian. This is because Georgian stress is weak [10,22] or not distinctive on the word level, and vowel harmony plays no role in the phonology of the language.

---

[16] In speech one could encounter the starred NN construct *čem saxl-s* where the modifier does not get case-marked. However, although the application of DAT case marking is not a very strong test for distinguishing phrases vs. compounds, any of the NOM, ERG, and VOC case marking is.

As mentioned above, in languages where stress is relevant on the word level, both components of syntactic phrases are characterized by stress, which is not true for compounds. Since the word level stress in Georgian is weak and the language relies mainly on phrasal prosody (see [6,7,19]), it might be helpful to consider the intonation contour, recognizable on the phrase and sentence level.

If we compare the compound *orγobe* (26a) to the phrase *or-i γobe* (26b), what we notice is that the compound is characterized by a HL[17] falling tone. As for the phrase, it is pronounced with a rise-fall contour, where the first component, the modifier, is characterized by a rising tone (LH) and the second component, the head noun, is realized as low (L). As indicated in the starred options in (26a) and (26b), the contours are not interchangeable: the compound cannot be characterized by the rise-fall contour and the phrase cannot be pronounced with a falling tone.

(26) a. Compound

      or + γobe-∅   / *or + γobe-∅
      H   L   LH   L
      two + fence-NOM   two + fence-NOM

      'path running between two fences'

   b. Phrase

      o· r-i   γobe-∅   / *or-i   γobe-∅
      L H   L   H   L
      two-NOM fence-NOM   two-NOM fence-NOM

      'two fences'

This is more visible when the first component of compounds consists of multiple syllables, like in the compound *vaxṭang-i+švil-i* (27a) with the three-syllable dependent component *vaxṭang-i*:

(27) a. vax· ṭan· g-i   + švil-i   / * vax· ṭan· g-i[18] + švil-i
      H   H   H   L   L   L   H   L
      Vakhtang-GEN + child-NOM   Vakhtang-GEN + child-NOM

      'Vakhtangishvili' (a Georgian family name)

   b. All-new contexts, no contrastive focus reading

      vax· ṭan· g-is   švil-i   / * vax· ṭan· g-is   švil-i
      L   L   H   L   H   H   H   L
      Vakhtang-GEN   child-NOM   Vakhtang-GEN   child-NOM

      'Vakhtang's child'

As in compounds with one-syllable dependent components (26a), the multiple-syllable dependents of compounds are similarly characterized by a H pitch and the head component by a L one (27a), resulting in a falling tone for the whole compound. Similarly, as in phrases with two-syllable modifiers (see, for instance, (26b)), the multiple-syllable modifiers of phrases are pronounced with a rise-fall contour as well, where the first component, the modifier, is characterized by a rising tone LH) and the second component, the head noun, is pronounced as low (L).

---

[17] H stands for high tone and L for low tone.

[18] The suffix -*i* in family names is a remnant of the GEN case marker -*is*.

As indicated in the starred options in (27a) and (27b), the contours are not interchangeable: the compound cannot be characterized by the rise-fall contour and the phrase cannot be pronounced with a falling tone (unless it has a contrastive focus reading as in "VAKHTANG's child (not somebody else's)").

However, before using the analysis of intonation contours as a decisive criterion in distinguishing between phrases and compounds, a thorough experimental investigation of word vs. phrasal intonation is needed. We will keep the experimental work as one of the future goals and refrain from using the intonation criterion in the present paper to make conclusions regarding the status of NN constructs in Georgian.

Morphological Tests. As we have mentioned in Sect. 3.2, morphological tests such as affixation would treat compounds as a whole word, as opposed to phrases, which will have affixation on both components – the head and its modifier.

The criterion works for case suffixes, like the ERG -*ma* in (28a) vs. (28b). Namely, the case marker -*ma* is applied to both components of the syntactic phrase in (28a) but only once in the case of the compound (28b).

(28) a. axal-ma kalak-ma
          new-ERG city-ERG
          'new city'

    b. axal+kalak-ma / *axal-ma+kalak-ma
          new+city-ERG   new-ERG+city-ERG
          'Akhalkalaki' (a toponym)

Thus, we can use affixation as a sound morphological criterion to decide about the syntactic status of linguistic forms in the case of Georgian.

Syntactic Tests. As mentioned in Sect. 3.3, one can differentiate compounds from phrases by the criteria of separability and expandability. The first one implies that new items can be inserted in between the components of phrases, but not between the components of compounds. As for expandability, dependent elements of phrases can be expanded by a modifier, while those of compounds cannot be.

Both criteria work for Georgian. Components of phrases can be separated (cf. (29a) vs. (29b)) but not components of compounds (cf. (22a) vs. (30)):

(29) Phrase

    a. or-i   γobe-∅
          two-NOM fence-NOM
          'two fences'

    b. or-i   ʒvel-i   γobe-∅
          two-NOM old-NOM fence-NOM
          'two old fences'

(30) Compound

    *or+ʒvel+γobe-∅
    two+old+fence-NOM

    'path running between two old fences'

The following examples illustrate the criterion of expandability. The noun phrase *axal-i kalak-i* (31a) can be expanded by a modifier, here, *sruliad* 'entirely'

(31b), while the same cannot be done in the case of a compound (cf. (32a) vs. (32b)):

(31) a. axal-i   kalak-i   b. sruliad axal-i   kalak-i
          new-NOM city-NOM   entirely new-NOM city-NOM
          'new city'   'entirely new city'

(32) a. axal+kalak-i   b. *sruliad-axal+kalak-i
          new-city-NOM   entirely-new-city-NOM
          'Akhalkalaki'

One more test discussed in the literature is to check whether the head element can be replaced by an anaphoric pronoun (see Sect. 3.4 above). The head of phrases may usually be replaced by anaphoric pronouns, while the same is not true for compounds.

This test does not work in Georgian, as the language usually does not replace the heads of phrases by anaphoric pronouns. Instead, we could use the test of head ellipsis. As is known, phrases in Georgian allow head ellipsis (cf. (33a) vs. (33b)), unlike compounds (cf. (34a) vs. (34b)):

(33) a. giorgi-m   iqida   vercxl-is   kamar-i, ilia-m   iqida   tqav-is
          Giorgi-ERG he.bought.it silver-GEN   belt-NOM Ilia-ERG he.bought.it leather-GEN
                                          DEPENDENT HEAD   DEPENDENT

          kamar-i.
          belt-NOM
          HEAD

          'Giorgi bought a silver belt, Ilia bought a leather belt.'

      b. giorgi-m   iqida   vercxl-is   kamar-i, ilia-m   iqida   tqav-is
          Giorgi-ERG he.bought.it silver-GEN   belt-NOM Ilia-ERG he.bought.it leather-GEN
                                          DEPENDENT HEAD   DEPENDENT

          ∅.
          ∅
          HEAD

          'Giorgi bought a silver belt, Ilia bought a leather one.'

(34) a. giorgi-m   dapaṭiža   dis+ċul-i   da zmis-ċul-i.
          Giorgi-ERG he.invited.him/her sister's+child-NOM and brother's+child-NOM
                                            DEPENDENT+HEAD   DEPENDENT+HEAD

          'Giorgi invited the child of his sister and the child of his brother.'

      b. *giorgi-m   dapaṭiža   dis+ċul-i   da zmis+∅.
          Giorgi-ERG he.invited.him/her sister's+child-NOM and brother's+∅
                                            DEPENDENT+HEAD   DEPENDENT+HEAD

          'Giorgi invited the child of his sister and the one of his brother.'

The examples given above show that when we apply head ellipsis to phrases, they remain grammatical (cf. (33a) vs. (33b)). However, if we apply the ellipsis test to compounds, we get an ungrammatical result (cf. (34a) vs. (34b)).

Therefore, the rule of head ellipsis can be used to check the syntactic status of phrases vs. compounds.

Semantic Tests. The semantic test of checking the referential status of the dependent element also works in Georgian. Namely, the rule that the dependent element of compounds is always generic also works in this language.

For instance, if the dependent element of the phrase *vir-is tav-i* (35a) refers to the head of a specific donkey, the GEN-marked dependent component *vir-i* of the compound *vir-is+tav-i* (35b) has a generic reading, while the compound itself can refer to a fool, an idiot.

(35)   a. vir-is   tav-i   b. vir-is+tav-i
           donkey-GEN head-NOM   donkey-GEN+head-NOM
           'a head of a donkey'   'a fool; an idiot'

5   Data Analysis: Applying the Tests to the Borrowed Pattern

Having reviewed the tests discussed in the literature for distinguishing phrases and compounds (see Sect. 3) and having considered the validity of those tests for Georgian (see Sect. 4.2), we can return to the borrowed pattern from English and determine its syntactic status in Georgian. Namely, we are interested in whether NN constructs form phrases or compounds in Georgian.

In order to find out what syntactic status the borrowed pattern has in Georgian, we need to check phonological, morphological, syntactic, and semantic tests determining phrase vs. compound status against the borrowed pattern. As we have seen above, some of the tests that work for English are not directly applicable for Georgian (see Sect. 4).

5.1   Phonological Tests

As we have seen in Sect. 4.2, phonological tests such as stress placement and vowel harmony do not work for Georgian. We have looked at the possibility of distinguishing phrases from compounds on the basis of phrasal intonation. Although with a number of examples it seems plausible to trust the differences in the intonation contours of particular words vs. phrases, in the absence of an experimental study (to be pursued in the future), we will not use the phonological test of intonation to make any general conclusions regarding the status of the NN constructs in Georgian.

5.2   Morphological Tests

In Sect. 3.2 we mentioned the morphological test of affixation, according to which compounds are treated as one word, as opposed to phrases, which are affixed on both the modifier and the head.

In Sect. 4.2, we looked at Georgian data, which also shows that both components of phrases get marked by case markers (24a), which is not true for compounds (cf. (28a) vs. (28b)).

If we apply the test of case marking here, for instance, ERG marking to a NN construct like *ioga mɣvime* (38a) or *ḳrċanis-i ṭqup-eb-i* (6a), we get the ungrammatical results in (36a) and (36b) respectively:

(36) a. *ioga-m mɣvime-m   b. *ḳrċanis-ma ṭqup-eb-ma
         yoga-ERG cave-ERG   Krtsanisi-ERG twin-PL-ERG

         'Yoga cave' (proper name)   'the Krtsanisi Twins'

The grammatically correct ERG marking for the NN construct would be as follows:

(37) a. ioga-∅   mɣvime-m   b. ḳrċanis-i   ṭqup-eb-ma
         yoga-NOM cave-ERG   Krtsanisi-NOM twin-PL-ERG

         'Yoga cave' (proper name)   'the Krtsanisi Twins'

The comparison of the examples (36a) vs. (37a) and (36b) vs. (37b) shows that the NN constructs *ioga mɣvime* (38a) and *ḳrċanis-i ṭqup-eb-i* (5a) behave as compounds. If they were phrases, they would have to be marked on both components – the modifier and the head, which is not the case in Georgian (36).

5.3 Syntactic Tests

In Sect. 3.3 we looked at two syntactic tests discussed in the literature that are helpful in distinguishing between syntactic phrases and compounds. These are the tests of separability and expandability and they both worked for Georgian (see Sect. 4.2). Let us apply the test of separability to the borrowed pattern below.

According to this test, the components of phrases can be separated by inserting new items, while the same is not possible for compounds. The example below illustrates a NN construct of the borrowed pattern type, *ioga mɣvime* (38a), and a possessive phrase, *ioga-s insṭruḳtor-i* (38b):

(38) a. ioga-∅   mɣvime-∅[19]
         yoga-NOM cave-NOM

         'Yoga cave' (proper name)

      b. ioga-s   insṭruḳtor-i[20]
         yoga-GEN instructor-NOM

         'yoga instructor'

Although both nouns, *ioga-* and *insṭruḳtor-*, in (38b) are matter borrowings (via Russian, an earlier donor language for Georgian), the structure is characteristic of possessive constructions in standard Georgian in having the dependent noun marked by GEN. As for the NN construct in (38a), it reflects the replicated pattern/rule of combining two nouns (characteristic to English), the dependent of which stands in NOM. None of the previous donor languages (including Russian) have contributed such a structural rule to Georgian. The only possible

---

[19] https://www.facebook.com/yogacavetbilisi/videos/1145438865569612/.
[20] http://yoga.ge/?show=yogajournal&Jid=1985.

way, from our point of view, is that it was replicated from the most recent donor language, English.

It seems that the borrowed pattern, reflected in the NN construct *ioga-∅ mɣvime-∅* (38a), behaves as a compound, as inserting a new item between its components yields an ungrammatical result (cf. (39a) vs. (38a)).

(39) Separability test

  a. *ioga-∅   axal-i   mɣvime-∅
      yoga-NOM new-NOM cave-NOM
      'new yoga cave'

  b. ioga-s   axal-i   instruktor-i
      yoga-GEN new-NOM instructor-NOM
      'new yoga instructor'

The behavior of the NN construct *ioga-∅ mɣvime-∅* (38a), which does not allow insertion of a new item (39a), certainly differs from that of the possessive phrase *ioga-s instruktor-i* (38b), the components of which can be separated (39b).

Therefore, the test of separability illustrated in this section shows that the pattern borrowed from English into Georgian can be categorized as a compound, not as a phrase.

The other test, according to which only the head of phrases (but not the head of compounds) can be replaced by anaphoric pronouns, did not work for Georgian. But we tried another syntactic test, namely, the test of head ellipsis. According to this test, deleting the head of a phrase works fine (33) but yields ungrammatical results when deletion is applied to the head of a compound (cf. (34a) vs. (34b)).

As the following examples illustrate, the test of head ellipsis works fine with the possessive phrase (cf. (40a) vs. (40b)) but gives an ungrammatical result when the head of the borrowed pattern gets deleted (cf. (41a) vs. (41b)):

(40) a. momvlel-ma   šeavso   mtred-is   barat-i   da ara zɣarb-is
          caretaker-ERG (s)he.filled.it.in pigeon-GEN card-NOM and not hedgehog-GEN
                                         DEPENDENT HEAD   DEPENDENT
          barat-i.
          card-NOM.
          HEAD
          'The caretaker filled in the pigeon's card but not the hedgehog's card.'

  b. momvlel-ma   šeavso   mtred-is   barat-i   da ara zɣarb-is   ∅.
      caretaker-ERG (s)he.filled.it.in pigeon-GEN card-NOM and not hedgehog-GEN ∅.
                                     DEPENDENT HEAD   DEPENDENT   HEAD
      'The caretaker filled in the pigeon's card but not the hedgehog's [card].'

(41) a. pacient-ma   dakarga   mtred-i   barat-i   da ara zɣarb-i   barat-i.
          patient-ERG (s)he.lost.it pigeon-NOM card-NOM and not hedgehog-NOM card-NOM
                                    DEPENDENT HEAD   DEPENDENT   HEAD
          'The patient lost the pigeon card but not the hedgehog card.'

  b. pacient-ma   dakarga   mtred-i   barat-i   da ara zɣarb-i   ∅.
      patient-ERG (s)he.lost.it pigeon-NOM card-NOM and not hedgehog-NOM ∅
                                DEPENDENT HEAD   DEPENDENT   HEAD
      (1) '*The patient lost the pigeon card but not the hedgehog one.'
      (2) 'The patient lost the pigeon card but not the hedgehog.'

If we compare the examples (40) vs. (41), and remember that the application of the head ellipsis test gives ungrammatical results in compounds, but not in phrases, we can conclude that the NN construct *zɣarb-i barat-i* (41a) behaves as a compound.

5.4 Semantic Tests

In Sect. 3.4, we mentioned the rule that the dependent element of compounds is always generic. And in Sect. 4.2 we determined that the rule worked in Georgian as well.

Let us compare the NN construct *mṭred-i barat-i* 'pigeon card' (8a) to the possessive phrase *mṭred-is barat-i* in (42). As we have already mentioned in Sect. 1 (in the paragraph immediately following example (9)), the construct does not refer to any particular pigeon. It refers to a type of card used by consumers to collect points and use them again when making a purchase in pharmacies of one of the Georgian pharmaceutical companies.

However, when used in a possessive phrase, the modifier *mṭred-is* can only get a referential reading (42), not a generic one:

(42) mṭred-is   barat-i
      pigeon-GEN card-NOM

      'pigeon's card' (card filled in with information on a particular pigeon in a zoo)

Therefore, according to the semantic test, NN constructs in Georgian behave as compounds.

6 Conclusion and Questions for Future Research

In this paper we have described a new development in modern Georgian spoken language that is spreading throughout the Georgian social media. The phenomenon under discussion is the use of NN constructs consisting of two consecutive nouns, the referents of which are in a possessive relationship. The main formal characteristic of the construct is that its dependent noun invariably stands in NOM, while the head gets marked according to the syntactic role of the construct in the sentence. Either one or both of the components of such constructs represent an English loan word. Even when both are represented by native Georgian lexemes (or earlier borrowings from other languages that already constitute a part of the Georgian vocabulary), the structure of the constructs resembles the structure of English NN constructs.

We have argued that the constructs are a result of pattern borrowing from English, which is a type of borrowing in which a recipient language replicates a pattern or a rule of the donor language. We have tried to analyze the borrowed pattern in Georgian and find out its status, whether the constructs represent phrases or are compounds.

Although the syntactic status of NN constructs in English is not always straightforward [3–5], for Georgian the status of the cognate borrowed pattern is clear: according to the morphological (Sect. 5.2), syntactic (Sect. 5.3), and semantic tests (Sect. 5.4), we can conclude that the borrowed pattern in Georgian behaves as a compound.

Depending on how one can interpret English NN constructs, one could have two possible developments in a contact-induced pattern borrowing:

(a) A compound item of a donor language is borrowed as a compound by a recipient language (*compound* ⇒ *compound*);
(b) A syntactic phrase structure of a donor language is borrowed as a compound by a recipient language (*phrase* ⇒ *compound*).

This result raises interesting questions: In general, is it possible that the category of the borrowed pattern (compound or phrase) changes between the source language and the recipient language? If yes, how? The English-Georgian case does not shed light on this problem, because the exact status of the source construction is not clear [3–5]. It would be interesting to study more pairs of languages in contact where pattern borrowing happens, but that goes beyond the scope of this paper and must be a subject of future work.

Is the borrowed NN construct in Georgian the (a) or (b) type of borrowing? If it is of type (b), what relevance does this have for the theories of borrowing? Is the (b)-type change attested elsewhere? Is it possible to have a syntactic category change in contact-induced language change? We believe that to answer these questions further research is necessary.

Acknowledgments. We would like to thank Alice C. Harris and the anonymous reviewers for helpful suggestions, and the editors for their support in the preparation of the final version of the paper. This work has been supported by the Shota Rustaveli National Science Foundation under the project 217500.

References

1. Amaghlobeli, N.: Language contacts in Georgian internet forums. STUF Lang. Typol. Universals **68**(1), 107–121 (2015)
2. Amiridze, N.: Accommodating loan verbs in Georgian: Observations and questions. J. Pragmatics **133**, 150–165 (2018)
3. Bauer, L., Lieber, R., Plag, I.: The Oxford Reference Guide to English Morphology. Oxford University Press, Oxford (2013)
4. Bauer, L.: When is a sequence of two nouns a compound in English? Engl. Lang. Linguist. **2**(1), 65–86 (1998)
5. Bell, M.J.: The English noun-noun construct: A morphological and syntactic object. In: Ralli, A., Booij, G., Scalise, S., Karasimos, A. (eds.) On-line Proceedings of the Eighth Mediterranean Morphology Meeting (MMM8), pp. 59–91 (2012)
6. Borise, L., Zientarski, X.: Word stress and phrase accent in Georgian. In: Proceedings of Tonal Aspects of Languages (TAL), vol. 6 (2018)
7. Butskhrikidze, M.: On the word level accentuation in Georgian. Presentation at The South Caucasian Chalk Circle, Paris, 24 September 2016
8. Giegerich, H.J.: Compound or phrase? English noun-plus-noun constructions and the stress criterion. Engl. Lang. Linguist. **8**, 1–24 (2004)

9. Haspelmath, M., Sims, A.D.: Understanding Morphology. Understanding Language Series, 2nd edn. Hodder Education, London (2010)
10. Hewitt, B.G.: Georgian: A Structural Reference Grammar, London Oriental and African Language Library, vol. 2. John Benjamins, Amsterdam/Philadelphia (1995)
11. Karlsson, F.: Finnish: An Essential Grammar. Understanding Language Series. Routledge, London and New York (1999)
12. Laalo, K.: Acquisition of compound nouns in Finnish. In: Dressler, W.U., Ketrez, F.N., Kilani-Schoch, M. (eds.) Nominal Compound Acquisition, pp. 191–207 (2017). Language Acquisition and Language Disorders 61, John Benjamins, Amsterdam/Philadelphia
13. Lewis, G.L.: Turkish Grammar, 2nd edn. Oxford University Press, Oxford (2000)
14. Matras, Y., Sakel, J.: Introduction. In: Matras, Y., Sakel, J. (eds.) Grammatical Borrowing in Cross-Linguistic Perspective, pp. 1–13. Mouton de Gruyter, Berlin and New York (2007)
15. Matras, Y., Sakel, J.: Investigating the mechanisms of pattern replication in language convergence. Stud. Lang. **31**(4), 829–865 (2007)
16. Newman, P.: The Hausa Language: An Encyclopedic Reference Grammar. Yale University Press, New Haven (2000)
17. Sakel, J.: Types of loan: Matter and pattern. In: Matras, Y., Sakel, J. (eds.) Grammatical Borrowing in Cross-Linguistic Perspective, pp. 15–29. Mouton de Gruyter, Berlin and New York (2007)
18. Shanidze, A.: Foundations of Georgian Grammar, I, Morphology, 2nd edn. Tbilisi University Press, Tbilisi (1973). (In Georgian)
19. Skopeteas, S., Féry, C.: Focus and intonation in Georgian: Constituent structure and prosodic realization (2016). https://pub.uni-bielefeld.de/download/2900490/2900491
20. Thomason, S.G.: Language Contact. An Introduction. Georgetown University Press, Washington (2001)
21. Thomason, S.G., Kaufman, T.: Language Contact, Creolization, and Genetic Linguistics. University of California Press, Berkeley, Los Angeles and London (1988)
22. Zhgenti, S.M.: The Rhythmical-Melodic Structure of Georgian. Codna, Tbilisi (1963). (In Georgian)

A Study of Subminimal Logics of Negation and Their Modal Companions

Nick Bezhanishvili[1], Almudena Colacito[2(✉)], and Dick de Jongh[1]

[1] Institute for Logic, Language and Computation, University of Amsterdam, Amsterdam, The Netherlands
{N.Bezhanishvili,D.H.J.deJongh}@uva.nl
[2] Mathematical Institute, University of Bern, Bern, Switzerland
almudena.colacito@math.unibe.ch

Abstract. We study propositional logical systems arising from the language of Johansson's minimal logic and obtained by weakening the requirements for the negation operator. We present their semantics as a variant of neighbourhood semantics. We use duality and completeness results to show that there are uncountably many subminimal logics. We also give model-theoretic and algebraic definitions of filtration for minimal logic and show that they are dual to each other. These constructions ensure that the propositional minimal logic has the finite model property. Finally, we define and investigate bi-modal companions with non-normal modal operators for some relevant subminimal systems, and give infinite axiomatizations for these bi-modal companions.

1 Introduction

Minimal propositional calculus (*Minimalkalkül*, denoted here as MPC) is the system obtained from the positive fragment of intuitionistic propositional calculus (equivalently, positive logic [29]) by adding a unary negation operator satisfying the so-called principle of contradiction (sometimes referred to as *reductio ad absurdum*, e.g., in [26]). This system was introduced in this form by Johansson [20] in 1937 by discarding *ex falso quodlibet* from the standard axioms for intuitionistic logic. The system proposed by Johansson has its roots in Kolmogorov's formalization of intuitionistic logic [22]. The axiomatization proposed by Johansson preserves the whole positive fragment and most of the negative fragment of Heyting's intuitionistic logic. As a matter of fact, many important properties of negation provable in Heyting's system remain provable (in some cases, in a slightly weakened form) in minimal logic.

In this work, we focus on propositional logical systems arising from the language of minimal logic and obtained by weakening the requirements for the negation operator in a 'maximal way'. More precisely, the bottom element of the bounded lattice of logics considered here is the system where the unary operator ¬ (that we still call 'negation') does not satisfy any conditions except for being functional. The top element of this lattice is minimal logic. We use the term

© Springer-Verlag GmbH Germany, part of Springer Nature 2019
A. Silva et al. (Eds.): TbiLLC 2018, LNCS 11456, pp. 21–41, 2019.
https://doi.org/10.1007/978-3-662-59565-7_2

N-logic to denote an arbitrary logical system in this lattice. This setting is para-consistent, in the sense that contradictory theories do not necessarily contain all formulas.

In this paper we continue the study of N-logics started in [12,13]. We investigate these logics from several different perspectives. In Sect. 2 we give an algebraic and model-theoretic presentation of N-logics, and provide a brief recap of the main duality and completeness results from [12]. We also review the semantics introduced for the N-logics in [12,13]—that we call N-semantics—and show that it is a variant of standard neighbourhood semantics.

In Sect. 3, we exploit these results to show that the lattice of N-logics has the cardinality of the continuum. The proofs from this section are obtained by exporting and adapting techniques of [5,19,21].

Section 4 is devoted to a study of the method of filtration for the basic N-logic. Model-theoretic and algebraic definitions of filtration are introduced and compared. This leads to the finite model property of the minimal logic.

Finally, Sect. 5 concludes the article by introducing (bi-)modal systems that are proved to play the role of modal companions for N-logics. More precisely, the language of these systems contains a normal (namely, S4) modality resulting from the positive fragment of intuitionistic logic, and a non-normal modality resulting from the negation operator. After characterizing these logics in terms of standard neighbourhood semantics, we continue using the equivalent N-semantics. We then give a proof that the standard Gödel translation of intuitionistic logic into S4 can be extended to translations of certain N-logics into these bi-modal systems. To the best of our knowledge this is the first use of non-normal modal operators in the context of the Gödel translation. On the other hand, modal systems using a mix of normal and non-normal modalities have been recently explored in the evidence-based semantics of epistemic logic [1,27].

2   Preliminaries

In this preliminary section we present the main technical tools that will be used throughout the paper. We start with a brief introduction to the Kripke semantics of minimal logic (here called *N-semantics*), in line with the tradition of intuitionistic logic. Later we present the algebraic semantics, and state some basic facts. In order to keep the structure of the paper as simple as possible, we skip the broader and introductory account of the topic and refer the interested reader to [12,13]. For a proof-theoretic account, see [7].

Let $\mathcal{L}(\mathsf{Prop})$ be the propositional language, where Prop is a countable set of propositional variables, generated by the following grammar:

$$p \mid \top \mid \varphi \wedge \varphi \mid \varphi \vee \varphi \mid \varphi \rightarrow \varphi \mid \neg\varphi$$

where $p \in \mathsf{Prop}$. We omit $\bot$ from the language. We call a formula *positive* if it contains only connectives from $\{\wedge, \vee, \rightarrow, \top\}$, and we refer to the positive fragment of intuitionistic logic as *positive logic*. We start by considering a system defined by the axioms of positive logic, with the additional axiom $(p \leftrightarrow q) \rightarrow$

$(\neg p \leftrightarrow \neg q)$ defining the behaviour of $\neg$. We call the resulting system N. We fix the positive logical fragment, and we strengthen the negation operator up to reaching minimal propositional logic, which can be seen as the system obtained by adding the axiom $(p \rightarrow q) \wedge (p \rightarrow \neg q) \rightarrow \neg p$ to positive logic [29]. An alternative axiomatization of minimal logic is obtained by extending N with the axiom $(p \rightarrow \neg p) \rightarrow \neg p$ [12, Proposition 1.2.5].

If we interpret $\neg$ as a 'modality', and disregard the fact that we consider extensions of positive logic, the basic system N can be seen as an extension of classical modal logic (see [28]) which is based on the rule $p \leftrightarrow q \,/\, \Box p \leftrightarrow \Box q$. Thus, N can be regarded as a weak intuitionistic modal logic that—as far as we know—has not been previously studied. Note that extending it with more of the properties of a negation would lead to 'very non-standard' modal logics. This relationship with modal logic will be further clarified towards the end of the paper, where the negation will be interpreted by a full-fledged modal operator.

A first algebraic account of Johansson's logic can be found in Rasiowa's work on non-classical logic [29], where the algebraic counterpart of minimal logic is identified as the variety of contrapositionally complemented lattices. A contrapositionally complemented lattice is an algebraic structure $\langle A, \wedge, \vee, \rightarrow, \neg, 1\rangle$, where $\langle A, \wedge, \vee, \rightarrow, 1\rangle$ is a relatively pseudo-complemented lattice (which algebraically characterizes positive logic [29]) and the unary fundamental operation $\neg$ satisfies the identity $(x \rightarrow \neg y) \approx (y \rightarrow \neg x)$. The variety presented by Rasiowa is term-equivalent to the variety of relatively pseudo-complemented lattices with a negation operator defined by the algebraic version of the principle of contradiction $(x \rightarrow y) \wedge (x \rightarrow \neg y) \rightarrow \neg x \approx 1$, originally employed in Johansson's axiomatization. Observe that Heyting algebras can be seen as contrapositionally complemented lattices where $\neg 1$ is a distinguished bottom element 0.

We further generalize the notion of Heyting algebra to that of an N-algebra. An *N-algebra* is an algebraic structure $\mathbf{A} = \langle A, \wedge, \vee, \rightarrow, \neg, 1\rangle$, where $\langle A, \wedge, \vee, \rightarrow, 1\rangle$ is a relatively pseudo-complemented lattice and $\neg$ is a unary operator satisfying the identity $(x \leftrightarrow y) \rightarrow (\neg x \leftrightarrow \neg y) \approx 1$. The latter can be equivalently formulated as $x \wedge \neg y \approx x \wedge \neg(x \wedge y)$. Note that this variety plays a fundamental role in the attempts of defining a connective over positive logic. In fact, the considered equation states that the function $\neg$ is a *compatible function* (or *compatible connective*), in the sense that every congruence of $\langle A, \wedge, \vee, \rightarrow, 1\rangle$ is a congruence of $\langle A, \wedge, \vee, \rightarrow, \neg, 1\rangle$. This is somehow considered a minimal requirement when introducing a new connective over a fixed setting (see, e.g., [9,14]). Clearly, with every N-logic L we can associate a variety of N-algebras. Each of these logics is complete with respect to its algebraic semantics [12]. Contrapositionally complemented lattices are the strongest structures that we consider here, and can be seen as the variety of N-algebras defined by the equation $(x \rightarrow \neg x) \rightarrow \neg x \approx 1$, or $x \rightarrow \neg x \approx \neg x$. We also consider the two varieties of N-algebras defined, respectively, by the identity $(x \wedge \neg x) \rightarrow \neg y \approx 1$, and by the identity $(x \rightarrow y) \rightarrow (\neg y \rightarrow \neg x) \approx 1$. They were studied in detail in [12], and we shall refer to the corresponding logics as *negative ex falso* logic (NeF) and *contraposition* logic (CoPC). We point out that the logic CoPC has appeared before under the name 'Subminimal Logic' with a completely different semantics ([16,17], [18, Section 8.33]). It was proved in [12,13] that the following relations hold between the considered logical systems:

$$\mathsf{N} \subset \mathsf{NeF} \subset \mathsf{CoPC} \subset \mathsf{MPC},$$

where the strict inclusion $\mathsf{L}_1 \subset \mathsf{L}_2$ means that every theorem of L_1 is a theorem of L_2, but there is at least one theorem of L_2 that is not provable in L_1. Note the strict inclusion $\mathsf{NeF} \subset \mathsf{CoPC}$ can be seen semantically (through the semantics we introduce below), by taking $\langle W, \leq, \mathrm{N} \rangle$, with $W := \{w, v\}$, $\leq := \{(w, v)\}$, and $\mathrm{N}(\emptyset) = \mathrm{N}(W) := \{v\}$, $\mathrm{N}(\{v\}) := W$.

Recall that an *intuitionistic Kripke frame* $\mathfrak{F}$ is a partially ordered set (briefly, poset) $\langle W, \leq \rangle$, and a Kripke model is a frame $\mathfrak{F}$ equipped with a valuation V assigning to every propositional variable $p \in \mathsf{Prop}$ an upward closed subset (*upset*) $V(p) \in \mathcal{U}(W, \leq)$ of $\mathfrak{F}$ where p is true. In the subminimal setting, we call an *N-frame* (sometimes we may call it just a *frame*) a triple $\mathfrak{F} = \langle W, \leq, \mathrm{N} \rangle$, where $\langle W, \leq \rangle$ is a poset and the function $\mathrm{N} \colon \mathcal{U}(W, \leq) \to \mathcal{U}(W, \leq)$ satisfies, for every $X, Y \in \mathcal{U}(W, \leq)$,[1]

$$\mathrm{N}(X) \cap Y = \mathrm{N}(X \cap Y) \cap Y. \tag{1}$$

An N-model (sometimes we may call it just a *model*) is again a pair $\langle \mathfrak{F}, V \rangle$, where $\mathfrak{F}$ is a frame and V a valuation from the set of propositional variables to $\mathcal{U}(W, \leq)$. Given a model $\langle \mathfrak{F}, V \rangle$, we define truth of positive formulas inductively as in the intuitionistic setting, and say that a negative formula $\neg \varphi$ is true at a node w, written $\langle \mathfrak{F}, V \rangle, w \models \neg \varphi$, if $w \in \mathrm{N}(V(\varphi))$, where $V(\varphi) := \{w \in W \colon \mathfrak{M}, w \models \varphi\}$. We use the customary notation and write $\mathfrak{F} \models \varphi$ if $\langle \mathfrak{F}, V \rangle, w \models \varphi$ holds for every $w \in W$ and every valuation V. Note that $\mathfrak{F} \models (p \wedge \neg p) \to \neg q$ holds on a frame satisfying property (1) if and only if $X \cap \mathrm{N}(X) \subseteq \mathrm{N}(Y)$ for arbitrary upsets X, Y, and $\mathfrak{F} \models (p \to q) \to (\neg q \to \neg p)$ holds if and only if the function N is antitone (i.e., $X \subseteq Y$ implies $\mathrm{N}(Y) \subseteq \mathrm{N}(X)$). The least element of a poset, when it exists, is called a *root*, and we say that a frame is *rooted* if its underlying poset has a root. Given a poset $\langle W, \leq \rangle$, we use $R(w)$ to denote the set $\{v \in W \colon w \leq v\}$ of successors of w in W. In the rest of the paper, we refer to this 'Kripke style' semantics for subminimal logics as N-semantics.

For a frame $\mathfrak{F} = \langle W, \leq, \mathrm{N} \rangle$, we define the corresponding neighbourhood frame $\mathfrak{F}_{\mathfrak{n}} = \langle W, \leq, \mathfrak{n} \rangle$ by setting a neighbourhood function $\mathfrak{n} \colon W \to \mathcal{P}(\mathcal{U}(W, \leq))$ to be $\mathfrak{n}(w) := \{X \subseteq W \mid w \in \mathrm{N}(X)\}$. A model is again a pair $\langle \mathfrak{F}, V \rangle$ consisting of a neighbourhood frame, and a valuation V defined in the same way as for the N-semantics. Then, $\langle \mathfrak{F}_{\mathfrak{n}}, V \rangle, w \models \neg \varphi$ if and only if $V(\varphi) \in \mathfrak{n}(w)$ or, equivalently, $w \in \mathrm{N}(V(\varphi))$, that is, $\langle \mathfrak{F}, V \rangle, w \models \neg \varphi$. Conversely, take a neighbourhood frame $\mathfrak{F}_{\mathfrak{n}} = \langle W, \leq, \mathfrak{n} \rangle$ such that $\mathfrak{n} \colon W \to \mathcal{P}(\mathcal{U}(W, \leq))$ is a monotone function (i.e., if $w \leq v$ then $\mathfrak{n}(w) \subseteq \mathfrak{n}(v)$) satisfying $X \in \mathfrak{n}(w)$ if and only if $X \cap R(w) \in \mathfrak{n}(w)$ for every upset X. The triple $\mathfrak{F} = \langle W, \leq, \mathrm{N} \rangle$ with $\mathrm{N}(X) := \{w \in W \mid X \in \mathfrak{n}(w)\}$

---

[1] This property is equivalent to $w \in \mathrm{N}(X) \iff w \in \mathrm{N}(X \cap R(w))$ for every $w \in W$ and $X \in \mathcal{U}(W, \leq)$; see, e.g., [12, Lemma 4.3.1].

is an N-frame, and the two definitions are mutually inverse. It will become clear in Sect. 5 that an approach using the N-semantics is helpful in practice. Further, the N-semantics is synergic to the algebraic approach, as the next paragraph shows.

For an N-algebra $\mathbf{A}$, we consider the set $W_{\mathbf{A}}$ of prime filters of $\mathbf{A}$, and let $\widehat{a} := \{w \in W_{\mathbf{A}} \mid a \in w\}$. Then the triple $\mathfrak{F}_{\mathbf{A}} = \langle W_{\mathbf{A}}, \subseteq, N_{\mathbf{A}} \rangle$ is a frame, where

$$N_{\mathbf{A}}(X) := \{w \in W_{\mathbf{A}} \mid (\exists \neg a \in w)(R(w) \cap \widehat{a} = R(w) \cap X)\},$$

for any upset $X \in \mathcal{U}(W_{\mathbf{A}}, \subseteq)$. This definition makes sure that $N_{\mathbf{A}}(\widehat{b})$ for $b \in \mathbf{A}$ includes every filter w which contains $\neg a$, for any $a \in \mathbf{A}$ that is equivalent to b with respect to w (i.e., $R(w) \cap \widehat{a} = R(w) \cap \widehat{b}$). Observe that the notion of prime filter in this context does not require the filter to be proper, i.e., the whole algebra A is always a prime filter. On the other hand, starting from a frame $\mathfrak{F}$, we obtain an N-algebra $\mathbf{A}_{\mathfrak{F}} = \langle \mathcal{U}(W, \leq), \cap, \cup, \rightarrow, N, W \rangle$ whose universe is the set of upsets of $\mathfrak{F}$ equipped with the usual intersection, union, Heyting implication, and unit W, and with the unary operator given by the function N. Note that the N-algebra $\mathbf{A}$ embeds into the N-algebra $\mathbf{A}_{\mathfrak{F}_{\mathbf{A}}}$ via the map $\alpha \colon a \mapsto \widehat{a}$. A consequence of this is that each valuation $\mu \colon \mathsf{Prop} \rightarrow \mathbf{A}$ on $\mathbf{A}$ gives rise to a valuation $V = \alpha \circ \mu$ on $\mathfrak{F}_{\mathbf{A}}$.

In order to obtain a full duality result between algebraic and frame-theoretic structures, in analogy with intuitionistic logic we introduce general frames. A *general frame* is a quadruple $\mathfrak{F} = \langle W, \leq, \mathcal{P}, N \rangle$, where $\langle W, \leq \rangle$ is a partially ordered set, $\mathcal{P} \subseteq \mathcal{U}(W, \leq)$ contains W and is closed under $\cup$, $\cap$, Heyting implication $\rightarrow$, and $N \colon \mathcal{P} \rightarrow \mathcal{P}$ satisfies for all $X, Y \in \mathcal{P}$, $N(X) \cap Y = N(X \cap Y) \cap Y$. Elements of $\mathcal{P}$ are called *admissible sets*. Note that $\emptyset$ need not be admissible. We next recall the definition of *refined* and *compact* general frames, see e.g., [10] for analogous definitions in the intuitionistic setting. If a general frame $\mathfrak{F}$ is refined and compact, we call it a *descriptive frame*. Finally, a *top descriptive frame* $\mathfrak{F} = \langle W, \leq, \mathcal{P}, N, t \rangle$ is a descriptive frame whose partially ordered underlying set $\langle W, \leq \rangle$ has a greatest element t included in every admissible upset $X \in \mathcal{P}$. Following [6], we call *top model* a pair $\langle \mathfrak{F}, V \rangle$ where $\mathfrak{F}$ is a top descriptive frame and V is a valuation on $\mathfrak{F}$ which makes every propositional variable true at the top node. Every finite frame with a top node is a top descriptive frame with a set of admissible upsets $\{X \in \mathcal{U}(W, \leq) \colon X \neq \emptyset\}$.

It can be proved [4,12] that the correspondence between $\mathfrak{F}$ and $\mathbf{A}_{\mathfrak{F}}$, and between $\mathbf{A}$ and $\mathfrak{F}_{\mathbf{A}}$ that holds for every frame and N-algebra, in the case of top descriptive frames gives rise to a proper duality. More precisely, given a top descriptive frame $\mathfrak{F} = \langle W, \leq, \mathcal{P}, N, t \rangle$, the structure $\mathbf{A}_{\mathfrak{F}} = \langle \mathcal{P}, \cap, \cup, \rightarrow, N, W \rangle$ is the N-algebra dual to $\mathfrak{F}$, and the set of prime filters of any N-algebra $\mathbf{A} = \langle A, \wedge, \vee, \rightarrow, \neg, 1 \rangle$ induces a dual top descriptive frame $\mathfrak{F}_{\mathbf{A}} = \langle W_{\mathbf{A}}, \subseteq, \mathcal{P}_{\mathbf{A}}, N_{\mathbf{A}}, A \rangle$, where $\mathcal{P}_{\mathbf{A}}$ is the set $\{\widehat{a} \colon a \in A\}$, and the map $N_{\mathbf{A}}$ is defined as $N_{\mathbf{A}}(\widehat{a}) = \widehat{(\neg a)}$. Moreover, we have $\mathfrak{F} \cong \mathfrak{F}_{\mathbf{A}_{\mathbf{F}}}$ and $\mathbf{A} \cong \mathbf{A}_{\mathfrak{F}_{\mathbf{A}}}$. Observe that the fact that prime filters of $\mathbf{A}$ do not need to be proper ensures that the corresponding frame structure has a top element.

As in the case of Heyting algebras, for N-algebras there exists a one-to-one correspondence between congruences and filters. We can therefore characterize subdirectly irreducible N-algebras as those N-algebras containing a second greatest element, or equivalently, as those N-algebras **A** whose dual frame $\mathfrak{F}_{\mathbf{A}}$ is a rooted top descriptive frame. We call a top descriptive frame $\mathfrak{F}$ finitely generated if $\mathbf{A}_{\mathfrak{F}}$ is a finitely generated N-algebra, thereby obtaining a completeness result for every N-logic with respect to the class of its finitely generated rooted top descriptive frames.

3   Continuum Many Logics

In this section, we construct continuum many N-logics by exporting and adapting techniques of [5,19,21]. We start from a countable family of positive formulas that we adapt so to define uncountably many independent subsystems of minimal logic containing the basic logic N.

Given two finite rooted posets $\mathfrak{F}, \mathfrak{G}$, we write $\mathfrak{F} \leq \mathfrak{G}$ if $\mathfrak{F}$ is an order-preserving image of $\mathfrak{G}$, that is, if there is an onto map $f: \mathfrak{G} \twoheadrightarrow \mathfrak{F}$ which preserves the order. This relation can be proved to be a partial order on the set of finite (rooted) posets [5]. Consider the sequence $\Delta = \{\mathfrak{F}_n \mid n \in \mathbb{N}\}$ of finite rooted posets with a top node (Fig. 1). The sequence Δ is obtained from the antichain presented in [3, Figure 2] by adding three nodes on the top. An argument resembling the one in [3, Lemma 6.12] shows that the sequence Δ is a $\leq$-antichain. We quickly sketch it here. If $f: \mathfrak{F}_n \twoheadrightarrow \mathfrak{F}_m$ with $n \geq m$ is order-preserving, then the upset of $\mathfrak{F}_m$ consisting of points of depth $\leq (m+1)$ must be the image of the upset of $\mathfrak{F}_n$ consisting of points of depth $\leq (m+1)$. Thus, for $x_{m+2} \in \mathfrak{F}_m$, there is $y \in \mathfrak{F}_n$ such that $f(y) = x_{m+2}$. But then, there is a $z \in \mathfrak{F}_n$ such that $y \leq z$ and $f(z) = y_{m+1}$. Since $x_{m+2} \not\leq y_{m+1}$, we get a contradiction.

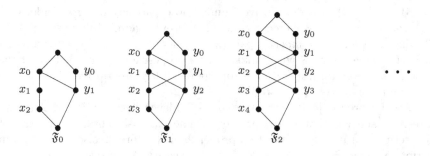

Fig. 1. The sequence Δ

Lemma 1. *The sequence Δ forms a $\leq$-antichain.*

We call a map $f: \mathfrak{F} \to \mathfrak{G}$ between posets $\mathfrak{F} = \langle W, \leq \rangle$ and $\mathfrak{G} = \langle W', \leq' \rangle$ a p-morphism if f is order-preserving, and $f(w) \leq' v$ implies the existence of $w' \in W$

such that $w \leq w'$ and $f(w') = v$. A partial p-morphism such that $\mathrm{dom}(f)$ is a downward closed set is called a *positive morphism*. We write $\mathfrak{F} \preceq \mathfrak{G}$ if $\mathfrak{F}$ is a positive morphic image of $\mathfrak{G}$. Assume $\mathfrak{F} \preceq \mathfrak{G}$ and let f be a positive morphism from $\mathfrak{G}$ onto $\mathfrak{F}$. Extending f by mapping all the points of $\mathfrak{G} \setminus \mathrm{dom}(f)$ to the top node of $\mathfrak{F}$, we obtain a total order-preserving map, yielding $\mathfrak{F} \leq \mathfrak{G}$. This ensures the $\leq$-antichain Δ to be also a $\preceq$-antichain.

Lemma 2. *If $\mathfrak{F} \preceq \mathfrak{G}$ and $\mathfrak{F}$ is a finite rooted posets with a top node, then $\mathfrak{F} \leq \mathfrak{G}$.*

Having constructed the desired $\preceq$-antichain, we now proceed by adjusting the technique of Jankov-de Jongh formulas. Recall that the universal model $\mathcal{U}(n)$ can be roughly thought of as the upper generated submodel of the descriptive model dual to the free Heyting algebra on $n \in \mathbb{N}$ generators. The universal model $(\mathcal{U}^\star(n))^+$ for positive logic is isomorphic to a generated submodel of $\mathcal{U}(n)$, and is a top model [5,6]. It was shown in [6] that every finite rooted poset with a top node, equipped with an appropriate valuation, is isomorphic to a generated submodel of $(\mathcal{U}^\star(m))^+$ for some m. Note that, for the frames $\mathfrak{F}_n$ in Fig. 1, it suffices to take $m = 3$. Hence, we call $w_n \in (\mathcal{U}^\star(3))^+$ the node in the positive universal model corresponding to the root of $\mathfrak{F}_n$. We can assume without loss of generality the positive Jankov-de Jongh formula of $\mathfrak{F}_n$ to be defined as follows [6]:

$$\chi^\star(\mathfrak{F}_n) = \psi^\star_{w_n} := \varphi^\star_{w_n} \to \bigvee_{i=1}^{r} \varphi^\star_{w_{n_i}} ,$$

where $\varphi^\star_{w_n}, \varphi^\star_{w_{n_i}}$ are defined as in [6] and $w_n \prec \{w_{n_1}, \ldots, w_{n_r}\}$. So, a rooted poset $\mathfrak{G}$ refutes $\chi^\star(\mathfrak{F}_n)$ if and only if $\mathfrak{F}_n \preceq \mathfrak{G}$. As Δ is a $\preceq$-antichain, this means that, for every $n, m \in \mathbb{N}$, the formula $\chi^\star(\mathfrak{F}_m)$ is valid on the poset $\mathfrak{F}_n$ if and only if $n \neq m$. In fact, the construction of $\varphi^\star_{w_n}, \varphi^\star_{w_{n_i}}$ ensures that the formula $\varphi^\star_{w_n}$ is satisfied at the root w_n in $(\mathcal{U}^\star(2))^+$, while none of the formulas $\varphi^\star_{w_{n_i}}$ is.

Lemma 3. *For $n, m \in \mathbb{N}$, the formula $\mathfrak{F}_n \models \chi^\star(\mathfrak{F}_m)$ if and only if $n \neq m$.*

At this point, we are going to equip each rooted poset $\mathfrak{F}_n$ with an appropriate function N_n to make it a top descriptive frame. Moreover, we enhance the definition of $\chi^\star(\mathfrak{F}_n)$ in order to obtain formulas $\theta(\mathfrak{F}_n)$ with the same defining property of the Jankov-de Jongh formulas for the signature of positive logic, but with the extra addition that the formulas $\theta(\mathfrak{F}_n)$ are theorems of MPC. Given the rooted poset $\mathfrak{F}_n$, we consider the top descriptive frame $\langle \mathfrak{F}_n, N_n, t_n \rangle$ where N_n has the property $N_n(\{t_n\}) = \{t_n\}$, the element t_n being the top node of $\mathfrak{F}_n$. We denote the new family of top descriptive frames $\langle \mathfrak{F}_n, N_n, t_n \rangle$ by Δ_N. Now we consider a fresh propositional variable p, and define:

$$\theta(\mathfrak{F}_n) := (p \to \neg p) \wedge \varphi^\star_{w_n} \to \neg p \vee \bigvee_{i=1}^{r} \varphi^\star_{w_{n_i}} .$$

We note that the formulas $\theta(\mathfrak{F}_n)$ are not the Jankov-de Jongh formulas for the considered signature. It is nonetheless easy to see that, if $n \neq m$, the formula

$\theta(\mathfrak{F}_n)$ is valid on the frame $\langle \mathfrak{F}_m, N_m, t_m \rangle$. On the other hand, for checking that $\langle \mathfrak{F}_n, N_n, t_n \rangle \not\vDash \theta(\mathfrak{F}_n)$ it is enough to consider a valuation $\tilde{V}_n$ extending V_n by $\tilde{V}_n(p) = \{t_n\}$. This way, the root of $\mathfrak{F}_n$ under the considered valuation makes $(p \to \neg p) \land \varphi^*_{w_n}$ true, while neither $\neg p$ nor $\bigvee_{i=1}^r \varphi^*_{w_{n_i}}$ is true at w_n.

As a consequence, we obtain the following result.

Theorem 1. *There are continuum many N-logics.*

Proof. The fact that each $\theta(\mathfrak{F}_n)$ is a theorem of MPC ensures that for each subset $\Gamma \subseteq \Delta_N$, the logic

$$L(\Gamma) := \mathsf{N} + \{\theta(\mathfrak{F}) \mid \mathfrak{F} \in \Gamma\}$$

belongs to the interval $[\mathsf{N}, \mathsf{MPC}]$. Finally, observe that for each pair of different subsets $\Gamma_1 \neq \Gamma_2$ of Δ_N, we have $L(\Gamma_1) \neq L(\Gamma_2)$. Indeed, without loss of generality we may assume that there is $\mathfrak{F} \in \Gamma_1$ such that $\mathfrak{F} \notin \Gamma_2$. Moreover, we have $\mathfrak{F} \not\vDash \theta(\mathfrak{F})$ and $\mathfrak{F} \vDash \theta(\mathfrak{G})$, for each $\mathfrak{G}$ in Γ_2. Therefore, there is a top descriptive frame $\mathfrak{F}$ which is an $L(\Gamma_2)$-frame and not an $L(\Gamma_1)$-frame. Since every N-logic is complete with respect to top descriptive frames, the latter entails that $L(\Gamma_1) \neq L(\Gamma_2)$. $\square$

Observe that the choices of N_n and of the formula $(p \to \neg p) \to \neg p$ that were made above are arbitrary among the ones that ensure a proof of Theorem 1. This suggests the question what would have happened if we had made different choices. As a matter of fact, the result of Theorem 1 can be refined as follows.

Proposition 1.

(1) *There are continuum many N-logics between* NeF *and* CoPC.
(2) *There are continuum many N-logics below* NeF.
(3) *There are continuum many N-logics above* CoPC.

Proof. The idea is to slightly modify the previous proof by choosing different theorems of minimal propositional logic, and different functions N_n. Changing the former allows us to refine the result from above, while changing the latter from below.

To prove (1), we consider posets $\mathfrak{F}_n$, but this time we consider the function N_n^1 defined by

$$N_n^1(X) := \begin{cases} W_n \setminus \{w_n\}, & \text{if } X \neq W_n \setminus \{w_n\} \\ W_n, & \text{if } X = W_n \setminus \{w_n\}. \end{cases}$$

Let p, q be fresh propositional variables. Consider the formulas

$$\theta^1(\mathfrak{F}_n) := (p \to q) \land \varphi^*_{w_n} \to (\neg q \to \neg p) \lor \bigvee_{i=1}^r \varphi^*_{w_{n_i}}.$$

Observe that the frame $\langle \mathfrak{F}_n, N_n^1, t_n \rangle$ is a frame of the logic NeF. In fact, if we consider an arbitrary $w \in W_n$ such that $w \in X \cap N_n^1(X)$ (this amounts to every

$w \in W_n \setminus \{w_n\}$), then $w \in \mathrm{N}_n^1(Y)$ for every admissible upset Y. Moreover, it is easy to see that, if $n \neq m$, the formula $\theta^1(\mathfrak{F}_n)$ is valid on the frame $\langle \mathfrak{F}_m, \mathrm{N}_m^1, t_m \rangle$. On the other hand, for checking that $\langle \mathfrak{F}_n, \mathrm{N}_n^1, t_n \rangle \nVdash \theta^1(\mathfrak{F}_n)$ it is enough to consider a valuation $\tilde{V}_n$ extending V_n by $\tilde{V}_n(p) = \{t_n\}$ and $\tilde{V}_n(q) = W_n \setminus \{w_n\}$. This way, the root of $\mathfrak{F}_n$ under the considered valuation makes the whole antecedent of $\theta^1(\mathfrak{F}_n)$ true (since $\tilde{V}_n(p) \subseteq \tilde{V}_n(q)$), while the consequent is not true at w_n (since $w_n \in \mathrm{N}_n^1(\tilde{V}_n(q))$, while $w_n \notin \mathrm{N}_n^1(\tilde{V}_n(p))$).

A proof of (2) is obtained by considering functions:

$$\mathrm{N}_n^2(X) := \begin{cases} W_n \setminus \{w_n\}, & \text{if } X \subset W, \\ W_n, & \text{if } X = W, \end{cases}$$

and formulas

$$\theta^2(\mathfrak{F}_n) := (p \wedge \neg p) \wedge \varphi_{w_n}^* \to \neg q \vee \bigvee_{i=1}^{r} \varphi_{w_{n_i}}^*,$$

with fresh propositional variables p, q.

For (3), it is enough to define $\mathrm{N}_n^3(X) := \{t_n\}$, for every upset X, and to use the formulas $\theta(\mathfrak{F}_n)$. □

4 Filtration

The method of filtration is among the oldest and most used techniques for proving the finite model property for modal and superintuitionistic logics. If a model refutes a formula φ, then the idea is to filter it out (identify the points that agree on the subformulas of φ) to obtain a finite model that refutes φ. Thus the new model is a quotient of the original one. The method of filtration was originally developed algebraically [24,25], and later model-theoretically [23,30], see also [10]. In this section, we complete the results and constructions from [12], where the finite model property was proved via model-theoretic filtrations applied to the canonical models only. A general model-theoretic definition of filtration was missing in [12]. We give one here, which is a combination of the standard filtration for intuitionistic logic [10] and the one for non-normal logic [11,28]. Building on the ideas of [2,3], we define algebraic filtrations for N-algebras. As in the modal and intuitionistic cases, we show, similarly to [2], that the algebraic and model-theoretic methods turn out to be "two sides of the same coin".

Let Σ be a finite set of formulas closed under subformulas. Given a model $\mathfrak{M}$, we define an equivalence relation $\sim$ on W by considering, for $w, v \in W$:

$$w \sim v \text{ if } (\forall \varphi \in \Sigma)(w \in V(\varphi) \Longleftrightarrow v \in V(\varphi)).$$

Consider $\mathfrak{F}^* = \langle W^*, \leq^*, \mathrm{N}^* \rangle$, where $W^* = W/\sim$, and $\leq^*$ and N^* are, respectively, a partial order on W^* and a function $\mathrm{N}^* \colon \mathcal{U}(W^*, \leq^*) \to \mathcal{U}(W^*, \leq^*)$ satisfying the following conditions for all $w, v \in W$, $X \in \mathcal{U}(W^*, \leq^*)$ and $\varphi, \neg\psi \in \Sigma$:

(a) $w \leq v$ implies $[w] \leq^* [v]$;

(b) $[w] \leq^* [v]$ and $w \in V(\varphi)$ imply $v \in V(\varphi)$;

(c) $[w] \in N^*(X)$ implies $w \in N(\pi^{-1}(X))$,

where $\pi^{-1}(X) := \{w \in W \colon [w] \in X\}$; and

(d) $w \in N(V(\psi))$ implies $[w] \in N^*(\pi[V(\psi)])$,

where $\pi[V(\psi)] := \{[w] \in W^* \colon w \in V(\psi)\}$. Finally, let V^* be a valuation on $\mathfrak{F}^*$ such that
$$V^*(p) := \{[w] \in W^* \colon w \in V(p)\} = \pi[V(p)],$$
for each $p \in \Sigma$. We call $\mathfrak{M}^* = \langle \mathfrak{F}^*, V^* \rangle$ a *filtration* of the model $\mathfrak{M}$ through Σ.

Theorem 2 (Filtration Theorem). *Let* $\mathfrak{M} = \langle \mathfrak{F}, V \rangle$ *be a model. Consider the filtration* $\mathfrak{M}^* = \langle \mathfrak{F}^*, V^* \rangle$ *of* $\mathfrak{M}$ *through a finite set* Σ *of formulas closed under subformulas. Then, for each* $\varphi \in \Sigma$ *and* $w \in W$ *we have*
$$w \in V(\varphi) \ \textit{iff} \ [w] \in V^*(\varphi).$$

Proof. We focus on the case $\neg\varphi \in \Sigma$. Consider $w \in V(\neg\varphi)$, i.e., $w \in N(V(\varphi))$. By induction hypothesis, $V^*(\varphi) = \{[w] \colon w \in V(\varphi)\} = \pi[V(\varphi)]$. Hence, by (d), we are able to conclude $[w] \in N^*(V^*(\varphi))$. On the other hand, assume $w \notin N(V(\varphi))$. By induction hypothesis, $V(\varphi) = \{w \colon [w] \in V'(\varphi)\} = \pi^{-1}(V^*(\varphi))$. Therefore, by (c), $[w] \notin N^*(V^*(\varphi))$. □

Among the filtrations of $\mathfrak{M} = \langle W, \leq, N, V \rangle$ through Σ, there always exists a greatest filtration. Consider the partial order on W^* defined by:
$$[w] \leq^g [v] \ \text{if} \ (\forall\varphi \in \Sigma)(w \in V(\varphi) \Rightarrow v \in V(\varphi)), \tag{2}$$

and the map $N^g(X) := \{[w] \in W^* \colon w \in N(\pi^{-1}(X))\}$. It is easy to check that $\mathfrak{F}^g = \langle W^*, \leq^g, N^g \rangle$ is a frame and $\mathfrak{M}^g = \langle \mathfrak{F}^g, V^* \rangle$ is a filtration of $\mathfrak{M}$. Consider now an arbitrary filtration $\mathfrak{M}^* = \langle \mathfrak{F}^*, V^* \rangle$. Suppose $[w] \leq^* [v]$. Then $w \in V(\varphi)$ implies $v \in V(\varphi)$. Moreover, observe that every upset of $\langle W^*, \leq^g \rangle$ is an upset of $\langle W^*, \leq^* \rangle$. Furthermore, for a given $X \in \mathcal{U}(W^*, \leq^g)$ such that $[w] \in N^*(X)$ we have then $[w] \in N^g(X)$. To see this, assume that $[w] \in N^*(X)$. Then, by (c), $w \in N(\pi^{-1}(X))$ and hence, $[w] \in N^g(X)$.

The rest of this section will be devoted to an algebraic development of the notion of filtration. Let Σ be a finite set of formulas closed under subformulas, and consider an N-algebra with a valuation $\langle \mathbf{A}, \mu \rangle$, where $\mu \colon \mathsf{Prop} \to \mathbf{A}$. Since Σ is finite, so is $\mu[\Sigma]$ as a subset of $\mathbf{A}$. Let $\mathbf{S}$ be the $(\wedge, \vee, 1)$-reduct of $\mathbf{A}$ generated by $\mu[\Sigma]$. Observe that $(\wedge, \vee, 1)$-reducts are locally finite, i.e., every finitely generated $(\wedge, \vee, 1)$-structure is finite. Hence, the resulting $\mathbf{S}$ is finite and we can make $\mathbf{S}$ an N-algebra by equipping it with operations $\to_{\mathbf{S}}$ and $\neg_{\mathbf{S}}$ defined as follows:
$$a \to_{\mathbf{S}} b := \bigvee \{s \in S \mid a \wedge s \leq b\}, \tag{3}$$

and

$$\neg_S a := \bigvee \{s \in S \mid s \leq \neg a\}, \tag{4}$$

for every $a, b \in S$. Note that $\neg_S a \leq \neg a$, and $\neg_S a = \neg a$ whenever $\neg a \in \mu[\Sigma]$, and $a \rightarrow_S b \leq a \rightarrow b$ and $a \rightarrow_S b = a \rightarrow b$ whenever $a \rightarrow b \in \mu[\Sigma]$. We call V the valuation $V = \alpha \circ \mu$ on $\mathfrak{F}_\mathbf{A}$, where α is defined as in Sect. 2.

Lemma 4. *The subset $S \subseteq A$ gives rise to a filtration $\mathfrak{M}_\mathbf{A}^* = \langle \mathfrak{F}_\mathbf{A}, V^* \rangle$ of the model $\mathfrak{M}_\mathbf{A} = \langle \mathfrak{F}_\mathbf{A}, V \rangle$ through Σ.*

Proof. Define $\sim$ on $W_\mathbf{A}$ by $w \sim v$ if and only if $S \cap w = S \cap v$. The first step amounts to proving that $w \sim v$ is equivalent to $w \cap \mu[\Sigma] = v \cap \mu[\Sigma]$. The proof follows the one of [2, Lemma 2.5]. Observe that this first step is clearly equivalent to $w \sim v$ if and only if $(\forall \varphi \in \Sigma)(\mu(\varphi) \in w$ if and only if $\mu(\varphi) \in v)$, which in turn is equivalent to $(\forall \varphi \in \Sigma)(w \in V(\varphi)$ if and only if $v \in V(\varphi))$.

Now, let $W_\mathbf{A}^* = W_\mathbf{A}/\sim$ and define $[w] \leq^* [v] \iff w \cap S \subseteq v \cap S$, and

$$N^*(X) := \{[w] \in W_\mathbf{A}^* : w \in N(\pi^{-1}(X))\},$$

for $X \in \mathcal{U}(W_\mathbf{A}^*, \leq^*)$. It is straightforward to see that $\mathfrak{F}_\mathbf{A}^* = \langle W_\mathbf{A}^*, \leq^*, N^* \rangle$ is a frame. Let now V^* be a valuation on $\mathfrak{F}_\mathbf{A}^*$ such that:

$$V^*(p) := \{[w] \in W_\mathbf{A}^* : w \in V(p)\}.$$

The structure $\langle \mathfrak{F}_\mathbf{A}^*, V^* \rangle$ is a model, and it satisfies conditions (a)–(d).   □

The construction presented in Lemma 4 can be generalized as follows. Let $\langle \mathbf{A}, \mu \rangle$ be an N-algebra with a valuation and Σ be a finite set of formulas closed under subformulas. Suppose that L is the universe of a finite sublattice of $\mathbf{A}$ with unit $1_\mathbf{A}$ such that $\mu[\Sigma] \subseteq L$. Consider the N-algebra $\mathbf{L} = \langle L, \wedge, \vee, \rightarrow_\mathbf{L}, \neg_\mathbf{L}, 1_\mathbf{A} \rangle$, where $\rightarrow_\mathbf{L}$ and $\neg_\mathbf{L}$ are defined as in (3) and (4). Let $\mu_\mathbf{L}$ be a valuation on $\mathbf{L}$ such that $\mu_\mathbf{L}(p) = \mu(p)$ for each variable $p \in \Sigma$. We call the pair $(\mathbf{L}, \mu_\mathbf{L})$ a *filtration* of $(\mathbf{A}, \mu)$ through Σ.

Theorem 3 (Filtration Theorem). *If $(\mathbf{L}, \mu_\mathbf{L})$ is a filtration of $(\mathbf{A}, \mu)$ through Σ, then $\mu_\mathbf{L}(\varphi) = \mu(\varphi)$ for each $\varphi \in \Sigma$.*

Proof. The proof goes again by induction, and we focus on the case $\neg \varphi \in \Sigma$. But then, $\neg \mu(\varphi) = \mu(\neg \varphi) \in \mu[\Sigma] \subseteq L$. Therefore, $\neg \mu(\varphi) = \neg_\mathbf{L} \mu(\varphi)$. Thus, $\mu_\mathbf{L}(\neg \varphi) = \neg_\mathbf{L} \mu_\mathbf{L}(\varphi) = \neg \mu(\varphi) = \mu(\neg \varphi)$.   □

Among the filtrations $(\mathbf{L}, \mu_\mathbf{L})$ of $(\mathbf{A}, \mu)$, the filtration $(\mathbf{S}, \mu_\mathbf{S})$ is the least one. The following shows that $(\mathbf{S}, \mu_\mathbf{S})$ corresponds to the greatest filtration $\mathfrak{M}_\mathbf{A}^g$ of $\mathfrak{M}_\mathbf{A}$ through Σ. Consider $[w], [v] \in W_\mathbf{A}^*$ such that $[w] \leq^g [v]$. But then, by (2), this means that $(\forall \varphi \in \Sigma)(w \in V(\varphi) \Rightarrow v \in V(\varphi))$. This is again equivalent to $w \cap S \subseteq v \cap S$. Hence, we conclude $[w] \leq^* [v]$.

5   Modal Companions

A normal extension M of S4 is a *modal companion* of an intermediate logic L if for each propositional formula φ we have $L \vdash \varphi$ iff $M \vdash \varphi^\square$, where $\varphi^\square$ is the Gödel translation of φ. In this section we define (bi-)modal companions of subminimal logics. The negation operator in N and its extensions behave like a non-normal modal operator which suggests that bi-modal companions of subminimal logics must have a non-normal modal operator. For the theory of non-normal modal logics and neighbourhood semantics we refer the reader to [11,15,28].

Consider the bi-modal language $\mathcal{L}_\square(\mathsf{Prop})$, where Prop is a countable set of propositional variables, generated by the following grammar:

$$p \mid \bot \mid \top \mid \varphi \wedge \varphi \mid \varphi \vee \varphi \mid \varphi \to \varphi \mid \square\varphi \mid \blacksquare\varphi$$

where $p \in \mathsf{Prop}$. We write $\neg\varphi$ to denote the implication $\varphi \to \bot$. Recall that the axioms and rules for the modal logic S4 are:

(K) $\square(p \to q) \to (\square p \to \square q)$;

(T) $\square p \to p$;

(4) $\square p \to \square\square p$,

in addition to all the classical tautologies, and the rules of *modus ponens*, uniform substitution, and necessitation $(p/\square p)$. Now, consider the following additional bi-modal axioms:

$$\square(p \leftrightarrow q) \to (\blacksquare p \leftrightarrow \blacksquare q), \tag{5}$$

$$\blacksquare p \to \square\blacksquare p. \tag{6}$$

Note that the $\blacksquare$ modality is not a normal modality, since we do not have the rule of necessitation for it. As for the notation, we denote the new modality as a *box* modality since it behaves as a universal modality from the point of view of neighbourhood semantics. We denote this bi-modal system by NS4.

A neighbourhood frame for NS4 is a triple $\mathfrak{F} = \langle W, \leq, \mathfrak{n} \rangle$, where $\langle W, \leq \rangle$ is a set equipped with a reflexive transitive relation $\leq$, and $\mathfrak{n}$ a monotone function $\mathfrak{n} \colon W \to \mathcal{P}(\mathcal{P}(W))$ (i.e., if $w \leq v$ then $\mathfrak{n}(w) \subseteq \mathfrak{n}(v)$) such that:

$$X \in \mathfrak{n}(w) \iff X \cap R(w) \in \mathfrak{n}(w). \tag{7}$$

Using the fact that a neighbourhood function $\mathfrak{n}$ induces the existence of a function $N \colon \mathcal{P}(W) \to \mathcal{U}(W, \leq)$ via the equivalence $X \in \mathfrak{n}(w)$ if and only if $w \in N(X)$, we consider the following generalization of N-semantics.

An *N-frame for* NS4 is a triple $\mathfrak{F} = \langle W, \leq, N \rangle$, where $\langle W, \leq \rangle$ is a set equipped with a reflexive transitive relation, and $N \colon \mathcal{P}(W) \to \mathcal{U}(W, \leq)$ is a map such that:

$$w \in N(X) \iff w \in N(X \cap R(w)). \tag{8}$$

An *N-model for* NS4 is a pair $\mathfrak{M} = \langle \mathfrak{F}, V \rangle$, where $\mathfrak{F}$ is an N-frame and V is a valuation $V \colon \mathsf{Prop} \to \mathcal{P}(W)$. Truth of a formula is defined as follows:

$$\mathfrak{M}, w \models p \qquad \Longleftrightarrow w \in V(p);$$

$$\mathfrak{M}, w \models \varphi \wedge \psi \Longleftrightarrow \mathfrak{M}, w \models \varphi \text{ and } \mathfrak{M}, w \models \psi;$$

$$\mathfrak{M}, w \models \varphi \vee \psi \Longleftrightarrow \mathfrak{M}, w \models \varphi \text{ or } \mathfrak{M}, w \models \psi;$$

$$\mathfrak{M}, w \models \varphi \rightarrow \psi \Longleftrightarrow \mathfrak{M}, w \models \varphi \text{ implies } \mathfrak{M}, w \models \psi;$$

$$\mathfrak{M}, w \models \Box\varphi \qquad \Longleftrightarrow R(w) \subseteq V(\varphi);$$

$$\mathfrak{M}, w \models \blacksquare\varphi \qquad \Longleftrightarrow w \in \mathrm{N}(V(\varphi)),$$

where $V(\varphi) := \{w \in W : \mathfrak{M}, w \models \varphi\}$. A formula φ is said to be valid in a model $\mathfrak{M}$ if $\mathfrak{M}, w \models \varphi$ for every $w \in W$, and φ is valid in a frame $\mathfrak{F}$ if every model on $\mathfrak{F}$ validates φ. We let $\mathcal{F}$ denote the class of all N-frames for NS4. We will now show that NS4 is sound with respect to $\mathcal{F}$.

Theorem 4 (Soundness). *The system* NS4 *is sound with respect to* $\mathcal{F}$.

Proof. We only check the cases for the axioms (5) and (6), by proving that they are valid on each frame. The validity of (6) follows immediately from the fact that $\mathrm{N}(X)$ is upward closed for every $X \in \mathcal{P}(W)$. For (5), suppose that $\langle \mathfrak{F}, V \rangle, w \models \Box(p \leftrightarrow q)$. This means that $R(w) \subseteq V(p \leftrightarrow q)$, which is in turn equivalent to $V(p) \cap R(w) = V(q) \cap R(w)$. In order to conclude $\langle \mathfrak{F}, V \rangle, w \models \blacksquare p \leftrightarrow \blacksquare q$, we assume without loss of generality that $w \in \mathrm{N}(V(p))$. By (8), this means that $w \in \mathrm{N}(V(p) \cap R(w))$. But then, $w \in \mathrm{N}(V(q) \cap R(w))$, since $V(p) \cap R(w) = V(q) \cap R(w)$. Thus, again by (8), we conclude $w \in \mathrm{N}(V(q))$. $\square$

We now focus on showing that NS4 is complete with respect to $\mathcal{F}$. To do this, we follow the standard approach for normal modal logics (see, e.g., [8,10]), and combine it with the standard approach for neighbourhood semantics [28]. Recall that a set Γ of formulas is said to be maximally consistent if it is consistent (in the sense that it does not contain both φ and $\neg\varphi$, for any φ) and, for every formula φ, either $\varphi \in \Gamma$ or $\neg\varphi \in \Gamma$. Every consistent set of formulas can be extended to a maximally consistent set of formulas (Lindenbaum's Lemma). As a consequence, it can be proved that every formula that is contained in all maximally consistent sets of formulas is a theorem. Consider now the set $\mathcal{W}$ of all maximally consistent sets of NS4 formulas. We denote by $|\varphi|$ the set of all maximally consistent sets of formulas containing φ. We define the canonical relation $\leq$ by: $\Gamma \leq \Delta$ if, whenever $\Box\varphi \in \Gamma$, then $\varphi \in \Delta$. A map $\mathcal{N} : \mathcal{P}(\mathcal{W}) \rightarrow \mathcal{U}(\mathcal{W}, \leq)$ is a canonical N-function provided that for all $\varphi \in \mathcal{L}_\Box(\mathsf{Prop})$,

$$\Gamma \in \mathcal{N}(|\varphi|) \Longleftrightarrow \blacksquare\varphi \in \Gamma.$$

We consider the canonical valuation $\mathcal{V} : \mathrm{Prop} \rightarrow \mathcal{P}(\mathcal{W})$ defined by

$$\mathcal{V}(p) = |p| := \{\Gamma \in \mathcal{W} : p \in \Gamma\}.$$

The quadruple $\langle \mathcal{W}, \leq, \mathcal{N}, \mathcal{V} \rangle$ is called a canonical model. To prove completeness of NS4, we use the following function

$$\mathcal{N}^{loc}(X) := \{\Gamma \in \mathcal{W} \mid (\exists \blacksquare \psi \in \Gamma)(X \cap R(\Gamma) = |\psi| \cap R(\Gamma))\}.$$

which makes $\mathcal{M} = \langle \mathcal{W}, \leq, \mathcal{N}^{loc}, \mathcal{V} \rangle$ into a canonical model, and its underlying frame into an NS4-frame. First of all, it is easy to see that $\mathcal{N}^{loc}$ satisfies (8). Further, the set $\mathcal{N}^{loc}(X)$ is upward closed for every $X \in \mathcal{P}(\mathcal{W})$. In fact, if $\blacksquare \psi \in \Gamma$, then also $\square \blacksquare \psi \in \Gamma$. But then, $\Gamma \leq \Delta$ implies $\blacksquare \psi \in \Delta$. Since $X \cap R(\Delta) = |\psi| \cap R(\Delta)$, we conclude. Moreover, $\mathcal{M}$ is canonical. In fact, $\Gamma \in \mathcal{N}^{loc}(|\varphi|)$ implies that $|\psi| \cap R(\Gamma) = |\varphi| \cap R(\Gamma)$ for some $\blacksquare \psi \in \Gamma$, which means that for every successor Δ of Γ, $\varphi \leftrightarrow \psi \in \Delta$ and, therefore, $\square(\varphi \leftrightarrow \psi) \in \Gamma$. By the axiom (5) and the fact that $\blacksquare \psi \in \Gamma$, we conclude $\blacksquare \varphi \in \Gamma$. A canonical function $\mathcal{N}$ is what ensures that membership and truth coincide in the canonical model (Truth Lemma); the cases for the other formulas proceed by the standard induction.

Theorem 5 (Completeness). *The system* NS4 *is complete with respect to* $\mathcal{F}$.

The next step is to show that NS4 is the modal companion of N. We start by recalling that it is possible to translate formulas from the intuitionistic language into the modal language via the Gödel translation:

$$\bot^{\square} = \bot;$$

$$p^{\square} = \square p;$$

$$(\varphi \circ \psi)^{\square} = \varphi^{\square} \circ \psi^{\square}, \text{ where } \circ \in \{\wedge, \vee\};$$

$$(\varphi \to \psi)^{\square} = \square(\varphi^{\square} \to \psi^{\square}).$$

The celebrated Gödel-McKinsey-Tarski theorem states that for every intuitionistic formula φ,

$$\mathsf{IPC} \vdash \varphi \iff \mathsf{S4} \vdash \varphi^{\square}.$$

The logic S4 is not the only modal companion of IPC (e.g., Grzegorczyk's logic Grz is another well-known example). A fundamental fact to be used in the proof is that any intuitionistic Kripke model $\mathfrak{M}$ can be seen as a model for S4, and for every node in the model the following holds:

$$\mathfrak{M}, w \models \varphi \iff \mathfrak{M}, w \models_{\square} \varphi^{\square},$$

where $\models$ denotes the intuitionistic and $\models_{\square}$ the modal notion of truth. Similarly, every N-model has an equivalent NS4-model, as any N-frame $\langle W, \leq, N \rangle$ induces a corresponding NS4-frame $\langle W, \leq, N^* \rangle$, with $N^* \colon \mathcal{P}(W) \to \mathcal{U}(W, \leq)$ defined by:

$$N^*(X) := \{w \in W \mid (\exists Y \in \mathcal{U}(W, \leq))(X \cap R(w) = Y \cap R(w) \text{ and } w \in N(Y))\}.$$

We start by adapting the Gödel translation in order to obtain a translation of N into NS4. It is sufficient to add a translation clause for the negation in the language of N:

$$(\neg \varphi)^{\square} = \blacksquare \varphi^{\square},$$

and to discard the clause for $\bot$. The following preliminary result is easy to prove.

Lemma 5. *Let* $\mathfrak{M} = \langle W, \leq, N, V \rangle$ *be a model for* N, *and let* $\mathfrak{M}^* = \langle W, \leq, N^*, V \rangle$ *be the corresponding N-model for* NS4 *defined in the way explained above. Then, for every formula* φ,

$$\mathfrak{M}, w \models \varphi \Longleftrightarrow \mathfrak{M}^*, w \models_\Box \varphi^\Box,$$

where $\models$ *and* $\models_\Box$ *are, respectively, the* N *and the modal notion of truth.*

In what follows, we allow ourselves to identify $\Box\varphi \wedge \Box\psi$ and $\Box(\varphi \wedge \psi)$ for formulas φ and ψ, omitting thereby some trivial steps in the derivations.

Theorem 6. *For every formula* $\varphi \in \mathcal{L}(\mathsf{Prop})$,

$$\mathsf{N} \vdash \varphi \Longleftrightarrow \mathsf{NS4} \vdash \varphi^\Box.$$

Proof. From left to right, we show that $\mathsf{NS4} \vdash \big((p \leftrightarrow q) \to (\neg p \leftrightarrow \neg q)\big)^\Box$, that is, $\mathsf{NS4} \vdash \Box\big(\Box(\Box p \leftrightarrow \Box q) \to \Box(\blacksquare\Box p \leftrightarrow \blacksquare\Box q)\big)$. Indeed:

$\vdash \Box(\Box p \leftrightarrow \Box q) \to (\blacksquare\Box p \leftrightarrow \blacksquare\Box q)$	By axiom (5)
$\vdash \Box\Box(\Box p \leftrightarrow \Box q) \to \Box(\blacksquare\Box p \leftrightarrow \blacksquare\Box q)$	By necessitation
$\vdash \Box(\Box p \leftrightarrow \Box q) \to \Box\Box(\Box p \leftrightarrow \Box q)$	By (4)
$\vdash \Box(\Box p \leftrightarrow \Box q) \to \Box(\blacksquare\Box p \leftrightarrow \blacksquare\Box q)$	
$\vdash \Box\big(\Box(\Box p \leftrightarrow \Box q) \to \Box(\blacksquare\Box p \leftrightarrow \blacksquare\Box q)\big)$	By necessitation

For the other direction, suppose that there exists an N-countermodel of a formula φ. By Lemma 5, this leads to a countermodel for $\varphi^\Box$ in NS4, as desired. $\square$

The next part of the section aims for a better understanding of the bi-modal logic NS4 by studying the behaviour of the modality $\blacksquare$ resulting from the subminimal negation. We hence focus now on proving the following main result.

Theorem 7. *The* $\{\wedge, \vee, \to, \blacksquare\}$-*fragment of* NS4 *is axiomatized by the following countably many rules:*

$$\frac{(\blacksquare p_1 \wedge \cdots \wedge \blacksquare p_n) \to (q \leftrightarrow r)}{(\blacksquare p_1 \wedge \cdots \wedge \blacksquare p_n) \to (\blacksquare q \leftrightarrow \blacksquare r)} \; R_n$$

for each $n \in \mathbb{N}$.

In what follows, we call the logic axiomatized by this countable set of rules $\mathsf{E}_\mathbb{N}$. This notation is justified by the fact that *classical modal logic* is the non-normal modal logic axiomatized by:

$$\frac{p \leftrightarrow q}{\Box p \leftrightarrow \Box q}$$

and is denoted by E. From the perspective of the N-semantics, the defining rule of E is expressed by the fact that N lifts to a well-defined function on sets.

First, we give a proof that each of these rules is derivable in NS4. For each rule, we provide a derivation of the conclusion from the premise in NS4.

Lemma 6. *The rule*

$$\frac{(\blacksquare p_1 \wedge \cdots \wedge \blacksquare p_n) \to (q \leftrightarrow r)}{(\blacksquare p_1 \wedge \cdots \wedge \blacksquare p_n) \to (\blacksquare q \leftrightarrow \blacksquare r)} \; R_n$$

is derivable in NS4, for each $n \in \mathbb{N}$.

Proof. The case $n = 0$ is clear. For $n > 0$, assume $(\blacksquare p_1 \wedge \cdots \wedge \blacksquare p_n) \to (q \leftrightarrow r)$. But then:

$\vdash \blacksquare p_i \to \square \blacksquare p_i$, for $i \in \{1, \ldots, n\}$ By axiom (6)

$\vdash (\blacksquare p_1 \wedge \cdots \wedge \blacksquare p_n) \to (\square \blacksquare p_1 \wedge \cdots \wedge \square \blacksquare p_n)$

$\vdash (\blacksquare p_1 \wedge \cdots \wedge \blacksquare p_n) \to \square(\blacksquare p_1 \wedge \cdots \wedge \blacksquare p_n)$ (a)

$\vdash \square(\blacksquare p_1 \wedge \cdots \wedge \blacksquare p_n) \to \square(q \leftrightarrow r)$ By necessitation

$\vdash \square(q \leftrightarrow r) \to (\blacksquare q \leftrightarrow \blacksquare r)$ By axiom (5)

$\vdash \square(\blacksquare p_1 \wedge \cdots \wedge \blacksquare p_n) \to (\blacksquare q \leftrightarrow \blacksquare r)$ (b)

$\vdash (\blacksquare p_1 \wedge \cdots \wedge \blacksquare p_n) \to (\blacksquare q \leftrightarrow \blacksquare r)$ By (a) and (b)

$\square$

As a consequence, we obtain the following partial result:

Lemma 7. *Given a $\{\wedge, \vee, \to, \blacksquare\}$-formula φ,*

$$\mathsf{E}_\mathbb{N} \vdash \varphi \implies \mathsf{NS4} \vdash \varphi.$$

In order to conclude the proof of Theorem 7 we take an $\mathsf{E}_\mathbb{N}$-countermodel for a $\{\wedge, \vee, \to, \blacksquare\}$-formula φ, and build from it an NS4-countermodel for φ. Therefore, we first provide a completeness result for the logic $\mathsf{E}_\mathbb{N}$.

Consider the modal N-frames $\langle W, N \rangle$ where, for every $X, Y, Z_1, \ldots, Z_n \in \mathcal{P}(W)$ and $n \in \mathbb{N}$, the function $N \colon \mathcal{P}(W) \to \mathcal{P}(W)$ satisfies (E_n):

$$N(X) \cap N(Z_1) \cap \cdots \cap N(Z_n) = N(X \cap N(Z_1) \cap \cdots \cap N(Z_n)) \cap N(Z_1) \cap \cdots \cap N(Z_n).$$

The corresponding notion of model is defined in the standard way. We next show that R_n is characterized by E_n.

Proposition 2. *For any $n \in \mathbb{N}$, a modal N-frame $\mathfrak{F} = \langle W, N \rangle$ satisfies E_n if and only if $\mathfrak{F} \models R_n$.*

Proof. From left to right, consider a frame $\mathfrak{F} = \langle W, \mathrm{N} \rangle$ satisfying E_n, and assume that for a given valuation V,

$$\langle \mathfrak{F}, V \rangle \models (\blacksquare p_1 \wedge \cdots \wedge \blacksquare p_n) \to (q \leftrightarrow r).$$

This means that:

$$V(q) \cap \mathrm{N}(V(\blacksquare p_1)) \cap \cdots \cap \mathrm{N}(V(\blacksquare p_n)) = V(r) \cap \mathrm{N}(V(\blacksquare p_1)) \cap \cdots \cap \mathrm{N}(V(\blacksquare p_n)).$$

Now, suppose $\langle \mathfrak{F}, V \rangle, w \models \blacksquare p_1 \wedge \cdots \wedge \blacksquare p_n$, that is,

$$w \in \mathrm{N}(V(p_1)) \cap \cdots \cap \mathrm{N}(V(p_n)).$$

It is now sufficient to prove that $w \models \blacksquare q$ if and only if $w \models \blacksquare r$. Indeed, without loss of generality, $w \in \mathrm{N}(V(q))$ entails

$$w \in \mathrm{N}(V(q)) \cap \mathrm{N}(V(p_1)) \cap \cdots \cap \mathrm{N}(V(p_n)).$$

Further, by E_n we get

$$w \in \mathrm{N}(V(q) \cap \mathrm{N}(V(p_1)) \cap \cdots \cap \mathrm{N}(V(p_n))) \cap \mathrm{N}(V(p_1)) \cap \cdots \cap \mathrm{N}(V(p_n))$$

that is equivalent to

$$w \in \mathrm{N}(V(r) \cap \mathrm{N}(V(p_1)) \cap \cdots \cap \mathrm{N}(V(p_n))) \cap \mathrm{N}(V(p_1)) \cap \cdots \cap \mathrm{N}(V(p_n))$$

by our assumption. Therefore, by E_n again, $w \in \mathrm{N}(V(r))$. It follows that $\mathfrak{F} \models \mathrm{R}_n$.

For the right-to-left direction, we assume $\mathfrak{F} \models \mathrm{R}_n$, and suppose that there exist subsets $X, Y, Z_1, \ldots, Z_n$ such that

$$\mathrm{N}(X) \cap \mathrm{N}(Z_1) \cap \cdots \cap \mathrm{N}(Z_n) \neq \mathrm{N}(X \cap \mathrm{N}(Z_1) \cap \cdots \cap \mathrm{N}(Z_n)) \cap \mathrm{N}(Z_1) \cap \cdots \cap \mathrm{N}(Z_n). \quad (9)$$

Set $V(p_i) = Z_i$ for $i \in \{1, \ldots, n\}$, $V(q) = X$, and $V(r) = X \cap \mathrm{N}(Z_1) \cap \cdots \cap \mathrm{N}(Z_n)$, for propositional variables p_i, q, r. But then, $V(\blacksquare p_i) = \mathrm{N}(Z_i)$, $V(\blacksquare q) = \mathrm{N}(X)$ and

$$V(\blacksquare r) = \mathrm{N}(X \cap \mathrm{N}(Z_1) \cap \cdots \cap \mathrm{N}(Z_n)).$$

Take any $w \in \mathrm{N}(Z_1) \cap \cdots \cap \mathrm{N}(Z_n)$. Obviously then $w \in X$ is equivalent to $w \in X \cap \mathrm{N}(Z_1) \cap \cdots \cap \mathrm{N}(Z_n)$, and hence, $(\blacksquare p_1 \wedge \cdots \wedge \blacksquare p_n) \to (q \leftrightarrow r)$ holds in $\mathfrak{F}$. But then, $(\blacksquare p_1 \wedge \cdots \wedge \blacksquare p_n) \to (\blacksquare q \leftrightarrow \blacksquare r)$ holds in $\mathfrak{F}$ as well by R_n, which contradicts (9). □

We conclude the completeness proof by means of a canonical model construction. Consider again the set $\mathcal{W}$ of maximally consistent sets of $\{\wedge, \vee, \to, \blacksquare\}$-formulas. For any $\Gamma \in \mathcal{W}$, set $\mathcal{R}(\Gamma) := \{\Delta \in \mathcal{W} \mid \forall \chi (\blacksquare \chi \in \Gamma \Rightarrow \blacksquare \chi \in \Delta)\}$ and define

$$\mathcal{N}(X) := \{\Gamma \in \mathcal{W} \mid (\exists \blacksquare \psi \in \Gamma)(X \cap \mathcal{R}(\Gamma) = |\psi| \cap \mathcal{R}(\Gamma))\}, \quad (10)$$

for any $X \in \mathcal{P}(\mathcal{W})$. We need to prove that $\langle \mathcal{W}, \mathcal{N} \rangle$ is an N-frame satisfying E_n for each $n \in \mathbb{N}$. Consider a sequence of subsets $X, Z_1, \ldots, Z_n \subseteq \mathcal{W}$, and take $\Gamma \in \mathcal{N}(Z_1) \cap \ldots \cap \mathcal{N}(Z_n)$. To show that E_n holds it will be sufficient to prove

$$\Gamma \in \mathcal{N}(X) \text{ if and only if } \Gamma \in \mathcal{N}(X \cap \mathcal{N}(Z_1) \cap \cdots \cap \mathcal{N}(Z_n)). \quad (11)$$

By (10), $\Gamma \in \mathcal{N}(Z_1) \cap \ldots \cap \mathcal{N}(Z_n)$ means that, for each $i \in \{1, \ldots, n\}$, there exists $\blacksquare \psi_i \in \Gamma$ such that

$$Z_i \cap \mathcal{R}(\Gamma) = |\psi_i| \cap \mathcal{R}(\Gamma). \tag{12}$$

To prove (11), it suffices to show that $\Delta \in X \cap \mathcal{R}(\Gamma)$ implies $\Delta \in \mathcal{N}(Z_i)$ for every $\Delta \in \mathcal{W}$ and every $i \in \{1, \ldots, n\}$. So, assume $\Delta \in X \cap \mathcal{R}(\Gamma)$. But $\Delta \in \mathcal{R}(\Gamma)$ means, by (12), that for each $i \in \{1, \ldots, n\}$, $\blacksquare \psi_i \in \Delta$. Moreover, $\mathcal{R}(\Delta) \subseteq \mathcal{R}(\Gamma)$. So

$$Z_i \cap \mathcal{R}(\Delta) = Z_i \cap \mathcal{R}(\Gamma) \cap \mathcal{R}(\Delta) = |\psi_i| \cap \mathcal{R}(\Gamma) \cap \mathcal{R}(\Delta) = |\psi_i| \cap \mathcal{R}(\Delta),$$

and hence, $\Delta \in \mathcal{N}(Z_i)$ for each $i \in \{1, \ldots, n\}$ by (12) again. By defining now the canonical valuation as $\mathcal{V}(p) = |p|$, it is standard to prove that $\varphi \in \Gamma$ if and only if $\mathcal{M}, \Gamma \models \varphi$, with $\mathcal{M} = \langle \mathcal{W}, \mathcal{N}, \mathcal{V} \rangle$. Therefore:

Proposition 3. *The logic* $\mathsf{E}_\mathbb{N}$ *is sound and complete with respect to the* N*-models satisfying* $\{\mathrm{E}_n : n \in \mathbb{N}\}$.

Now, we endow the canonical model of $\mathsf{E}_\mathbb{N}$ with a reflexive transitive relation so as to obtain an NS4-model. Set $\Gamma \leq \Delta$ if and only if $\Delta \in \mathcal{R}(\Gamma)$. It is clear that $\mathcal{N}(X) \in \mathcal{U}(\mathcal{W}, \leq)$, for any $X \in \mathcal{P}(\mathcal{W})$. It suffices to show that $\Gamma \in \mathcal{N}(X)$ is equivalent to $\Gamma \in \mathcal{N}(X \cap \mathcal{R}(\Gamma))$, for $\Gamma \in \mathcal{W}$. Indeed, we have $\Gamma \in \mathcal{N}(X \cap \mathcal{R}(\Gamma))$ if and only if $X \cap \mathcal{R}(\Gamma) \cap \mathcal{R}(\Gamma) = |\psi| \cap \mathcal{R}(\Gamma)$ for some $\blacksquare \psi \in \Gamma$, that is, $X \cap \mathcal{R}(\Gamma) = |\psi| \cap \mathcal{R}(\Gamma)$. Therefore, $\Gamma \in \mathcal{N}(X)$. The following is now immediate:

Lemma 8. *Given a* $\{\wedge, \vee, \rightarrow, \blacksquare\}$*-formula* φ,

$$\mathsf{NS4} \vdash \varphi \implies \mathsf{E}_\mathbb{N} \vdash \varphi.$$

At this point, Theorem 7 follows from Lemmas 7 and 8.

Theorem 7. *The* $\{\wedge, \vee, \rightarrow, \blacksquare\}$*-fragment of* NS4 *is axiomatized by the following countably many rules:*

$$\frac{(\blacksquare p_1 \wedge \cdots \wedge \blacksquare p_n) \rightarrow (q \leftrightarrow r)}{(\blacksquare p_1 \wedge \cdots \wedge \blacksquare p_n) \rightarrow (\blacksquare q \leftrightarrow \blacksquare r)} \ \mathrm{R}_n$$

for each $n \in \mathbb{N}$.

We conclude the section, and the whole article, with a brief sketch of how to obtain a modal companion for contraposition logic CoPC. We recall that it is defined by the axiom $(p \rightarrow q) \rightarrow (\neg q \rightarrow \neg p)$. We consider the bi-modal logic CoS4 obtained by replacing the axiom (5) by

$$\Box(p \rightarrow q) \rightarrow (\blacksquare q \rightarrow \blacksquare p). \tag{13}$$

The corresponding N-semantics is given by frames $\mathfrak{F} = \langle W, \leq, \mathrm{N} \rangle$, where $\langle W, \leq \rangle$ is a poset, and N is antitone, i.e., for $X, Y \in \mathcal{P}(W)$,

$$X \subseteq Y \implies \mathrm{N}(Y) \subseteq \mathrm{N}(X).$$

The logic CoPC can be translated into the system obtained by replacing (5) by (13) via the translation defined at the beginning of the section.

Theorem 8. *For every formula φ we have*

$$\mathsf{N} \vdash \varphi \Longleftrightarrow \mathsf{CoS4} \vdash (\varphi)^{\square}.$$

Proof. For the left-to-right direction, consider the translation of the axiom $(p \to q) \to (\neg q \to \neg p)$, i.e., $\square\big(\square(\square p \to \square q) \to \square(\blacksquare\square q \to \blacksquare\square p)\big)$. We have:

$\vdash \square(\square p \to \square q) \to (\blacksquare\square q \to \blacksquare\square p)$ By axiom (13)

$\vdash \square\square(\square p \to \square q) \to \square(\blacksquare\square q \to \blacksquare\square p)$ By necessitation

$\vdash \square(\square p \to \square q) \to \square\square(\square q \to \square p)$ By (4)

$\vdash \square(\square p \to \square q) \to \square(\blacksquare\square q \to \blacksquare\square p)$

$\vdash \square\big(\square(\square p \to \square q) \to \square(\blacksquare\square q \to \blacksquare\square p)\big)$ By necessitation

For the reverse direction, it suffices again to observe that from any N-frame for CoPC we can obtain an N-frame for CoS4. $\qquad\square$

We finish by highlighting a few further questions and research directions. First, it is reasonable to expect the $\{\wedge, \vee, \to, \blacksquare\}$-fragment of CoS4 to be axiomatized by the following countably many rules

$$\frac{(\blacksquare p_1 \wedge \cdots \wedge \blacksquare p_n) \to (q \to r)}{(\blacksquare p_1 \wedge \cdots \wedge \blacksquare p_n) \to (\blacksquare r \to \blacksquare q)}$$

for $n \in \mathbb{N}$, the argument being similar to the one for Theorem 7. This raises the question whether there exist finite axiomatizations for the $\{\wedge, \vee, \to, \blacksquare\}$-fragments of the considered bi-modal systems.

More generally, the notion and properties of bi-modal companions of subminimal logics naturally lead to the question whether the theory of modal companions of intermediate logics ([10, Section 9.6]) can be paralleled in this case. For example, do there always exist the least and greatest bi-modal companions of subminimal logics? Can one prove an analogue of the Blok-Esakia theorem in this setting? Also the study of the interplay between the two modalities could be taken further through a comparison with widely studied non-normal modal logics, such as classical modal logic E and monotonic modal logic M.

Finally, it will be interesting to investigate the relations between our semantics for the logic CoPC and the one proposed by Hazen [17].

Acknowledgments. We would like to thank the referees for their careful reading, which has helped us to clarify a number of issues and improve the paper.

References

1. van Benthem, J., Pacuit, E.: Dynamic logics of evidence-based beliefs. Studia Logica **99**(1–3), 61–92 (2011)

2. Bezhanishvili, G., Bezhanishvili, N.: An algebraic approach to filtrations for super-intuitionistic logics. In: Liber Amicorum Albert Visser. Tributes Series, vol. 30, pp. 47–56 (2016)
3. Bezhanishvili, G., Bezhanishvili, N.: Locally finite reducts of Heyting algebras and canonical formulas. Notre Dame J. Form. Log. **58**(1), 21–45 (2017)
4. Bezhanishvili, G., Moraschini, T., Raftery, J.G.: Epimorphisms in varieties of residuated structures. J. Algebra **492**, 185–211 (2017)
5. Bezhanishvili, N.: Lattices of intermediate and cylindric modal logics. Ph.D. thesis, ILLC Dissertations Series DS-2006-02, Universiteit van Amsterdam (2006)
6. Bezhanishvili, N., de Jongh, D., Tzimoulis, A., Zhao, Z.: Universal models for the positive fragment of intuitionistic logic. In: Hansen, H.H., Murray, S.E., Sadrzadeh, M., Zeevat, H. (eds.) TbiLLC 2015. LNCS, vol. 10148, pp. 229–250. Springer, Heidelberg (2017). https://doi.org/10.1007/978-3-662-54332-0_13
7. Bílková, M., Colacito, A.: Proof theory for positive logic with weak negation. Studia Logica (2019, to appear)
8. Blackburn, P., De Rijke, M., Venema, Y.: Modal Logic. Cambridge Tracts in Theoretical Computer Science, vol. 53. Cambridge University Press, Cambridge (2002)
9. Caicedo, X., Cignoli, R.: An algebraic approach to intuitionistic connectives. J. Symb. Log. **66**(04), 1620–1636 (2001)
10. Chagrov, A., Zakharyaschev, M.: Modal Logic. Oxford Logic Guides, vol. 35. The Clarendon Press, New York (1997)
11. Chellas, B.F.: Modal Logic: An Introduction. Cambridge University Press, Cambridge (1980)
12. Colacito, A.: Minimal and subminimal logic of negation. Master's thesis, ILLC Master of Logic Series Series MoL-2016-14, Universiteit van Amsterdam (2016)
13. Colacito, A., de Jongh, D., Vargas, A.L.: Subminimal negation. Soft Comput. **21**(1), 165–174 (2017)
14. Ertola, R.C., Galli, A., Sagastume, M.: Compatible functions in algebras associated to extensions of positive logic. Log. J. IGPL **15**(1), 109–119 (2007)
15. Hansen, H.H.: Monotonic modal logics. Master's thesis, ILLC Prepublication Series PP-2003-23, Universiteit van Amsterdam (2003)
16. Hazen, A.P.: Subminimal negation. University of Melbourne Philosophy Department, Preprint 1/92 (1992)
17. Hazen, A.P.: Is even minimal negation constructive? Analysis **55**, 105–107 (1995)
18. Humberstone, L.: The Connectives. MIT Press, Cambridge (2011)
19. Jankov, V.A.: The construction of a sequence of strongly independent superintuitionistic propositional calculi. Sov. Math. Dokl. **9**, 806–807 (1968)
20. Johansson, I.: Der Minimalkalkül, ein reduzierter intuitionistischer Formalismus. Compositio mathematica **4**, 119–136 (1937)
21. de Jongh, D.: Investigations on the intuitionistic propositional calculus. Ph.D. thesis, ILLC Historical Dissertation Series HDS-5, Universiteit van Amsterdam (1968)
22. Kolmogorov, A.N.: On the principle of excluded middle. Matematicheskii Sbornik **32**(24), 646–667 (1925)
23. Lemmon, E.J.: An introduction to modal logic: the lemmon notes. In: Collaboration with Scott, D., Segerberg, K. (eds.) American Philosophical Quarterly. Monograph Series, vol. 11. Basil Blackwell, Oxford (1977)
24. McKinsey, J.C.C.: A solution of the decision problem for the Lewis systems S2 and S4, with an application to topology. J. Symb. Logic **6**, 117–134 (1941)
25. McKinsey, J.C.C., Tarski, A.: The algebra of topology. Ann. Math. **54**, 141–191 (1944)

26. Odintsov, S.: Constructive Negations and Paraconsistency. Trends in Logic, vol. 26. Springer, Dordrecht (2008). https://doi.org/10.1007/978-1-4020-6867-6
27. Özgün, A.: Evidence in epistemic logic: a topological perspective. Ph.D. thesis, ILLC Dissertations Series DS-2017-07, Universiteit van Amsterdam (2017)
28. Pacuit, E.: Neighborhood Semantics for Modal Logic. Springer, Cham (2017). https://doi.org/10.1007/978-3-319-67149-9
29. Rasiowa, H.: An Algebraic Approach to Non-classical Logics. Studies in Logic and the Foundations of Mathematics. North-Holland Publishing Company, Amsterdam (1974)
30. Segerberg, K.: An Essay in Classical Modal Logic, vols. 1–3. Filosofiska Föreningen och Filosofiska Institutionen vid Uppsala Universitet (1971)

Finite Identification with Positive
and with Complete Data

Dick de Jongh and Ana Lucia Vargas-Sandoval[(✉)]

Institute for Logic, Language and Computation, University of Amsterdam,
Amsterdam, The Netherlands
D.H.J.deJongh@uva.nl, ana.varsa@gmail.com

Abstract. We study the differences between finite identifiability of recursive languages with positive and with complete data. In finite families the difference lies exactly in the fact that for positive identification the families need to be anti-chains, while in the infinite case it is less simple, being an anti-chain is no longer a sufficient condition. We also study maximal learnable families, identifiable families with no proper extension which can be identified. We show that these often though not always exist with positive identification, but that with complete data there are no maximal learnable families at all. We also investigate a conjecture of ours, namely that each positively identifiable family has either finitely many or uncountably many maximal noneffectively positively identifiable extensions. We verify this conjecture for the restricted case of families of equinumerous finite languages.

Keywords: Formal learning theory · Finite identification ·
Positive data · Complete data · Indexed family · Anti-chains

1 Introduction

The groundbreaking work of Gold [7] in 1967 started a new era for developing mathematical and computational frameworks for studying the formal process of *learning*. Gold's model, *identification in the limit*, has been studied for learning recursive functions, recursively enumerable languages, and recursive languages with *positive data* and with *complete data*. The learning task consists of identifying a language amidst a family of languages on the basis of an infinite stream of inputs concerning the language. The stream consists of either positive information: an enumeration of all members of the language, or complete information labeling all sentences as belonging to the language or not. The learning function will output infinitely many conjectures, and for a successful learning function these are required to stabilize into one permanent right one. In Gold's model, a huge difference in power between learning with positive data and with complete data is exposed. With positive data

The original version of this chapter was revised: One information in the acknowledgment was missing. This has now been rectified. The correction to this chapter is available at
https://doi.org/10.1007/978-3-662-59565-7_18

a family of languages containing all finite languages and at least one infinite one is not learnable. With complete data the learning task becomes almost trivial.

Based on Gold's model and results, Angluin's [1] work focuses on indexed families of recursive languages, i.e., families of decidable languages with a uniform decision procedure for membership. Such families naturally occur as the sets of languages generated by many types of grammars. In particular, Angluin [1] gave a characterization of the cases in which Gold's learning task can be executed. Her work shows that many non-trivial families of recursive languages can be learned by means of positive data only. A few years later, Mukouchi [12] (and simultaneously Lange and Zeugmann [10]) introduced the framework of *finite identification* in the Angluin style for both positive and complete data. The learning task is as in Gold's model with the difference that the learning function can only output a single conjecture. Mukouchi presents an Angluin style characterization theorem for positive and complete finite identification. As expected, finite identification with complete data is more powerful than with positive data only. However, the distinction is much less huge than in Gold's framework. Mukouchi's work didn't draw much attention until recently, when Gierasimczuk and de Jongh [5] further developed the theory of finite identification.

The difference between finite identification with positive and with complete data, if not as huge as in the limit case, is as we shown in this article, considerable not only in power but also in character. The motivation for studying learning in this formal way is no longer predominantly first language learning by children as it was for Gold. His motivation for concentrating on positive data was because of indication that children do not use negative data when they learn their native language [7], but this is no longer believed in general. A large amount of theoretical and experimental work in computational linguistics (see e.g. [11]) has been conducted to analyze and test the intuition that there is a powerful contribution of "negative" data for improving and speeding up children language acquisition (see e.g. [8,15]). Formal learning theory goes beyond its linguistic purpose for studying children's language acquisition. The proximity between dynamic epistemic (and doxastic) logic and formal learning theory has recently gained attention in attempts to jointly analyse learnability, scientific inquiry and information processing (see [6] for a detailed summary of this connection). In particular, the connection between finite identification and dynamic epistemic logic originates from the common interests that concern the process of *learning leading to knowledge*, such as solvability, success, Ockham's razor, learner, learning situations and learning methods, just to mention some. Dégremont and Gierasimczuk in [4] show that finite identification of languages (sets) can be modelled in dynamic epistemic logic via a suitable translation of finite identification's basic concepts (data stream, class of languages, and languages) into the semantics of dynamic epistemic logic and alternatively of epistemic temporal logic. The purpose for this translation is to comprehend more deeply the semantics of learning, as in formal learning theory, and to analyse its multiple dimensions, including ways of formalisation (see e.g. [2,3] for more results on this). By bringing both frameworks together we have the zooming-in step by step information changes from dynamic epistemic logic, together with the zooming-out of their long-term learning horizon as in finite identification. This combined study gives us for learning, criteria for choosing appropriate rules for updates, adequate learning conditions and a generic reasoning procedure. Moreover, regarding the step by step information changes of dynamic epistemic logic, the study of fastest learning in [5] gives us the indication that the effectiveness of the procedure of retrieving specific information, needs to be studied seriously. All these studies have been done with positive data only. It is less clear how the relation is with respect to complete data. Approaching various perspectives for the analysis of learning is crucial not only for understanding better these phenomena but for enriching learning procedures in artificial intelligence and automata theory. In order to develop further this connection, we first need a complete picture and a clear intuition of finite identification with positive data and with complete data.

In this work, we focus on a more fine-grained theoretical analysis of the distinction between finite identification with positive and with complete data in the Angluin style. Our aim is to formally study the issue of the difference in learning power stemming from, on the one hand, positive and, on the other hand, positive and negative data.

After a section with preliminaries we start, in the small Sect. 3, with finite identification of finite families. Here, the distinction between positive and complete data comes out very clearly: the difference is exactly described by the fact that with positive data families can only be identified if they are anti-chains with respect to the subset relation $\subset$.

Then, in Sect. 4, we investigate whether any finitely identifiable family is contained in a maximal finitely identifiable one. Maximal learnable families are of special interest because any learner which can learn a maximal learnable family can also learn any of its subfamilies. First, in Subsect. 4.1, we address this in the setting of positive data. Simple examples of positively identifiable families are often maximal, like the set of all sets of exactly n elements for a fixed natural number n. If we widen the question to the existence of a *non-effectively finitely identifiable* maximal extension for positively identifiable families, we get a positive answer for families containing only finite languages. We point out obstacles to generalising this result to arbitrary families containing also infinite languages, this wider question remains an open problem in general. Then we present a family of finite sets coded standardly by canonical codes which does not have an *effectively finitely identifiable* maximal extension, i.e., answering the above question negatively. In Subsect. 4.2, we come to study the complete data setting. Surprisingly, we provide a strong negative result concerning maximal learnable families for effective or non-effective finite identification with complete data: any finitely identifiable family can be extended to a larger one which is also finitely identifiable, *ergo* maximal identifiable families do not exist in the case of complete data.

After this, in Sect. 5, we address the question: how many maximal extensions a positively identifiable family has. We prove that any family containing only pairs of natural numbers has either only finitely many or uncountably many maximal non-efficient pfi extensions. We are able to extend this result to families containing only n-tuples of natural numbers for a fixed n, equinumerous families. With this we solve the conjecture we made that families of finite languages always have finitely many or uncountably many maximal non-efficient pfi extensions for this restricted case. The conjecture remains open for the general case, even for families of finite languages with bounded cardinality.

Finally, in Sect. 6, we focus on families which are anti-chains. We show that there exist infinite anti-chains of infinite languages which can be identified with complete information but not with positive information only. For infinite anti-chains of finite languages such an example cannot exist if the indexing of the languages is by canonical indices because such families are always positively identifiable. The case of arbitrary indexing is investigated, but not fully solved. We do exhibit an example of an indexable family of finite sets that cannot be given a canonical indexing.

2 Preliminaries

We use standard notions used in recursion theory and learning theory (see e.g., [9,13,14]), and in Angluin style identification in the limit (see [1,12]).

Since we can represent strings of symbols by natural numbers, we will always refer to $\mathbb{N}$ as our universal set. Thus, *languages* are sets of natural numbers, i.e., $L \subseteq \mathbb{N}$. A *family* $\mathcal{L} = \{L_i : i \in \mathbb{N}\}$ will be *an indexed family of non-empty recursive languages*, i.e., the two-placed predicate $y \in L_i$ is recursive. In case all languages are finite and there is a recursive function f such that for each i, $f(i)$ is a canonical index[1] for L_i i.e., $L_i = F_{f(i)}$, then we call $\mathcal{L}$ a *canonical family*.

In finite identification a *learner* will be a total recursive function that takes its values in $\mathbb{N} \cup \{\uparrow\}$ where $\uparrow$ stands for *undefined*.

A *positive data presentation* of a language L is an infinite sequence $\sigma^+ := x_0, x_1, \ldots$ of elements of $\mathbb{N}$ such that $\{x_0, x_1, \ldots\} = L$. A *complete data presentation* of a language L is an infinite sequence of pairs $\sigma := (x_0, t_0), (x_1, t_1), \ldots$ of $\mathbb{N} \times \{0, 1\}$ such that $\{x_n \in \mathbb{N} : t_n = 1, n \geq 0\} = L$ and $\{x_m \in \mathbb{N} : t_m = 0, m \geq 0\} = \mathbb{N} \setminus L$. An initial segment of length n of σ and σ^+ is indicated by $\sigma[n]$ and $\sigma^+[n]$, respectively.

A family $\mathcal{L}$ of languages is said to be *finitely identifiable from positive data (pfi)* (or *finitely identifiable from complete data (cfi)*) if there exists a recursive learner φ which satisfies the following: for any language L_i of $\mathcal{L}$ and for any positive data sequence σ^+ (or complete data sequence σ) of L_i as input to φ, φ produces on exactly one initial segment $\sigma^+[n]$ a guess $\varphi(\sigma^+[n]) = j$ such that $L_j = L_i$, and stops. We occasionally relax the condition of the recursivity of the learner φ, in such cases φ is said to be a *non-effective learner* and $\mathcal{L}$ is said to be *non-effectively finitely identifiable from positive data (non-effectively pfi)* (or *non-effectively finitely identifiable from complete data (non-effectively cfi)*). Clearly a family that is pfi is also *non-effectively pfi*. Similarly for cfi. For readability, we will use nepfi to refer to the notion of *non-effectively pfi*.

---

[1] We write F_n for the finite set with canonical index n.

Let $\mathcal{L}$ be a family of languages, and let L be a language in $\mathcal{L}$. A finite set D_L is a *definite tell-tale set* (DFTT) for L if $D_L \subseteq L$ and for all $L' \in \mathcal{L}$ if $D_L \subseteq L'$, then $L' = L$.

A language L' is said to be *consistent* with a pair of finite sets (B, C) if $B \subseteq L'$ and $C \subseteq \mathbb{N} \backslash L'$. A pair of finite sets $(D_L, \overline{D}_L)$ is a *definite tell-tale pair* for L if L is consistent with $(D_L, \overline{D}_L)$, and for all $L' \in \mathcal{L}$, if L' is consistent with $(D_L, \overline{D}_L)$, then $L' = L$. We will refer to D_L as the positive member of the tell-tale pair and to $\overline{D}_L$ as its negative member.

Theorem 1 (Mukouchi's Characterization Theorem [10,12]). *A family $\mathcal{L}$ of languages is finitely identifiable with positive data (pfi) iff for every $L \in \mathcal{L}$ there is a DFTT set D_L obtainable in a uniformly computable way. That is, there exists an effective procedure Φ that on input i, index of L, produces the canonical index $\Phi(i)$ of some definite finite tell-tale set of L.*

A family $\mathcal{L}$ of languages is finitely identifiable with complete data (cfi) iff for every $L \in \mathcal{L}$ there is a tell-tale pair $(D_L, \overline{D}_L)$ in a uniformly computable way.

Corollary 1. *If a family $\mathcal{L}$ has two languages such that $L_i \subset L_j$, then $\mathcal{L}$ is not pfi.*

Clearly if a family is pfi then it is cfi. A completely analogous theorem holds for non-effective learners and non-effective procedures for pfi and cfi.

Theorem 2. *If $\mathcal{L}$ is a canonical family where no $L_i \in \mathcal{L}$ is a proper subset of any other $L_j \in \mathcal{L}$, then $\mathcal{L}$ is pfi.*

Proof. For every $L_i \in \mathcal{L}$. Simply take $D_i = L_i$ as the DFTT.

Similarly, if $\mathcal{L}$ is any family of finite languages at all that is an anti-chain with respect to $\subset$, then $\mathcal{L}$ is non-effectively pfi.

3   Finite Families of Languages

This section is dedicated to finite families of languages. A pair of simple but striking results already provides a good insight in a feature underlying the difference between finite identification on positive and on complete data.

Theorem 3. *A finite family of languages $\mathcal{L}$ is finitely identifiable from positive data iff no language $L \in \mathcal{L}$ is a proper subset of another $L' \in \mathcal{L}$.*

Proof. From left to right follows straightforwardly by contraposition of Corollary 1. From right to left take any language L_i in $\mathcal{L}$. Since $L_i \nsubseteq L_j$ for any $j \neq i$, choose $n_{ij} = \mu n \{ n \in L_i \backslash L_j \}$ and let $D_i = \{ n_{ij} : j \neq i \}$. Let us verify that D_i is a DFTT for L_i: Clearly it is finite because the family is finite, so $\{ n_{ij} : j \neq i \}$ is finite and $D_i \subseteq L_i$. By construction, if $D_i \subseteq L_k \in \mathcal{L}$ then $i = k$.   □

Theorem 4. *Any finite collection of languages $\mathcal{L} = \{ L_1, \ldots, L_n \}$ is finitely identifiable with complete data.*

Proof. Let $\mathcal{L}$ be any finite family of languages and let L_i any language in $\mathcal{L}$. Take any $j \neq i$, then $L_i \setminus L_j \neq \emptyset$ or $L_j \setminus L_i \neq \emptyset$. If $L_i \setminus L_j \neq \emptyset$, take the smallest $n_{ij} \in L_i \setminus L_j$ to be in D_i. If $L_j \setminus L_i \neq \emptyset$, take the smallest $m_{ij} \in L_j \setminus L_i$ to be in $\overline{D}_i$. Repeat this for all $j \leq n$. The pair of sets obtained in that manner is consistent with L_i by construction, in fact they form a tell-tale pair for L_i. Note that this pair cannot be consistent with any other language $L_k \in \mathcal{L}$ such that $L_k \neq L_i$ simply by construction. Since i was arbitrary, by Mukouchi's characterization theorem for complete data we have that $\mathcal{L}$ is cfi.   □

4   Looking for Maximal Learnable Families

In this section we investigate whether any finitely identifiable family is contained in a finitely identifiable family which is maximal with respect to inclusion. We first address the question for maximal nepfi families and later for maximal pfi. We provide a positive result for maximal nepfi extensions of families with finite languages. For families containing infinite languages this question remains open. For the more common pfi families, maximal pfi extensions exist, but we do give an example of a canonical family which does not have a maximal pfi extension. The case of cfi is rather different, as we will show in Subsect. 4.2 that maximal extensions for cfi families never exist.

Theorem 5. *Every indexed family of finite languages which is pfi is contained in a maximal family of finite languages which is nepfi.*

Proving Theorem 5 is by a classical Zorn lemma argument because the union of a chain of pfi families of finite languages is again a family of finite languages. Moreover, if each family in the chain is pfi then it has to be an anti-chain, and therefore its union is an anti-chain as well. Then it will be nepfi because each language functions as its own DFTT. If infinite languages are present in the family, such an argument cannot be applied since not every family which is an anti-chain is nepfi (see Proposition 1 in Sect. 6).

4.1   The Existence of Maximal pfi Families

The following example shows that not every pfi family can be extended into a maximal pfi family. First we present the following standard definition of *recursively inseparable sets.*

Definition 1. We say that $A, B \subseteq \mathbb{N}$, $A \cap B = \emptyset$, are recursively inseparable iff there is no recursive set $C \subseteq \mathbb{N}$ such that ($C \supseteq A$ and $C \cap B = \emptyset$).

It is well-known that r.e. sets A, B exist which are recursively inseparable [16].

Theorem 6. *Let $A \subseteq \mathbb{N}$ and $B \subset \mathbb{N} \setminus A$ be two recursively inseparable r.e. sets. Let $\mathcal{L} := \{\{a\} : a \in A\} \cup \{\{b, c\} : b, c \in B \text{ with } b \neq c\}$. The family $\mathcal{L}$ is pfi and there is no canonical maximal pfi extension of $\mathcal{L}$.*

Proof. First note that since both A and B are r.e., $\mathcal{L}$ is a canonical indexed family. It is easy to see that it is pfi since any language serves as its own DFTT (as in the proof of Theorem 2). Now, towards a contradiction, suppose there is a maximal canonical pfi family extending $\mathcal{L}$, say $\mathcal{L}'$. Because of maximality and canonicity of $\mathcal{L}'$, we can decide for each finite set $Y \subseteq \mathbb{N}$, whether $Y \in \mathcal{L}'$ or $Y \notin \mathcal{L}'$. This can be done in the following manner. We first note that if $Y \notin \mathcal{L}'$ then $Y \subset L_i$ or $L_i \subset Y$ for some $L_i \in \mathcal{L}'$, otherwise Y can be added to $\mathcal{L}'$ as a new element without impairing positive identifiability, which would make $\mathcal{L}'$ non-maximal. To decide whether $Y \in \mathcal{L}'$ we simply run through $L_0, L_1, L_2, \ldots$ until we meet L_i which is Y itself, or a sub- or superset of Y. Since $\mathcal{L}'$ is decidable the set $A' \supseteq A$ of singletons in $\mathcal{L}'$ is recursive and $A' \cap B = \emptyset$, so A' separates A from B. This contradicts the inseparability of A and B. $\square$

We can strengthen Theorem 6 to conclude that $\mathcal{L}$ has no maximal pfi extension at all.

Theorem 7. *The family $\mathcal{L}$ of Theorem 6 has no maximal pfi extension at all.*

Proof. Let $\mathcal{L}' \supseteq \mathcal{L}$ be a maximal pfi family. We define recursive A', B', such that $A \subseteq A'$ and $B \subseteq B'$, $A' \cap B' = \emptyset$, $A' \cup B' = \mathbb{N}$. Let $\mathcal{L}' = \{L_n : n \in \mathbb{N}\}$. For each i determine whether $i \in A'$ or $i \in B'$ as follows. Since $\mathcal{L}'$ is indexed and maximal, we can find the first n such that $i \in L_n$. Now consider D_n the DFTT of L_n. We distinguish two possibilities: (1) $D_n = \{i\}$, and (2) $D_n \neq \{i\}$. In case (1) put $i \in A'$. Note that $A \subseteq A'$, because if $i \in A$ then $\{i\} \in \mathcal{L}$ and thus $\{i\} \in \mathcal{L}'$. Note that $\{i\} \subset L_n$ is impossible since $\mathcal{L}'$ is pfi. Thus $\{i\} = L_n$ and so $D_n = \{i\}$. In case (2) put $i \in B'$. Note that $B \subseteq B'$, because if $i \in B$ then $\{i, j\} \in \mathcal{L} \subseteq \mathcal{L}'$ for some $j \neq i$. So $D_n \neq \{i\}$ because $\mathcal{L}'$ is pfi. Therefore A', B' have been constructed as required, contradiction. $\square$

This theorem applies not only to families with infinite members but also to non-canonical families of only finite languages. In Theorem 11 from Sect. 6 we show that such families exist.

4.2   Do Maximal cfi Families Exist?

In this section we address the question whether every cfi family is contained in a maximal one. Surprisingly, we show that the answer is negative.

First observe the following. Let $\overline{\mathcal{L}}$ be the complement family of any cfi family $\mathcal{L}$, i.e., $\overline{\mathcal{L}} = \{\overline{L} : L \in \mathcal{L}\}$ where $\overline{L} = \mathbb{N} \setminus L$. Note that for every sequence σ of complete data for a family $\overline{\mathcal{L}}$ there is mirror image of σ, say sequence $\overline{\sigma}$ (presented in exactly the same order), for the cfi family $\mathcal{L}$ with inverted values of 0's and 1's. So $(k, 1)_j$ is in σ iff $(k, 0)_j$ is in $\overline{\sigma}$, for any $j \in \mathbb{N}$. We obtain the following result.

Proposition 1. *If a family $\mathcal{L}$ is cfi then $\overline{\mathcal{L}}$ is cfi as well.*

Proof. Let φ be a learner for $\mathcal{L}$ we can define a learner $\overline{\varphi}$ for $\overline{\mathcal{L}}$ as follows:

$$\overline{\varphi}(\sigma[n]) = \overline{L} \text{ iff } \varphi(\overline{\sigma}[n]) = L$$

Clearly $\overline{\varphi}$ is a recursive learner for $\overline{\mathcal{L}}$. □

Corollary 2. *If either $\mathcal{L}$ or $\overline{\mathcal{L}}$ is cfi then $\mathcal{L}$ and $\overline{\mathcal{L}}$ are cfi.*

This is not the case for pfi families, since for instance the family of all singletons $\mathcal{L}^s$ is pfi but its complement family, namely the family of all co-singletons, is clearly not pfi. This is because no finite subset of a co-singleton can determine which co-singleton it is.

Consider any language L, then a *direct successor* of L is $L \cup \{n\}$ with n not in L. For every non-cofinite language $L_i \subseteq \mathbb{N}$ let $Suc(L_i)$ be the set of all direct successors of L_i and $Suc_{\mathcal{L}}(L_i) = Suc(L_i) \cap \mathcal{L}$.

Proposition 2. *If $\mathcal{L}$ is cfi then $Suc_{\mathcal{L}}(L_i)$ is finite for every L_i in $\mathcal{L}$.*

Proof. Let $\mathcal{L}$ be a cfi family and $L \in \mathcal{L}$. Towards a contradiction, suppose there are infinitely many direct successors of L in $\mathcal{L}$. Thus, we assume that $Suc_{\mathcal{L}}(L)$ has infinitely many elements.

Since $\mathcal{L}$ is cfi we have a tell-tale pair for L, namely $(D_L, \overline{D}_L)$. First note that $D_L \subseteq L_i$ for any $L_i \in Suc(L)$. Since $\overline{D}_L$ is finite, the contradiction will follow by showing that $\overline{D}_L$ only serves to disambiguate between a finite number of direct successors of L in $\mathcal{L}$. We prove the following: $\overline{D}_L \cap L_j \neq \emptyset$ only for finitely many $L_j \in Suc_{\mathcal{L}}(L)$.

First note that $\overline{D}_L$ is finite and for all distinct $L_j, L_i \in Suc_{\mathcal{L}}(L)$, $L_j = L \cup \{k_j\} \neq L \cup \{k_i\} = L_i$ for some $k_j, k_i \in \mathbb{N}$. Since $\overline{D}_L$ is the negative member of the tell-tale pair for L, $\overline{D}_L \cap L = \emptyset$. Thus, if $L_j \in Suc_{\mathcal{L}}(L)$ and $\overline{D}_L \cap L_j \neq \emptyset$ then $\overline{D}_L \cap L_j = \{k_j\}$. Since for each $L_i, L_j \in Suc_{\mathcal{L}}(L)$ we have that $k_j \neq k_i$ and $\overline{D}_L$ is finite, $\overline{D}_L$ can only intersect finitely many $L_j \in Suc_{\mathcal{L}}(L)$.

Continuing with the general proof, take $L_i \in Suc_{\mathcal{L}}(L)$ such that $L_i \cap \overline{D}_L = \emptyset$. We can take such a language $L_i \in Suc_{\mathcal{L}}(L)$ because of the previous claim and our initial assumption that the set $Suc_{\mathcal{L}}(L)$ is infinite. Then L_i is a witness for showing that $(D_L, \overline{D}_L)$ is not a tell-tale pair for L which is a contradiction. This is because $D_L \subseteq L_i$ and $\overline{D}_L \cap L_i = \emptyset$, so disambiguation between L and L_i is not possible.

Since the choice of the tell-tale pair $(D_L, \overline{D}_L)$ was arbitrary it follows that $Suc_{\mathcal{L}}(L)$ must be finite. □

Next comes the crucial result, the non-existence of maximal cfi families.

Theorem 8. *Maximal cfi extensions do not exist for any cfi family $\mathcal{L}$.*

Proof. Let $\mathcal{L}$ be a cfi family. The strategy is to find a proper extension of $\mathcal{L}$ which is cfi and is such that it can always be extended in a similar way remaining cfi. W.l.o.g. we can assume that $\mathcal{L}$ has a non-cofinite language, and such language we fix as L. This is simply because of the following: if all languages in $\mathcal{L}$ are cofinite, then all languages in the complement family $\overline{\mathcal{L}}$ are finite. Thus we can find a cfi extension $\mathcal{L}'$ for $\overline{\mathcal{L}}$ by proving the theorem for $\overline{\mathcal{L}}$. By Proposition 1 we know that $\overline{\mathcal{L}'}$ is cfi. Note that $\overline{\mathcal{L}'}$ is an extension of our original family $\mathcal{L}$. Therefore, we can assume that $L \in \mathcal{L}$ is not cofinite. We proceed with the core of our proof. We start with proving the following claim.

Claim 1: For any tell-tale pair $(D_L, \overline{D}_L)$ of L, there exists $n \in \mathbb{N}$ such that $n \notin \overline{D}_L \cup L$.

Let $\overline{D}_L$ be the negative member in the tell-tale pair for L obtained by the tell-tale pair function f on $\mathcal{L}$. Since L is not cofinite we know that L has infinitely many direct successors. By Proposition 2 we have that $\mathcal{L}$ contains only finitely many direct successors of L. Note that since $\overline{D}_L$ is finite, we can choose infinitely many $m \in \mathbb{N}$ such that $m \notin \overline{D}_L$ and $m \notin L' \in Succ_{\mathcal{L}}(L)$. Take $n \in \mathbb{N}$ satisfying these characteristics and $D_L \cup \{n\}$ the respective direct successor for D_L. The set $D_L \cup \{n\}$ is precisely the candidate for extending $\mathcal{L}$.

Now, if n satisfies the claim, we show that the family $\mathcal{L}' = \mathcal{L} \cup \{D_L \cup \{n\}\}$ is cfi. In fact we claim that the finite sets $D'_{D_L \cup \{n\}} = D_L \cup \{n\}$, $\overline{D}'_{D_L \cup \{n\}} = \overline{D}_L$, and $D'_L = D_L$, $\overline{D}'_L = \overline{D}_L \cup \{n\}$ are tell-tale pairs for $D_L \cup \{n\}$ and L respectively in $\mathcal{L}'$. If $D_L \cup \{n\} = D_j \cup \{n'\}$ happens to be the case for some $L_j \in \mathcal{L}$, then fix $D'_j = D_j$ and $\overline{D}'_j = \overline{D}_j \cup \{m \in D_L \cup \{n\} : m \notin L_j\}$. For the rest of the languages in $\mathcal{L}'$ the tell-tale pairs will be exactly the ones chosen by the function f initially for $\mathcal{L}$, i.e, for all the rest of $L_i \in \mathcal{L}'$, $(D'_i, \overline{D}'_i) = (D_i, \overline{D}_i)$. To see this, note that by construction of $(D'_L, \overline{D}'_L)$, and $D_L \cup \{n\}$, the pair $(D'_L, \overline{D}'_L)$ cannot be consistent with $D_L \cup \{n\}$. It cannot be consistent with any other $L_j \in \mathcal{L}'$ either because that will contradict that $\mathcal{L}$ is cfi. Now we need to show that $D'_{D_L \cup \{n\}}$ and $\overline{D}'_{D_L \cup \{n\}}$ are not consistent with any other $L_j \in \mathcal{L}'$. By contradiction, suppose that the pair $(D'_{D_L \cup \{n\}}, \overline{D}'_{D_L \cup \{n\}})$ is consistent with a language $L_j \in \mathcal{L}'$ and $L_j \neq D_L \cup \{n\}$. Since $L_j \neq D_L \cup \{n\}$, by definition of $\mathcal{L}'$ we obtain $L_j \in \mathcal{L}$. By definition of $\mathcal{L}$ and since $L, L_j \in \mathcal{L}$ we have $L_j \neq L$. Thus,

$$D_L \subseteq D'_{D_L \cup \{n\}} \subseteq L_j$$

and

$$\overline{D}'_{D_L \cup \{n\}} = \overline{D}_L \subseteq \mathbb{N} \setminus L_j.$$

This implies that $(D_L, \overline{D}_L)$ was not a tell-tale pair for L w.r.t. $\mathcal{L}$, contradicting that $\mathcal{L}$ is cfi. Similarly we cannot have any pair $(D_i, \overline{D}_i)$ chosen by the function f consistent with any other $L_j \neq L_i \in \mathcal{L}'$. Therefore $\mathcal{L} \cup \{D_L \cup \{n\}\}$ is cfi. $\square$

There are other ways of extending a cfi family than the one described in Theorem 8 as the following example shows.

Example 1. *Take the family* $\mathcal{L} = \{\{0\}, \{0, 1\}, \{0, 1, 2\}, \ldots, \{0, 1, 2, 3, \ldots, n\}, \ldots\}$. *This family is cfi. Note that for* $L = \{0\}$ *we can extend* $\mathcal{L}$ *with* $L \cup \{2\}$ *and preserve cfi even though a tell-tale pair for* L *is* $(\{0\}, \{1, 2\})$. *Moreover we can extend it with* $L \cup \{3\}$, $L \cup \{4\}$ *and so on, and preserve cfi.*

5   Counting Maximal Extensions

We are also interested in the follow-up question: How many maximal nepfi extensions can a pfi family have? It is a conjecture of ours that every pfi or nepfi family has finitely many or uncountably many maximal nepfi extensions. In this section our work is more purely combinatorial, we are after structural properties only, so we ignore whether a family of languages is or can be represented as an indexed family, and similarly we are indifferent to whether we have a pfi or nepfi maximal extension of such a family.

First consider the following example: Let $\mathcal{L}^s$ be the family of all singletons. Clearly it is maximal with respect to nepfi. However, if we take out one of the singletons, say $\{0\}$, we obtain a nepfi subfamily $\mathcal{L}_0^s$ which is no longer maximal and its only nepfi extension is $\mathcal{L}^s$. If we remove $\{1\}$ from $\mathcal{L}_0^s$, we can maximally extend this family in two different ways, either adding $\{0,1\}$ or adding $\{0\}$ and $\{1\}$. Thus we have two independent maximal nepfi extensions for $\mathcal{L}_1^s$. We can repeat this effective deletion-procedure finitely many times and still obtain finitely many extensions. For regaining maximality, we are indeed "restricted" in the structural sense. The following lemma illustrates this.

Lemma 1. *Let $\mathcal{L}$ be a maximal nepfi family and let $\mathcal{L}'$ be a maximal nepfi extension of $\mathcal{L} \setminus \{\{x\}\}$ where $\{x\} \in \mathcal{L}$. Then for all $L \in \mathcal{L}'$ which are not in $\mathcal{L} \setminus \{\{x\}\}$, L is of the form $\{x\} \cup A$, for some $A \subseteq L_i \in \mathcal{L} \setminus \{\{x\}\}$.*

Proof. Note first that in order to achieve nepfi maximality in an extension $\mathcal{L}'$ of $\mathcal{L} \setminus \{\{x\}\}$, any new language $L \in \mathcal{L}'$ in the extension needs to have x as an element. Thus any $L \in \mathcal{L}' \setminus (\mathcal{L} \setminus \{\{x\}\})$ is such that $x \in L$. Let $A = L \setminus \{\{x\}\}$. We will prove that $A \subset L_i$ for some $i \in \mathbb{N}$. By maximality of $\mathcal{L}$, A itself could not be added to $\mathcal{L}$ and preserve nepfi. Thus, either $A \subset L_i$ or $L_i \subseteq A$ for some $L_i \in \mathcal{L}$. The latter cannot be since if $L_i \subseteq A$ then $L_i \subseteq A \cup \{x\}$ and $\mathcal{L}'$ should be an anti-chain. Therefore $A \subset L_i$.   □

However, in the following example we see that even when the languages are all finite, we may still regain uncountably many maximal pfi extensions.

Example 2. *Let $\mathcal{L} = \{\{0\} \cup \mathcal{L}'\}$ where $\mathcal{L}' = \{\{i,j,k\} : i,j,k \in \mathbb{N} \setminus \{0\}\}$. Clearly $\mathcal{L}$ is a maximal nepfi family. Consider $\mathcal{L}'$, by Lemma 1 in order to regain maximality, the languages to add must be of the form $\{0\} \cup A$ for some $A \subseteq L_i$ and some $L_i \in \mathcal{L}'$. Therefore we have the following procedure for constructing uncountably many maximal nepfi extensions of $\mathcal{L}'$: For each $B \subseteq \mathbb{N} \setminus \{0\}$ add the triples of the form $\{0,n,m\}$ with $n \neq m$ and $n,m \in B$ and all the pairs of the form $\{0,c\}$ with $c \notin B$. This construction is for all $B \subseteq \mathbb{N} \setminus \{0\}$, thus $\mathcal{L}'$ has uncountably many maximal nepfi extensions.*

5.1   Uncountably Many Maximal nepfi Extensions

We dedicate this subsection to study cases in which we can recover uncountably many maximal nepfi extensions of a given family. We first address some cases of families with finite languages similar to Example 2. After studying these cases, we exhibit some sufficient conditions for a family in order to have uncountably many maximal extensions.

In this section we name *pairs, triplets* and *n-tuples* the unordered sets with 2, 3 and n elements respectively. Consider the following example that uses the easy fact discussed in the next subsection that the family of all pairs, $\mathcal{L}^2$, is maximal nepfi.

Example 3. *Let $\mathcal{L}$ be the family $\{\{0\}, \{1\}\} \cup \{\{i,n\} : i,n \in \mathbb{N} \setminus \{0,1\}\}$. This is clearly a nepfi family because every language is mutually incomparable with any other language in the family. Moreover it is maximal (w.r.t. nepfi) precisely because any other subset of $\mathbb{N}$ is either a subset or a superset of $\{\{i,n\} : i,n \in \mathbb{N} \setminus \{0,1\}\}$, or a superset of $\{0\}, \{1\}$. Now consider the subfamily $\mathcal{L}' = \mathcal{L} \setminus \{\{0\}, \{1\}\}$, $\mathcal{L}'$ has uncountably many maximal nepfi extensions. Clearly $\mathcal{L}$ is one, and for every $B \subseteq \mathbb{N}$, the family $\mathcal{L}' \cup \{\{0,1,b\} : b \in B\} \cup \{\{0,c\}, \{1,c\} : c \notin B\}$ is a maximal nepfi extension of $\mathcal{L}'$. So we have uncountably many maximal nepfi extensions of $\mathcal{L}'$.*

However, if we take the similar maximal nepfi family $\{\{0\}\} \cup \{\{i,n\} : i,n \in \mathbb{N} \setminus \{0\}\}$ and consider the nepfi subfamily $\{\{i,n\} : i,n \in \mathbb{N} \setminus \{0\}\}$ it turns out that it has only two maximal nepfi extensions, namely $\mathcal{L}^2$ and $\{\{0\}\} \cup \{\{i,n\} : i,n \in \mathbb{N} \setminus \{0\}\}$ itself.

By a similar combinatorial argument as in Example 3, we straightforwardly obtain the following result.

Proposition 3. *For every finite set $\{0,1,\ldots,m\}$ with $m > 0$, the subfamily $\mathcal{L}^2_{\mathbb{N} \setminus \{0,1,\ldots,m\}} = \{\{i,n\} : i,n \in \mathbb{N} \setminus \{0,1,\ldots,m\}\}$, of the family of all pairs $\mathcal{L}^2$, has uncountably many maximal nepfi extensions.*

Proof. Simply because $\mathcal{L}^2_{\mathbb{N} \setminus \{0,1,\ldots,m\}} \subset \mathcal{L}^2 \setminus \{\{0,a\}, \{1,b\} : a,b \in \mathbb{N}\}$ and, by Example 3, the latter has uncountably many maximal nepfi extensions.   □

By Example 2 we know that the subfamily $\mathcal{L}^3 \setminus \{\{0,a,b\} : a,b \in \mathbb{N}\}$ already has uncountably many maximal nepfi extensions. Therefore any subfamily $\mathcal{L}^3_{\mathbb{N} \setminus \{0,1,\ldots,m\}}$ obtained by removing all triplets of the form $\{i,a,b\}$ with $a,b \in \mathbb{N}$ and $i \in \{0,1,\ldots,m\}$ has uncountably many maximal nepfi extensions for any $m \in \mathbb{N}$. Since a similar combinatorial argument works for any subfamily of quadruples, quintuples etc, we can generalize this result to all $n \geq 3 \in \mathbb{N}$.

Proposition 4. *Let $n \geq 3 \in \mathbb{N}$ and $\mathcal{L}^n$ be the class of all n-tuples. Any subfamily $\mathcal{L}^n_{\mathbb{N} \setminus \{0,1,\ldots,m\}}$ obtained by removing all n-tuples of the form $\{i, x_1, \ldots, x_{n-1}\}$ with $x_j \in \mathbb{N}$ and $i \in \{0,1,\ldots,m\}$ has uncountably many maximal nepfi extensions for any $m \in \mathbb{N}$.*

5.2   The Class of All Pairs $\mathcal{L}^2$

In this subsection we study subfamilies of the family of all pairs. This will also bring some general insights for equinumerous families with more than two elements. First we provide the following definition.

Definition 2

– Let $\mathcal{Y} = \{Y_1, \ldots Y_n\}$ be any set of pairs in $\mathcal{L}^2$, let $NUM(\mathcal{Y})$ to be the set of all numbers which appear in the pairs $Y_1, \ldots Y_n$, and let $PAIRS(\mathcal{Y})$ the set of all pairs formed by elements in $NUM(\mathcal{Y})$. Finally, let $\mathcal{L}^{\mathcal{Y}}$ be the subfamily of all pairs which are not in $PAIRS(\mathcal{Y})$, i.e.,, $\mathcal{L}^{\mathcal{Y}} = \mathcal{L}^2 \setminus PAIRS(\mathcal{Y})$.
– We can easily generalize the definition above to the family $\mathcal{L}^n$ of all n-tuples for $n \in \mathbb{N}$. We denote as $nTUP(\mathcal{Y})$ the set of all n-tuples formed by elements in $NUM(\mathcal{Y})$ and $\mathcal{L}^{\mathcal{Y}} = \mathcal{L}^n \setminus nTUP(\mathcal{Y})$.

The combinatorial notion of Sperner family explains why for every finite set of pairs $\mathcal{Y} = \{Y_1, \ldots Y_n\}$, the subfamily $\mathcal{L}^{\mathcal{Y}}$ has finitely many maximal nepfi extensions.

Definition 3. A *Sperner family* (or Sperner system) on A is a family of subsets of A in which none of the sets is contained in any other. Equivalently, a Sperner family is an *anti-chain* in the inclusion lattice over the power set of A.

From here on we will refer to Sperner families as anti-chains. The number of different anti-chains on a set of n elements is counted by the so-called Dedekind numbers. Determining these numbers is known as the Dedekind problem. The number of anti-chains on sets of n elements for $n \in \mathbb{N}$ are $2, 3, 6, 20, 168, 7581, \ldots$ respectively. Concretely, we have for a set with 0 elements the anti-chains $\emptyset$ and $\{\emptyset\}$.

Lemma 2. *Let $\mathcal{L}^{\mathcal{Y}} \subseteq \mathcal{L}^2$ be the family corresponding to some finite set of pairs $\mathcal{Y} = \{Y_1, \ldots Y_n\}$. For every maximal nepfi extension $\mathcal{L}$ of $\mathcal{L}^{\mathcal{Y}}$ and every $L \in (\mathcal{L} \setminus \mathcal{L}^{\mathcal{Y}})$, $L \subseteq NUM(\mathcal{Y})$.*

Proof. To obtain a contradiction suppose there is a maximal nepfi extension $\mathcal{L} \neq \mathcal{L}^2$ of $\mathcal{L}^{\mathcal{Y}}$ such that for some $L \in (\mathcal{L} \setminus \mathcal{L}^{\mathcal{Y}})$, $L \not\subseteq NUM(\mathcal{Y})$. Thus, there is $z \in L$ such that $z \notin NUM(\mathcal{Y})$. Note that L cannot be a singleton simply because $\{z\}$ is contained in infinitely many $\{z, w\} \in \mathcal{L}^{\mathcal{Y}} \subseteq \mathcal{L}$. Thus a $w \neq z$ exists in L such that $\{w, z\} \notin PAIRS(\mathcal{Y})$. Therefore $\{w, z\} \in \mathcal{L}^{\mathcal{Y}}$ simply by definition of $\mathcal{L}^{\mathcal{Y}}$. But since $\{w, z\} \subseteq L \in \mathcal{L}$, $\mathcal{L}$ cannot be nepfi extension of $\mathcal{L}^{\mathcal{Y}}$ contradicting our initial assumption. □

Proposition 5. *For every finite set of pairs $\mathcal{Y} = \{Y_1, \ldots Y_n\}$, the number of maximal nepfi extensions of the subfamily $\mathcal{L}^{\mathcal{Y}}$ is bounded by the Dedekind number of the set $NUM(\mathcal{Y}) = \{y_1, \ldots, y_m\}$ or in other words, by the number of anti-chains in $NUM(\mathcal{Y}) = \{y_1, \ldots, y_m\}$. Moreover, the maximal nepfi extensions of $\mathcal{L}^{\mathcal{Y}}$ correspond to the maximal singleton-free anti-chains on $NUM(\mathcal{Y})$.*

Proof. Let $\mathcal{Y} = \{Y_1, \ldots Y_n\}$ be any finite set of pairs and $\mathcal{L}^{\mathcal{Y}} \subseteq \mathcal{L}^2$ the corresponding family. By Lemma 2 we know that for every $\mathcal{L}$ maximal nepfi extension of $\mathcal{L}^{\mathcal{Y}}$, if $L \in (\mathcal{L} \setminus \mathcal{L}^{\mathcal{Y}})$ then $L \subseteq NUM(\mathcal{Y}) = \{y_1, \ldots, y_m\}$. Therefore, for every $\mathcal{L}$ maximal nepfi extension of $\mathcal{L}^{\mathcal{Y}}$ we have that $(\mathcal{L} \setminus \mathcal{L}^{\mathcal{Y}}) \subseteq \mathcal{P}(NUM(\mathcal{Y}))$ which is finite. By Mukouchi's corollary (Corollary 1) we know that $(\mathcal{L} \setminus \mathcal{L}^{\mathcal{Y}})$ must be an anti-chain in $\mathcal{P}(NUM(\mathcal{Y}))$. Therefore, every maximal nepfi extension $\mathcal{L}$ of $\mathcal{L}^{\mathcal{Y}}$ corresponds to some anti-chain in $\mathcal{P}(NUM(\mathcal{Y}))$ without singletons. Moreover, since $\mathcal{L} \supseteq \mathcal{L}^{\mathcal{Y}}$ is maximal nepfi, $(\mathcal{L} \setminus \mathcal{L}^{\mathcal{Y}})$ is precisely a maximal singleton-free anti-chain in $\mathcal{P}(NUM(\mathcal{Y}))$. For the other direction, if we extend $\mathcal{L}^{\mathcal{Y}}$ with any maximal singleton-free anti-chain in $\mathcal{P}(NUM(\mathcal{Y}))$ then clearly the resulting family $\mathcal{L}$ is a maximal nepfi extension. Simply because any $L \in (\mathcal{L} \setminus \mathcal{L}^{\mathcal{Y}})$ has L itself as a DFTT set.   □

We can straightforwardly generalize Proposition 5 for a subfamily of the family of all n-tuples, $\mathcal{L}^n$, for every $n \in \mathbb{N}$.

Proposition 6. *For every finite set of n-tuples $\mathcal{Y} = \{Y_1, \ldots Y_n\}$, the number of maximal nepfi extensions of the subfamily $\mathcal{L}^{\mathcal{Y}}$ is bounded by the number of anti-chains in the finite set $NUM(\mathcal{Y})$. Moreover, the maximal nepfi extensions of $\mathcal{L}^{\mathcal{Y}}$ correspond to the maximal anti-chains in $NUM(\mathcal{Y})$ and such anti-chains contain no k-cardinality sets for any $k \leq n - 1$.*

So far we know the following about subfamilies of $\mathcal{L}^2$: (1) By Proposition 5, any subfamily of $\mathcal{L}^2$ obtained by removing finitely many pairs from $\mathcal{L}^2$ has only finitely many maximal nepfi extensions; (2) by Example 3 and Proposition 3, we know that any subfamily of $\mathcal{L}^2$ obtained by removing all pairs of the form $\{i, n\}$ with $n \in \mathbb{N}$ and $i \in \{0, 1, \ldots, m\}$ (of which there are infinitely many) has either 2 maximal nepfi extensions (when $0 = m$) or uncountably many (when $0 < m$). But what happens when we consider subfamilies obtained by removing finitely or infinitely many arbitrary pairs? The answer to this question will also clarify what happens to subfamilies of all n-tuples $\mathcal{L}^n$ for any $n \in \mathbb{N}$.

In this section we will first study what happens when we remove from $\mathcal{L}^2$ any finite group of pairs. Then we will study the case of removing infinitely many pairs. We provide a complete overview of our investigation on the number of maximal nepfi extensions of subfamilies of $\mathcal{L}^2$ and will be able to conclude that every subfamily of $\mathcal{L}^2$ has either finitely or uncountably many maximal nepfi extensions.

Definition 4. We say that $\mathcal{G} \subseteq \mathcal{L}^2$ is a *cluster* in $\mathcal{L}^2$ if $PAIRS(\mathcal{G}) = \mathcal{G}$ (see Definition 2) and $\|\mathcal{G}\| > 1$.

Clearly for every $\mathcal{Y} \subseteq \mathcal{L}^2$, $PAIRS(\mathcal{Y})$ is a cluster in $\mathcal{L}^2$. The minimal-in-size clusters of $\mathcal{L}^2$ are the ones that contain three pairs.

Lemma 3. *For any finite set $\mathcal{Y} \subseteq \mathcal{L}^2$, $\mathcal{Y} \subseteq PAIRS(\mathcal{Y})$ and this is the minimal cluster that contains $\mathcal{Y}$.*

To illustrate the lemma above, let $\mathcal{Y} = \{\{1,2\}, \{2,3\}\}$. Then the minimal cluster that contains $\mathcal{Y}$ is $PAIRS(\mathcal{Y}) = \{\{1,2\}, \{1,3\}, \{2,3\}\}$. We have many finite clusters that contain $PAIRS(\mathcal{Y})$ and therefore $\mathcal{Y}$. For instance the cluster $\mathcal{G} = \{\{1,2\}, \{1,3\}, \{2,3\}, \{1,4\}, \{2,4\}, \{3,4\}\}$.

Proposition 7. *Let $\mathcal{G}_1, \ldots, \mathcal{G}_n$ be a finite set of clusters. Then the family*

$$\mathcal{L}^2 \setminus (\mathcal{G}_1 \cup \ldots \cup \mathcal{G}_n)$$

has finitely many maximal nepfi extensions.

Proof. Take $\mathcal{G} \supseteq \mathcal{G}_1 \cup \ldots \cup \mathcal{G}_n$ the minimal cluster that contains all $\mathcal{G}_1, \ldots, \mathcal{G}_n$, which exists by Lemma 3. By Proposition 5, $\mathcal{L}^{\mathcal{G}}$ has finitely many maximal nepfi extensions, and $\mathcal{L}^2 \setminus (\mathcal{G}_1 \cup \ldots \cup \mathcal{G}_n)$ as well, since $\mathcal{L}^2 \setminus (\mathcal{G}_1 \cup \ldots \cup \mathcal{G}_n) \supseteq \mathcal{L}^{\mathcal{G}}$. □

Definition 5. We say that a cluster $\mathcal{G} \subset \mathcal{L}^2$ is a maximal cluster outside $\mathcal{L} \subset \mathcal{L}^2$ if $\mathcal{G} \cap \mathcal{L} = \emptyset$ and for any cluster $\mathcal{G}' \supseteq \mathcal{G}$ outside $\mathcal{L}$ it holds that $\mathcal{G}' = \mathcal{G}$. We say that $\{\mathcal{G}_1, \ldots, \mathcal{G}_n\}$ is the greatest set of clusters outside $\mathcal{L}$, if it is the set of all maximal clusters outside $\mathcal{L}$.

The three results that follow address the cases when a subfamily of $\mathcal{L}^2$ has uncountably many maximal extensions.

Lemma 4. *If $\mathcal{L} \subseteq \mathcal{L}^2$ and an infinite cluster outside $\mathcal{L}$ exists then $\mathcal{L}$ has uncountably many maximal nepfi extensions.*

Proof. Let $\mathcal{G}$ be an infinite cluster outside $\mathcal{L}$, let $\{\{a_i, b_i\}\}_{i \in \mathbb{N}}$ be an enumeration of $\mathcal{G}$ and let $\{a_0, b_0\}, \{a_1, b_1\}$ be the first two elements in the enumeration of $\mathcal{G}$. By definition of cluster (Definition 4) we have that $\{a_0, a_1\}, \{a_0, b_1\}, \{b_0, a_1\}, \{b_0, b_1\}$ are also in $\mathcal{G}$. Thus they are outside $\mathcal{L}$. Therefore $\mathcal{L}$ can be extended in at least two different ways and the extensions remain nepfi, by adding $\{a_0, b_0\}, \{a_1, b_1\}$ or by adding $\{a_0, b_0, a_1\}, \{a_1, b_1, a_0\}$. Note that these two ways of extending $\mathcal{L}$ are mutually exclusive. We repeat this procedure on the resulting extension from the step before applied to the couple of elements $\{a_{2k}, b_{2k}\}, \{a_{2k+1}, b_{2k+1}\}$ in $\mathcal{G}$. By a well-known combinatorial argument, we have uncountably many maximal nepfi extensions for $\mathcal{L}$. □

Proposition 8. *Let $\mathcal{L} \subseteq \mathcal{L}^2$ and let there be no infinite clusters outside $\mathcal{L}$. If $\{\mathcal{G}_i\}_{i \in \mathbb{N}}$ is a countable sequence of disjoint clusters such that $\bigcup_{i=1}^{\infty} \mathcal{G}_i \subseteq (\mathcal{L}^2 \setminus \mathcal{L})$, or if for more than one $k \in \mathbb{N}$ we have that $\{\{k, m\} : m \in \mathbb{N} \setminus \{k\}\} \cap \mathcal{L} = \emptyset$, then $\mathcal{L}$ has uncountably many maximal nepfi extensions.*

Proof. First we will study the case when for more than one $k \in \mathbb{N}$ we have that $\{\{k, m\} : m \in \mathbb{N}\} \cap \mathcal{L} = \emptyset$. Note that we already proved in Example 3 that the subfamily $\mathcal{L}' = \{\{i, n\} : i, n \in \mathbb{N} \setminus \{0, 1\}\}$ of $\mathcal{L}^2$ has uncountably many maximal nepfi extensions. Therefore any subfamily of $\mathcal{L}'$ has uncountably many as well. Clearly every $\mathcal{L}$ satisfying the condition just mentioned will be isomorphic to a subfamily of $\mathcal{L}'$. Therefore $\mathcal{L}$ has uncountably many maximal nepfi extensions.

For the remaining case, let $\mathcal{G}_1, \ldots, \mathcal{G}_n, \ldots$ be a countable sequence of finite clusters such that $\bigcup_{i=1}^{\infty} \mathcal{G}_i \subseteq (\mathcal{L}^2 \setminus \mathcal{L})$, and suppose the clusters are pair-wise disjoint. For each $\mathcal{G}_i$ consider $NUM(G_i)$, the set of all numbers that appear in each pair of the cluster $\mathcal{G}_i$. Then we can extend $\mathcal{L}$ in two different ways: (1) by adding the cluster $PAIRS(G_i) = G_i$, and (2) by adding the set $NUM(G_i)$. Note that these two ways of extending $\mathcal{L}$ are mutually exclusive. Therefore, and since we have countably many clusters G_i, by a well-known combinatorial argument, we have uncountably many maximal nepfi extensions for $\mathcal{L}$. □

In the following result we address when a subfamily of $\mathcal{L}^2$ has finitely many maximal nepfi extensions.

Proposition 9. *Let $\mathcal{L} \subseteq \mathcal{L}^2$ and let there be no infinite clusters outside $\mathcal{L}$. If the greatest set of clusters $\{\mathcal{G}_1, \ldots, \mathcal{G}_n\}$ outside $\mathcal{L}$ is a finite set and for at most one $k \in \mathbb{N}$ $\{\{k, m\} : m \in \mathbb{N} \setminus \{k\}\} \cap \mathcal{L} = \emptyset$, then $\mathcal{L}$ has finitely many maximal nepfi extensions.*

Proof. Let the greatest set of clusters outside $\mathcal{L}$ be the finite set $\{\mathcal{G}_1, \ldots, \mathcal{G}_n\}$ and for at most one $k \in \mathbb{N}$, $\{\{k, m\} : m \in \mathbb{N} \setminus \{k\}\} \cap \mathcal{L} = \emptyset$. By treating the following cases we exhaust all the possibilities.

(a) There are no clusters outside $\mathcal{L}$ contained in $(\mathcal{L}^2 \setminus \mathcal{L})$ and for exactly one $k \in \mathbb{N}$, $\{\{k, m\} : m \in \mathbb{N} \setminus \{k\}\} \cap \mathcal{L} = \emptyset$.
(b) There are no clusters contained in $(\mathcal{L}^2 \setminus \mathcal{L})$ and for no $k \in \mathbb{N}$, $\{\{k, m\} : m \in \mathbb{N} \setminus \{k\}\} \cap \mathcal{L} = \emptyset$, i.e., there are no clusters contained in $(\mathcal{L}^2 \setminus \mathcal{L})$ and for all $k \in \mathbb{N}$, $\{\{k, m\} : m \in \mathbb{N} \setminus \{k\}\} \cap \mathcal{L} \neq \emptyset$.
(c) $n \geq 1$ and for at most one $k \in \mathbb{N}$, $\{\{k, m\} : m \in \mathbb{N} \setminus \{k\}\} \cap \mathcal{L} = \emptyset$.

Case (a): Note that each pair $\{a, b\} \in (\mathcal{L}^2 \setminus \mathcal{L})$ is of the form $\{k, m\}$ for some $m \in \mathbb{N}$. This is because otherwise, the pairs $\{a, b\}, \{k, a\}, \{k, b\}$ would form a cluster contained in $(\mathcal{L}^2 \setminus \mathcal{L})$. Therefore the only possibility is that $(\mathcal{L}^2 \setminus \mathcal{L}) = \{\{k, m\} : m \in \mathbb{N} \setminus \{k\}\}$ and we already discussed this case in Example 3. So in this case $\mathcal{L}$ has only two possible maximal extensions.

Case (b): Note that since there are no clusters contained in the complement of $\mathcal{L}$, we cannot add any language larger than a pair. To see this, take any triple of elements in $\mathbb{N}$, $\{a, b, c\}$. By our assumption, the cluster $\mathcal{G} = \{\{a, b\}, \{b, c\}, \{a, c\}\}$ is not contained in $\mathcal{L}^2 \setminus \mathcal{L}$. Therefore one of these pairs is already in $\mathcal{L}$. Thus we cannot add the language $\{a, b, c\}$ to extend $\mathcal{L}$. The same reasoning applies to any set larger than a triple. Since for all $k \in \mathbb{N}$, $\{\{k, m\} : m \in \mathbb{N} \setminus \{k\}\} \cap \mathcal{L} \neq \emptyset$, we cannot add singletons either: for every $i \in \mathbb{N}$ there is a language $L_i := \{i, m\} \in \mathcal{L}$ such that $\{i\} \subseteq \{i, m\}$ which prevents nepfi. Therefore the only maximal nepfi extension is $\mathcal{L}^2$.

Case (c): Let $\mathcal{G} = \{\mathcal{G}_1, \ldots, \mathcal{G}_m\}$ be the greatest set of clusters outside $\mathcal{L}$, i.e., $(\bigcup_{i=1}^{n} \mathcal{G}_i) \subseteq (\mathcal{L}^2 \setminus \mathcal{L})$, and let $k \in \mathbb{N}$ be such that $\{\{k, m\} : m \in \mathbb{N} \setminus \{k\}\} \cap \mathcal{L} = \emptyset$. The strategy of the proof is to show that each maximal extension of $\mathcal{L}$ is uniquely characterized by some maximal anti-chain in $NUM(\bigcup_{i=1}^{m} \mathcal{G}_i) \cup \{k\}$. This is sufficient because the number of maximal anti-chains is bounded by the Dedekind number of $NUM(\bigcup_{i=1}^{m} \mathcal{G}_i) \cup \{k\}$, which is finite. In order to achieve this we need to prove the following claim:

Claim: For any maximal nepfi extension $\mathcal{L}_m$ of $\mathcal{L}$ and any $A \in \mathcal{L}_m \setminus \mathcal{L}$ we have that either $A \subseteq NUM(\bigcup \mathcal{G}) \cup \{k\}$ or $A = \{m, n\}$ for some $\{m, n\} \notin \mathcal{L}$.

We prove this by contradiction. Suppose there is $\mathcal{L}_m \supseteq \mathcal{L}$ such that $\mathcal{L}_m \neq \mathcal{L}^2$ and $A \in \mathcal{L}_m$ is such that $A \not\subseteq NUM(\bigcup \mathcal{G}) \cup \{k\}$ and $A \neq \{m, n\}$ for any $m, n \in \mathbb{N}$. Since $A \not\subseteq NUM(\bigcup \mathcal{G}) \cup \{k\}$, there is $y \in A$ such that $y \notin NUM(\bigcup \mathcal{G}) \cup \{k\}$. Therefore $y \neq k$. Note that A cannot be a singleton, say

$\{y\}$, simply because $\{y\} \subseteq \{y, m+1\} \in \mathcal{L}$ where $m = max\{NUM(\mathcal{G}) \cup \{k\}\}$. Thus, there is $z \neq y$ such that $\{z, y\} \subseteq A$. Note that $A \neq \{z, y\}$ since we are supposing $A \neq \{m, n\}$ for any $\{m, n\} \notin \mathcal{L}$. Thus we have that $x \in A$ exists with $x \neq z, y$. Note that if $PAIRS(\{x, z, y\}) \subseteq (\mathcal{L}^2 \setminus \mathcal{L})$ then $PAIRS(\{x, z, y\}) \subseteq \mathcal{G}_i$ for some $i \in \{1, \ldots, m\}$, but this cannot be since $y \notin NUM(\bigcup \mathcal{G}) \cup \{k\}$. Therefore $PAIRS(\{x, z, y\}) \cap \mathcal{L} \neq \emptyset$, i.e., there is a pair $\{a, b\} \subseteq \{x, z, y\} \subseteq A$ such that $\{a, b\} \in \mathcal{L}$, but this contradicts that $A \in \mathcal{L}_m \setminus \mathcal{L}$ where $\mathcal{L}_m$ is a maximal nepfi extension of $\mathcal{L}$.

We have proved the claim and have thereby shown that each maximal nepfi extension of $\mathcal{L}$ which is not $\mathcal{L}^2$ is characterised by some anti-chain of $NUM(\bigcup \mathcal{G}) \cup \{k\}$ of which there are just finitely many. $\square$

By Proposition 7, Lemma 4, Propositions 8 and 9 we have the following result.

Theorem 9. *Any subfamily $\mathcal{L}$ of $\mathcal{L}^2$ has either finitely many maximal nepfi extensions or uncountably many.*

Trivially, every $\mathcal{L} \subseteq \mathcal{L}^2$ has an indexable maximal effective nepfi extension, namely $\mathcal{L}^2$ itself, but one can also see that, if there are finitely many nepfi maximal extensions all of them are indexable maximal effective pfi extensions and if there are uncountably many, countably many of those are pfi.

Theorem 9 for $\mathcal{L}^2$ allows us to obtain rather straightforwardly a similar general result for subclasses of the family of all n-tuples $\mathcal{L}^n$ for any $n \in \mathbb{N}$. But there are some subtle details so that we need to treat carefully. Therefore we dedicate the following section to the class of all subfamilies of $\mathcal{L}^n$.

5.3   The Class of All n-Tuples $\mathcal{L}^n$

Here we generalize all the notions and results we obtained for $\mathcal{L}^2$.

Definition 6. We say that $\mathcal{G} \subseteq \mathcal{L}^n$ is an n-cluster in $\mathcal{L}^n$ if the set $nTUP(\mathcal{G})$ of all n-tuples formed by numbers in $NUM(\mathcal{G})$ (see Definition 2) is exactly $\mathcal{G}$, i.e., if $nTUP(\mathcal{G}) = \mathcal{G}$.

Clearly for every $\mathcal{Y} \subseteq \mathcal{L}^n$, $nTUP(\mathcal{Y})$ is an n-cluster in $\mathcal{L}^n$.

Lemma 5. *For any finite set $\mathcal{Y} \subseteq \mathcal{L}^n$, $\mathcal{Y} \subseteq nTUP(\mathcal{Y})$, and this is the minimal n-cluster that contains $\mathcal{Y}$.*

Proposition 10. *Let $\mathcal{G}_1, \ldots, \mathcal{G}_m$ be a finite set of n-clusters. Then the family $\mathcal{L}^n \setminus (\mathcal{G}_1 \cup \ldots \cup \mathcal{G}_m)$ has finitely many maximal nepfi extensions.*

Proof. The proof goes as in the case for $\mathcal{L}^2$, taking the minimal n-cluster that contains all $\mathcal{G}_1, \ldots, \mathcal{G}_n$, which by Lemma 5 we know exists. By Proposition 6 we know that $\mathcal{L}^\mathcal{G}$ has finitely many maximal nepfi extensions, and $\mathcal{L}^n \setminus (\mathcal{G}_1 \cup \ldots \cup \mathcal{G}_m) \supseteq \mathcal{L}^\mathcal{G}$, so $\mathcal{L}^n \setminus (\mathcal{G}_1 \cup \ldots \cup \mathcal{G}_m)$ has finitely many as well. $\square$

In what follows we will consider the subclasses of $\mathcal{L}^3$. This clarifies straight-forwardly what happens in the general case for every $n > 3$.

What is different in $\mathcal{L}^3 \setminus \{\{0, a, b\} : a, b \in \mathbb{N}\}$ that it allows for uncountably many maximal nepfi extensions, whereas $\mathcal{L}^3 \setminus \{\{0, 1, b\} : b \in \mathbb{N}\}$ does not? The difference lies in the elements that are fixed and the ones that remain 'free' in the triples that are discarded from the families. Whenever we fix two elements in the triples we are preventing the combinatorics to act out, since with only one 'free' element in the triple, there is not much that combinatorics can do. However, with two non-fixed entries we can build uncountably many nepfi extensions as in case (4) of Example 2.

The following proposition generalizes what happens in Example 2.

Proposition 11. Let $n \geq 3$ and let $\{a_0, a_1, \ldots, a_k\}$ be a fixed k-tuple of elements in $\mathbb{N}$ for some $k \leq n - 2$.

1. If $k \leq n - 3$, the family $\mathcal{L}^n \setminus \{\{a_0, a_1, \ldots, a_k, x_{k+1}, \ldots, x_{n-1}\} \in \mathcal{L}^n : x_i \in \mathbb{N} \setminus \{a_1, \ldots, a_k\}\}$ has uncountably many maximal nepfi extensions.
2. The family $\mathcal{L}^n \setminus \bigcup_{i=1}^m \{\{a_{i,0}, a_{i,1}, \ldots, a_{i,n-2}, b\} \in \mathcal{L}^m : b \in \mathbb{N} \setminus \{a_{i,0}, a_{i,1}, \ldots, a_{i,n-2}\}\}$ for finitely many $(n-2)$-tuples $\{a_{i,0}, a_{i,1}, \ldots, a_{i,n-2}\}$ in $\mathcal{L}^{n-2}$ (note that $k = n-2$) has finitely many maximal nepfi extensions.

As we mentioned before, the proof for the generalization of Theorem 9 needs to be treated carefully since there are cases that do not correspond exactly to the ones for $n = 2$. For instance in the proof of the following Lemma and Propositions for $n = 3$, there are more cases of a similar kind in which the subfamily has uncountably many maximal extensions. The general proof for $n \geq 3$ is basically the same as for $n = 3$. Due to lack of space we do not provide the proof of the following Lemma and Propositions here, the proof for $n = 3$ can be found in an extended version of our paper located in https://analuciavargassan.com/page/.

The following Lemma is the corresponding counterpart of Lemma 4 but for subfamilies of $\mathcal{L}^n$ for a fixed n.

Lemma 6. Let $\mathcal{L} \subseteq \mathcal{L}^n$ and let $\mathcal{L}^n \setminus \mathcal{L}$ contain an infinite n-cluster. Then $\mathcal{L}$ has uncountably many maximal nepfi extensions.

The following Propositions are the suitable generalisations of Propositions 8 and 9 respectively.

Proposition 12. Let $\mathcal{L} \subseteq \mathcal{L}^n$ and let there be no infinite n-cluster in $\mathcal{L}^n \setminus \mathcal{L}$. If $\mathcal{L}$ satisfies one of the following cases:

(i) there is an infinite sequence of n-clusters $\{\mathcal{G}_i\}_{i \in \mathbb{N}}$ such that $(\bigcup_{i=1}^\infty \mathcal{G}_i) \subseteq \mathcal{L}^n \setminus \mathcal{L}$, or

(ii) for infinitely many tuples $\{a_0, a_1, \ldots, a_{n-2}\} \in \mathcal{L}^{n-2}$ is that
$\{\{a_0, a_1, \ldots, a_{n-2}, x\} \in \mathcal{L}^n : x \in \mathbb{N} \setminus \{a_0, \ldots, a_{n-2}\}\} \cap \mathcal{L} = \emptyset$, or

(iii) for some $m \leq n-3$ and some tuple $\{a_0, a_1, \ldots, a_m\} \in \mathcal{L}^m$ it is the case that
$\{\{a_0, a_1, \ldots, a_m, x_{m+1}, \ldots, x_{n-1}\} \in \mathcal{L}^n : x \in \mathbb{N} \setminus \{a_0, \ldots, a_m\}\} \cap \mathcal{L} = \emptyset$,

then $\mathcal{L}$ has uncountably many maximal nepfi extensions.

Proposition 13. *Let $\mathcal{L} \subseteq \mathcal{L}^n$ and let there be no infinite n-cluster in $\mathcal{L}^n \setminus \mathcal{L}$. If $\mathcal{L}$ satisfies the following,*

(a) *there are at most finitely many n-clusters $\mathcal{G}_1, \ldots, \mathcal{G}_N$ such that $(\bigcup_{i=1}^{N} \mathcal{G}_i) \subseteq \mathcal{L}^n \setminus \mathcal{L}$, and*

(b) *for at most finitely many tuples $\{a_0, a_1, \ldots, a_{n-2}\} \in \mathcal{L}^{n-2}$ is that*
$$\{\{a_0, a_1, \ldots, a_{n-2}, x\} \in \mathcal{L}^n : x \in \mathbb{N} \setminus \{a_0, \ldots, a_{n-2}\}\} \cap \mathcal{L} = \emptyset, \text{ and}$$

(c) *for all $m \leq n - 3$ and tuples $\{a_0, a_1, \ldots, a_m\} \in \mathcal{L}^m$ it is the case that*
$$\{\{a_0, a_1, \ldots, a_m, x_{m+1}, \ldots, x_{n-1}\} \in \mathcal{L}^n : x \in \mathbb{N} \setminus \{a_0, \ldots, a_m\}\} \cap \mathcal{L} \neq \emptyset,$$

then $\mathcal{L}$ has finitely many maximal nepfi extensions.

By Propositions 10, 11, Lemma 6, Propositions 12 and 13 we have the following generalisation of Theorem 9.

Theorem 10. *Let $n \in \mathbb{N}$. Any subfamily $\mathcal{L}$ of the family of all n-tuples $\mathcal{L}^n$ has either finitely many maximal nepfi extensions or uncountably many.*

6   Infinite Anti-chains

Contrary to the results of Sect. 3, cfi identification is more powerful on infinite anti-chains of infinite languages than pfi identification.

Proposition 1. *The family of all co-singletons, $\{\mathbb{N} \setminus \{i\} : i \in \mathbb{N}\}$, is an anti-chain which is cfi but not pfi.*

The family of co-singletons is also not even nepfi, simply because DFTT's do not exist. The case of infinite anti-chains of finite languages is less clear. It is a trivial fact that canonical families which are anti-chains are always pfi. That families of finite sets which are not canonical do exist is already mentioned in [5] (a simple example is the family $\mathcal{L} = \{L_i : L_i = \{i\} \cup \{y : Tiiy\}\}$ where T is Kleene's T-predicate and which is not pfi simply because is not an anti-chain). By following a diagonalization strategy, we can construct a non-canonical family (but still indexable) of finite languages which is an anti-chain but is not pfi.

Theorem 11. *There is a family $\mathcal{L} = \{L_i : i \in \mathbb{N}\}$ of finite languages which is an anti-chain and for which there is no canonically indexed anti-chain $\{D_{f(n)} : n \in \omega\}$ such that $D_{f(i)} \subseteq L_i$ for all $i \in \mathbb{N}$ and $D_{f(i)} \not\subseteq L_j$ for all $j \neq i$, i.e., this family is not pfi.*

Proof. The strategy of the construction is by diagonalization. We diagonalize against all r.e. families of canonical finite sets, that is, families of the form break $\{F_e : e \in B\}$ with B r.e. Note that this includes all the families of the form $\{F_{f(n)} : n \in \mathbb{N}\}$ with f computable simply by definition of computable languages and r.e. languages. We will abuse our notation a bit and refer to the canonical families as $\{D_e : e \in B\}$ instead. Thus, we construct a uniformly computable family of finite languages $\{L_i : i \in \omega\}$ such that the following requirement is satisfied for each e:

(R_e): If $\{D_n : n \in W_e\}$ is an anti-chain

then the following does not hold: (1)

$D_i \subseteq L_i$ for all $i \in \mathbb{N}$, and $D_i \not\subseteq L_j$ for all $j \neq i$,

where, as usual, W_e denotes the e-th r.e. set. The strategy for meeting the requirement (R_e) is by making W_e fail on e. The idea is to let $L_e = \{e\}$ until, if ever, we see a canonical code n with $D_n = \{e\}$ appear in W_e, and then make sure that two incomparable extensions of $\{e\}$ occur in $\mathcal{L}$. To be able to construct these two incomparable extensions we use an infinite strictly increasing recursive sequence $\mathcal{A} = a_0, a_1 a_2, \ldots$ of indices of the empty set. The members of $\mathcal{A}$ itself are of course harmless because they produce no non-empty set at all.

We let $L_e = \{e\}$ until, if ever, we see a canonical code n with $D_n = \{e\}$ appear in W_e (R_e *requires attention*). In that stage s we change L_e to be $\{e, a_{2s}\}$. Moreover, we force $L_{a_{2s}}$ to be $\{a_{2s}, a_{2s+1}\}$ and $L_{a_{2s+1}}$ to be $\{a_{2s+1}, e\}$ in order to satisfy the anti-chain condition. In this manner we ensure that $L_i = \{e\}$ can only happen if $i = e$, and that for every stage $s' < s$, all $L_{a_{2s'}}, L_{a_{2s'+1}}$ have been already established for $i \leq a_{2s}$. Also L_e is changed in a way such that $D_e \subset L_e$ and $D_e \subset L_{a_{2s+1}}$, for $a_{2s+1} \in \mathcal{A}$. Thus (R_e) is guaranteed since in later stages we will not change L_e again.

Now the explicit stage-wise construction. We assume that at stage s, we have already determined whether $x \in L_i$ for every $i < a_{2s}$ and $x < a_{2s}$. Let $e < a_{2s}$ be the least number such that (R_e) requires attention. As indicated above e can never be a member of $\mathcal{A}$. We put $e, a_{2s} \in L_e$, $a_{2s}, a_{2s+1} \in L_{a_{2s}}$ and $a_{2s+1}, e \in L_{a_{2s+1}}$. Thereby, we can be sure (R_e) will be satisfied. For all other $j < a_{2s}$ we only put $j \in L_j$. For all $s' < s$, $L_{a_{2s'}}, L_{a_{2s'+1}}$ have already been established and for all $i < a_{2s}$ with $i \neq e$ we have $a_{2s}, a_{2s+1} \notin L_i$. We execute this procedure for any given e in one stage only. If there is no $e < a_{2s}$ such that (R_e) requires attention, then we do the latter for all $i < a_{2s+1}$. Indeed, we have now determined once and for all whether $x \in L_i$ for all $x, i \leq a_{2(s+1)}$. This construction defines a uniformly computable anti-chain of finite languages, since at every stage s, whether $x \in L_i$ is effectively determined for all $x, i \leq a_{2s+1}$ and every L_e for $e \notin \mathcal{A}$ is either $\{e\}$ or $\{e, a_{2s}\}$. Clearly it is an anti-chain because whenever L_e is a singleton, it has an empty intersection with the rest of the languages. And, whenever L_e is a pair, say $\{e, a_{2s}\}$, by construction there are only two other languages $L_{a_{2s}}, L_{a_{2s+1}}$ such that $L_e \cap L_{a_{2s}} \neq \emptyset \neq L_e \cap L_{a_{2s+1}}$, but these intersections are singletons.

To verify that our construction satisfies all requirements we follow a case-by-case procedure:

Case 1: There is no $n \in \mathbb{N}$ such that $D_n = \{e\}$ and $n \in W_e$. In this case (R_e) will never require attention. Then $L_e = \{e\}$, but there exists no $n \in W_e$ such that $D_n = \{e\}$. Thus (1) is satisfied.

Case 2: There is $n \in \mathbb{N}$ such that $D_n = \{e\}$ and $n \in W_e$. Then inevitably (R_e) will require attention, i.e., there exist $n, s' \in \mathbb{N}$ such that $D_n = \{e\}$ and

$n \in W_{e,s'}$. This is because there are only finitely many (R_i) with $i < e$ and each (R_i) receives attention at most once. It follows that there is a stage $s \geq s'$ at which (R_e) receives attention. We then have $L_e = \{e, a_{2s}\}$, $D_n = \{e\}$ for some $n \in W_e$ but also $e \in L_{a_{2s+1}}$, $a_{2s} \in L_{a_{2s}}$. This ensures that (R_e) gets satisfied for L_e because $D_n = \{e\} \subseteq L_e$, $D_n = \{e\} \subseteq L_{a_{2s+1}}$ and $L_e \neq L_{a_{2s+1}}$. Thus we have that (R_e) is satisfied. Note that the only possible DFTTs for L_e are $D_n = \{e\}$ or L_e itself. But by construction L_e cannot be canonically represented, i.e. it cannot be a DFTT (by definition of DFTT). Therefore $\mathcal{L}$ is an anti-chain of finite languages which is not pfi.   □

This particular example happens to be not cfi either. The question remains open whether there exists such a family which is cfi and not pfi.

7 Conclusion and Future Research

This paper provides a detailed analysis and novel results in finite identification with positive and with complete data. Our work solves questions that concern structural properties of certain families of languages such as maximality with respect to positive finite identification and the differences between pfi and cfi when studying anti-chains. Moreover, we also frame the computational features that are crucial for bringing pfi and cfi closer to each other.

Concerning the questions of structural properties with respect to positive data, on one hand we prove that there always exist a maximal non-effectively finitely identifiable family of an identifiable one that contains only finite languages. On the computational aspect, we provide a canonical family of languages which does not have a maximal effective pfi extension, answering negatively the posed question for maximal pfi extensions. When studying finite identifiable families with complete data, we prove that neither maximal cfi families nor maximal non-effective cfi families exist.

We also study the structural question that concerns the number of maximal finitely identifiable extensions with positive data for the cases of equinumerous families. We prove that these families have either finitely many or uncountably many maximal nepfi extensions.

Finally, we were concerned with families that are anti-chains. We present a cfi anti-chain of infinite languages which is not pfi. For the case of canonical anti-chains of only finite languages such an example cannot exist, simply because such families are always positively identifiable. We provide a first step in investigating the case of non-canonical anti-chains by presenting a, non-canonical but still computable, anti-chain of finite languages which is not pfi but also not cfi. Whether such an anti-chain exists which is cfi but not pfi is an intriguing open question.

Directions of future work involve studying non-canonical anti-chains in detail. In addition we want to investigate how our results stand up in the context of other types of learning, specifically the fastest learning of [5] and their consequences for the study of learning in dynamic epistemic logic. We are also interested in the general open question (with a combinatorial flavor) of how many maximal nepfi extensions a pfi family of finite languages has. In the light of our results, we conjecture that there are either finitely many or uncountably many. In ongoing work we investigate subfamilies containing only n-tuples and m-tuples for fixed numbers n and m, such subfamilies support our conjecture. The next step is to continue with the question for families of finite languages of bounded cardinality, we believe the answer will bring interesting insights, not only for finite identification but for discrete mathematics and combinatorics as well.

Acknowledgment. We want to thank Sebastiaan A. Terwijn for his assistance in the methods of the proof of Theorem 11. We thank the two anonymous referees that helped us to clarify a number of issues and improve the paper.

References

1. Angluin, D.: Inductive inference of formal languages from positive data. Inf. Control **45**(2), 117–135 (1980)
2. Bolander, T., Gierasimczuk, N.: Learning actions models: qualitative approach. In: van der Hoek, W., Holliday, W.H., Wang, W. (eds.) LORI 2015. LNCS, vol. 9394, pp. 40–52. Springer, Heidelberg (2015). https://doi.org/10.1007/978-3-662-48561-3_4
3. Bolander, T., Gierasimczuk, N.: Learning to act: qualitative learning of deterministic action models. J. Log. Comput. **28**(2), 337–365 (2017)
4. Dégremont, C., Gierasimczuk, N.: Finite identification from the viewpoint of epistemic update. Inf. Comput. **209**(3), 383–396 (2011)
5. Gierasimczuk, N., de Jongh, D.: On the complexity of conclusive update. Comput. J. **56**(3), 365–377 (2012)
6. Gierasimczuk, N., Hendricks, V.F., de Jongh, D.: Logic and learning. In: Baltag, A., Smets, S. (eds.) Johan van Benthem on Logic and Information Dynamics, pp. 267–288. Springer, Cham (2014). https://doi.org/10.1007/978-3-319-06025-5_10
7. Gold, M.E.: Language identification in the limit. Inf. Control **10**(5), 447–474 (1967)
8. Hiller, S., Fernández, R.: A data-driven investigation of corrective feedback on subject omission errors in first language acquisition. In: Proceedings of the 20th SIGNLL, Conference on Computational Natural Language Learning (CoNLL), pp. 105–114 (2016)
9. Jain, S., Osherson, D., Royer, J.S., Sharma, A.: Systems That Learn, vol. 2. MIT Press, Cambridge (1999)
10. Lange, S., Zeugmann, T.: Set-driven and rearrangement-independent learning of recursive languages. Theory Comput. Syst. **29**(6), 599–634 (1996)
11. Mitkov, R.: The Oxford Handbook of Computational Linguistics. Oxford University Press, Oxford (2005)
12. Mukouchi, Y.: Characterization of finite identification. In: Jantke, K.P. (ed.) AII 1992. LNCS, vol. 642, pp. 260–267. Springer, Heidelberg (1992). https://doi.org/10.1007/3-540-56004-1_18
13. Osherson, D.N., Stob, M., Weinstein, S.: Systems That Learn: An Introduction to Learning Theory for Cognitive and Computer Scientists. MIT Press, Cambridge (1986)

14. Rogers, H.: Theory of Recursive Functions and Effective Computability, vol. 5. McGraw-Hill, New York (1967)
15. Saxton, M., Backley, P., Gallaway, C.: Negative input for grammatical errors: effects after a lag of 12 weeks. J. Child Lang. **32**(3), 643–672 (2005)
16. Soare, R.I.: Recursively Enumerable Sets and Degrees: A Study of Computable Functions and Computably Generated Sets. Springer, Heidelberg (1999)

Two Neighborhood Semantics
for Subintuitionistic Logics

Dick de Jongh[1]([⊠]) and Fatemeh Shirmohammadzadeh Maleki[2]([⊠])

[1] Institute for Logic, Language and Computation, University van Amsterdam,
Amsterdam, The Netherlands
`D.H.J.deJongh@uva.nl`
[2] School of Mathematics, Statistics and Computer Science, College of Science,
University of Tehran, Tehran, Iran
`f.shmaleki2018@gmail.com`

Abstract. This investigation is concerned with weak subintuitionistic logics interpreted over neighborhood models introduced by the authors in 2016. The two types of neighborhood semantics introduced in that article are compared and their relationship is clarified. Thereby modal companions for various logics are recognized. Specifically, a logic is found which has basic monotonic logic with necessitation as its modal companion. Many of the extensions of the basic logics are discussed and characterized.

Keywords: Intuitionistic logic · Subintuitionistic logic ·
Modal companions · Neighborhood models

1 Introduction

In [7] we introduced two types of neighborhood models. In the NB-neighborhood frames the neighborhoods are pairs (X, Y) of sets of worlds. In N-neighborhood frames the neighborhoods are sets of worlds having the form $\overline{X} \cup Y$ for some choice of X and Y. The NB-neighborhood semantics is best suited to study the basic logic WF, the N-neighborhood semantics is more suitable for obtaining modal companions with respect to the Gödel type translation discovered for subintuitionistic logics with Kripke models by Corsi in [2]. The exact relationship between the two types of frames remained unclear in that article. In particular it was not sure whether the two types of frames define the same set of valid formulas.

In [4] we gave a partial solution of this problem without complete proofs, the main emphasis of that article was on conservativity results for sets of implications with regard to intuitionistic logic IPC. But we did introduce a rule N which when added to WF yields the system WF$_N$ which axiomatizes the validities of N-models. In the present article we give more complete proofs of this result, and make a

© Springer-Verlag GmbH Germany, part of Springer Nature 2019
A. Silva et al. (Eds.): TbiLLC 2018, LNCS 11456, pp. 64–85, 2019.
https://doi.org/10.1007/978-3-662-59565-7_4

finer analysis of the relationship between the two types of models. It turns out that N-neighborhood frames can be seen as a special type of NB-neighborhood frames.

Furthermore, we characterize the properties of many axiom schemata extending WF in both types of models and prove their completeness. In particular a new rule N_2 is introduced, related to N, which axiomatizes a stronger logic WF_{N_2} over WF. The modal companion of WF_{N_2} is the basic monotonic modal logic M_{Nec}. Finding such a logic was of course very desirable, but had escaped us so far.

The content of the different sections is as follows. In Sect. 2 we introduce the logic WF and its NB-neighborhood semantics. In Sect. 3 we introduce the logic WF_N and its N-neighborhood semantics. In Sect. 4 we discuss some formulas that highlight the difference between the two types of semantics, and in Sect. 5 some logics are discussed that extend WF. In Sect. 6 we extend Corsi's Gödel type translations of logics above F into modal logic to the logics above WF and provide modal companions for some of them. This is also where we introduce the new logic WF_{N_2}.

2 NB-Neighborhood Semantics

We first recall the NB-neighborhood frames introduced in [7], and further studied in [4]. The NB-neighborhood frames were created to deal with the very natural basic logical system WF introduced in [7], and recalled later in this section. The N-neighborhood frames of the next section seemed at first the natural choice to study weak subintuitionistic logics because they are a straightforward extension of the neighborhood frames used for modal logics. It turned out however that they were unsuitable for the study of WF. They are fitting for some stronger logics.

Here we choose general frames as the basic notion, restricting the valuations to a subset of the powerset of the set of worlds. We feel that in the case of neighborhood semantics this is a natural choice. The difference between general Kripke frames and full Kripke frames is very marked. In the case of full Kripke frames one deals with worlds only, not with sets of worlds, only in general frames one deals with sets. In the case of neighborhood frames one always deals with sets, hence the difference between full and general frames is much less marked. Of course in both cases the duality with algebras is concerned with general frames. It seems to us that completeness theorems with respect to full frames are less urgent in the case of neighborhood semantics than in the case of Kripke semantics. We have to confess that we were not very clear on this subject in [7]. In that article we dealt in the completeness proofs with general frames only, but we merely mentioned this in passing just above Definition 2.4.

So, if we use the word frame we mean general frame. If we intend to have a full frame we specifically say so. In the present paper, we will deal with results on full frames in subsections at the end of the Sects. 2 and 5. We prove completeness with respect to full frames for the most important logics. For a number of minor logics we have failed to do so for the NB-frames. In Sect. 3 in which we cover N-neighborhood semantics the completeness proofs for full frames are very straightforward and we deal with them in the section itself. In Sect. 6 it is important that we have completeness with respect to full neighborhood frames for the logics for which we are establishing modal companions, because the proofs rely on the fact that the models and their truth definitions obtained from the completeness proofs on both sides are the same.

We do not give all the proofs in the present section. The missing ones can be found in [7], and, anyhow, very similar ones can be found in the next section on N-neighborhoods.

Definition 1. *An* **NB-***Neighborhood Frame* $\mathfrak{F} = \langle W, NB, \mathcal{X} \rangle$ *consists of a non-empty set* W, *a non-empty collection* $\mathcal{X}$ *of subsets of* W, *and a function* $NB : W \to \mathcal{P}(\mathcal{X}^2)$ *such that:*

1. $\mathcal{X}$ *is closed under* $\cup, \cap$ *and* $\to$ *defined by* $U \to V := \{w \in W \mid (U, V) \in NB(w)\}$,
2. $\forall w \in W \; \forall X, Y \in \mathcal{X} \; (X \subseteq Y \Rightarrow (X, Y) \in NB(w))$.

A **full NB-neighborhood frame** *is an* NB-*neighborhood frame in which* $\mathcal{X} = \mathcal{P}(W)$.

In an **NB-Neighborhood Model** $\mathfrak{M} = \langle W, NB, \mathcal{X}, V \rangle$, $V : At \to \mathcal{X}$ *is a valuation function on the set of propositional variables* At.

In an NB-neighborhood frame, if there exists an element $g \in W$ such that

$$NB(g) = \{(X, Y) \in \mathcal{X}^2 \mid X \subseteq Y\},$$

then g is called **omniscient** and the frames are called **rooted NB-neighborhood frames**.

Definition 2. *Let* $\mathfrak{M} = \langle W, NB, \mathcal{X}, V \rangle$ *be an* NB-*neighborhood model.*

Truth *of a propositional formula in a world* w *is defined inductively as follows.*

1. $\mathfrak{M}, w \Vdash p \quad\quad \Leftrightarrow \quad w \in V(p)$;
2. $\mathfrak{M}, w \Vdash A \wedge B \quad \Leftrightarrow \quad \mathfrak{M}, w \Vdash A$ *and* $\mathfrak{M}, w \Vdash B$;
3. $\mathfrak{M}, w \Vdash A \vee B \quad \Leftrightarrow \quad \mathfrak{M}, w \Vdash A$ *or* $\mathfrak{M}, w \Vdash B$;
4. $\mathfrak{M}, w \Vdash A \to B \quad \Leftrightarrow \quad (A^{\mathfrak{M}}, B^{\mathfrak{M}}) \in NB(w)$;
5. $\mathfrak{M}, w \nVdash \bot$,

where $A^{\mathfrak{M}} := \{w \in W \mid \mathfrak{M}, w \Vdash A\}$. *Sets* $X \subseteq W$ *such that* $X = A^{\mathfrak{M}}$ *for some formula* A *are called* **definable**; A *is* **valid** *in* $\mathfrak{M}$, $\mathfrak{M} \Vdash A$, *if for all* $w \in W$, $\mathfrak{M}, w \Vdash A$, *and* A *is valid in* $\mathfrak{F}$, $\mathfrak{F} \Vdash A$ *if for all* $\mathfrak{M}$ *on* $\mathfrak{F}$, $\mathfrak{M} \Vdash A$. *We write* $\Vdash A$ *if* $\mathfrak{M} \Vdash A$ *for all* $\mathfrak{M}$. *Also we define* $\Gamma \Vdash A$ *iff for all* $\mathfrak{M}$, $w \in \mathfrak{M}$, *if* $\mathfrak{M}, w \Vdash \Gamma$ *then* $\mathfrak{M}, w \Vdash A$ *(local validity).*

In this paper we are not concerned with negation, so Clause 5 in the truth definition is just there for completeness' sake.

Proposition 1. *If* $\mathfrak{M} = \langle W, NB, \mathcal{X}, V\rangle$ *is an* NB-*neighborhood model then* $\mathfrak{M}$ *can be extended by adding an omniscient world to obtain a rooted* NB-*neighborhood model* $\mathfrak{M}'$ *such that for all formulas A and for all* $w \in W$,

$$\mathfrak{M}, w \Vdash A \text{ iff } \mathfrak{M}', w \Vdash A.$$

Proof. We add a world g to W and make a new model $\mathfrak{M}' = \langle W', NB', \mathcal{X}', V'\rangle$ with $W' = W \cup \{g\}$, $\mathcal{X}' = \{X, X \cup \{g\} \mid X \in \mathcal{X}\}$, for all propositional letters p, $(p)^{\mathfrak{M}'} = (p)^{\mathfrak{M}}$, and for all $w \in W$ and g:

$$NB'(g) = \left\{(X, Y) \in \mathcal{X}'^2 \mid X \subseteq Y\right\},$$

$$\begin{aligned} NB'(w) = NB(w) \ &\cup \{(X, Y \cup \{g\}) \mid (X, Y) \in NB(w)\} \\ &\cup \{(X \cup \{g\}, Y) \mid (X, Y) \in NB(w)\} \\ &\cup \{(X \cup \{g\}, Y \cup \{g\}) \mid (X, Y) \in NB(w)\}. \end{aligned}$$

The proof is by induction on A. The case where A is a proposition letter follows by definition. Conjunction and disjunction are easy. Now assume $\mathfrak{M}, w \Vdash B \to C$, then $(B^{\mathfrak{M}}, C^{\mathfrak{M}}) \in NB(w)$. By induction hypothesis we have $B^{\mathfrak{M}} = B^{\mathfrak{M}'} \cap W$ and $C^{\mathfrak{M}} = C^{\mathfrak{M}'} \cap W$. So by definition of NB':

$$(B^{\mathfrak{M}'}, C^{\mathfrak{M}'}) \in NB'(w).$$

That is $\mathfrak{M}', w \Vdash B \to C$.

Next assume $\mathfrak{M}', w \Vdash B \to C$. Then $(B^{\mathfrak{M}'}, C^{\mathfrak{M}'}) \in NB'(w)$. So, by induction hypothesis and definition of NB', $(B^{\mathfrak{M}}, C^{\mathfrak{M}}) \in NB(w)$. That is $\mathfrak{M}, w \Vdash B \to C$.□

Corollary 1. *Validity in* NB-*neighborhood frames and rooted* NB-*neighborhood frames is the same.*

The advantage of rooted models lies in the fact that they obey the next very convenient theorem.

Theorem 1 ([7], Theorem 2.13). *If* $\mathfrak{M}$ *is a rooted* NB-*neighborhood model, then:*

1. $\mathfrak{M} \Vdash A \to B$ *iff* $A^{\mathfrak{M}} \subseteq B^{\mathfrak{M}}$.
2. $\Vdash A \to B$ *iff for all models* $\mathfrak{M}$, $A^{\mathfrak{M}} \subseteq B^{\mathfrak{M}}$.

Proof. (1) If $\mathfrak{M} \Vdash A \to B$ then $\mathfrak{M}, g \Vdash A \to B$, so $A^{\mathfrak{M}} \subseteq B^{\mathfrak{M}}$. If $A^{\mathfrak{M}} \subseteq B^{\mathfrak{M}}$, then for all $w \in W$, $\mathfrak{M}, w \Vdash A \to B$, so $\mathfrak{M} \Vdash A \to B$. (2) is immediate from (1).   □

Definition 3. WF *is the logic given by the following axiom schemas and rules,*

1. $A \to A \vee B$ 2. $B \to A \vee B$ 3. $A \to A$

4. $A \wedge B \to A$ 5. $A \wedge B \to B$ 6. $\dfrac{A \quad A \to B}{B}$

7. $\dfrac{A \to B \quad A \to C}{A \to B \wedge C}$ 8. $\dfrac{A \to C \quad B \to C}{A \vee B \to C}$ 9. $\dfrac{A \to B \quad B \to C}{A \to C}$

10. $\frac{A}{B\to A}$ 11. $\frac{A\leftrightarrow B \quad C\leftrightarrow D}{(A\to C)\leftrightarrow(B\to D)}$ 12. $\frac{A \quad B}{A\wedge B}$

13. $A\wedge(B\vee C)\to(A\wedge B)\vee(A\wedge C)$ 14. $\bot\to A$

As stated before we are not concerned with negation. The results are independent of the inclusion of Axiom 14. The basic notion $\vdash_{WF}A$ means that A can be derived from the WF-axioms by means of its rules. But, as in modal logic, when one axiomatizes *local validity*, not all rules in a Hilbert type system have the same status when one considers deductions from assumptions. In modal logic the rule of necessitation can only be applied to prove theorems, not to derive conclusions from assumptions. Here similar restrictions apply to nearly all the rules. One may have some freedom in this respect, but we do like to have a notion of $\Gamma\vdash_{WF} A$ that satisfies some form of the deduction theorem. Of course, we cannot hope to get the full deduction theorem because that is equivalent to having the two implication axioms of IPC. Our choice in the case of deduction from assumptions is to restrict all the rules except the conjunction rule; the restriction on modus ponens is slightly weaker than on the other rules: when concluding B from $A, A\to B$ only the implication $A\to B$ need be a theorem. If we execute this in the definition of $\Gamma\vdash_{WF}A$ we will get a weak form of the deduction theorem with a single assumption. This will of course determine our notion of theory in WF. These considerations lead to the following definition.

Definition 4. *We define $\Gamma\vdash_{WF}A$ as, there is a derivation of A from Γ and theorems of WF using the rules $\frac{A \quad B}{A\wedge B}$ and $\frac{A \quad A\to B}{B}$ where in the latter case the restriction is that $A\to B$ has to be provable in WF.*

In this section we will write $\Gamma\vdash A$ for $\Gamma\vdash_{WF}A$.

Theorem 2 (Weak Deduction Theorem, [7] Theorem 2.19). $A\vdash B$ *iff* $\vdash A\to B$.

From Theorem 1 we can now obtain:

Theorem 3. *The logic WF is sound with respect to the class of rooted NB-neighborhood frames.*

Theorem 4. *The logic WF is sound with respect to the class of NB-neighborhood frames.*

Proof. Assume $\Gamma\vdash A$. We want to show that $\Gamma\Vdash A$. Let $\mathfrak{M}$ be an NB-neighborhood model such that $\mathfrak{M}, w\Vdash\Gamma$. Then by Proposition 1, there exists rooted NB-neighborhood model $\mathfrak{M}'$ such that $\mathfrak{M}', w\Vdash\Gamma$. So, by Theorem 3, $\mathfrak{M}', w\Vdash A$ and then by Proposition 1, we can conclude that $\mathfrak{M}, w\Vdash A$. That is $\Gamma\Vdash A$. $\square$

We will rely on [7] for the completeness proof of WF, but we state right now the definitions of prime theories and the canonical model because we will need them later anyway.

Definition 5. *A set of sentences Δ is a **theory** if and only if*

1. $A, B \in \Delta \Rightarrow A \wedge B \in \Delta$,
2. $\vdash A \to B$ and $A \in \Delta \Rightarrow B \in \Delta$,
3. $\vdash A \Rightarrow A \in \Delta$.

Δ *is a **prime theory** if Δ is a theory, and, if $A \vee B \in \Delta$, then $A \in \Delta$ or $B \in \Delta$.*

Definition 6 ([7] Definition 2.26). *Let W_{WF} be the set of all consistent prime theories of* WF. *Given a formula A, we define $[\![A]\!] = \{\Delta \mid \Delta \in W_{\mathsf{WF}}, A \in \Delta\}$. The **basic NB-canonical model** $\mathfrak{M}_{\mathsf{WF}} = \langle W, NB, V \rangle$ is defined by:*

1. $W = W_{\mathsf{WF}}$,
2. *For each $\Gamma \in W$, and all formulas A, B, $NB(\Gamma) = \{([\![A]\!], [\![B]\!]) \mid A \to B \in \Gamma\}$,*
3. *If $p \in At$, then $V(p) = [\![p]\!] = \{\Gamma \mid \Gamma \in W_{\mathsf{WF}}$ and $p \in \Gamma\}$.*

Lemma 1 ([7] Proposition 2.18). *Δ is a theory iff Δ is closed under $\vdash$.*

Theorem 5 ([7] Theorem 2.28). *The logic* WF *is strongly complete with respect to the class of rooted NB-neighborhood frames.*

2.1   Completeness with Respect to Full NB-Neighborhood Frames

The **full canonical model of** WF is defined as a small adaptation of Definition 6: all pairs (X, Y) in $(\mathcal{P}(W))^2$ with $X \subseteq Y$ are added to all $NB(\Gamma)$.

Theorem 6. *The logic* WF *is strongly complete with respect to the class of full NB-neighborhood frames.*

Proof. The proof of Theorem 2.28 in [7] only needs a small adaptation, in the proof of Lemma 2.25: if $([\![A]\!], [\![B]\!]) \in N(\Gamma)$ because $[\![A]\!] \subseteq [\![B]\!]$), then $\vdash A \to B$, so, also in this case, $\Gamma \vdash A \to B$. $\square$

3   N-Neighborhood Semantics

In this section we recall the N-neighborhood frames, also introduced in [7]. They are useful to study the connection to modal logic, specifically in pointing out modal companions to the various logics introduced here. In [4] the relationship with NB-neighborhood frames was clarified to a certain extent, but here we give a much fuller explanation of the connections and differences, and we give missing proofs. It is shown that the set of valid formulas on N-neighborhood frames is axiomatized by the system $\mathsf{WF_N}$ obtained by adding a rule N to the system WF. As in the previous section our basic notion is the one of general frames. Because they are more straightforward for N-frames, full frames will be covered here as we go along.

Definition 7. $\mathfrak{F} = \langle W, N, \mathcal{X} \rangle$ *is an* **N-Neighborhood Frame** *if W is a non-empty set, $\mathcal{X}$ is a non-empty collection of subsets of W such that $\emptyset$ and W belong to $\mathcal{X}$, and $\mathcal{X}$ is closed under $\cup$, $\cap$ and $\rightarrow$ defined by*

$$U \rightarrow V := \left\{ w \in W \mid \overline{U} \cup V \in N(w) \right\},$$

N is a function from W into $\mathcal{P}(\mathcal{X})$, and for each $w \in W$, $W \in N(w)$.
A **full N-neighborhood frame** *is an N-neighborhood frame in which $\mathcal{X} = \mathcal{P}(W)$.*

Valuation $V : At \rightarrow \mathcal{X}$ makes $\mathfrak{M} = \langle W, N, \mathcal{X}, V \rangle$ an **N-Neighborhood Model**. *Truth of a propositional formula in a world w is defined inductively as follows.*

1. $\mathfrak{M}, w \Vdash p \quad\quad\quad \Leftrightarrow \quad w \in V(p)$;
2. $\mathfrak{M}, w \Vdash A \wedge B \quad \Leftrightarrow \quad \mathfrak{M}, w \Vdash A$ and $\mathfrak{M}, w \Vdash B$;
3. $\mathfrak{M}, w \Vdash A \vee B \quad \Leftrightarrow \quad \mathfrak{M}, w \Vdash A$ or $\mathfrak{M}, w \Vdash B$;
4. $\mathfrak{M}, w \Vdash A \rightarrow B \quad \Leftrightarrow \quad \{v \mid v \Vdash A \Rightarrow v \Vdash B\} = \overline{A^{\mathfrak{M}}} \cup B^{\mathfrak{M}} \in N(w)$;
5. $\mathfrak{M}, w \not\Vdash \bot$,

where $A^{\mathfrak{M}} := \{w \in W \mid \mathfrak{M}, w \Vdash A\}$. Sets $X \subseteq W$ such that $X = A^{\mathfrak{M}}$ for some formula A are called **definable**. *A formula A is* **valid** *in $\mathfrak{M}$, $\mathfrak{M} \Vdash A$, if for all $w \in W$, $\mathfrak{M}, w \Vdash A$, and A is valid in $\mathfrak{F}$, $\mathfrak{F} \Vdash A$ if for all $\mathfrak{M}$ on $\mathfrak{F}$, $\mathfrak{M} \Vdash A$. We write $\Vdash A$ if $\mathfrak{M} \Vdash A$ for all $\mathfrak{M}$. Also we define $\Gamma \Vdash A$ iff for all $\mathfrak{M}, w \in \mathfrak{M}$, if $\mathfrak{M}, w \Vdash \Gamma$ then $\mathfrak{M}, w \Vdash A$, i.e., we have* **local validity**.

If in an N-neighborhood frame there exists an element $g \in W$ such that $N(g) = \{W\}$, then g is called **omniscient** and the frame a **rooted N-neighborhood frame**.

Proposition 2 ([7], Theorem 6.2). *If $\mathfrak{M} = \langle W, N, \mathcal{X}, V \rangle$ is an N-neighborhood model then $\mathfrak{M}$ can be extended by adding an omniscient world to obtain a rooted N-neighborhood model $\mathfrak{M}'$ such that for all formulas A and for all $w \in W$,*

$$\mathfrak{M}, w \Vdash A \text{ iff } \mathfrak{M}', w \Vdash A.$$

Corollary 2. *Validity in N-neighborhood frames and in rooted N-neighborhood frames is the same.*

Theorem 7. *If $\mathfrak{M}$ is a rooted N-neighborhood model, then:*

1. $\mathfrak{M} \Vdash A \rightarrow B$ iff $A^{\mathfrak{M}} \subseteq B^{\mathfrak{M}}$.
2. $\Vdash A \rightarrow B$ iff for all models $\mathfrak{M}$, $A^{\mathfrak{M}} \subseteq B^{\mathfrak{M}}$.

Proof. (1) Assume $\mathfrak{M} \Vdash A \rightarrow B$. Then, for all $w \in \mathfrak{M}$, $\overline{A^{\mathfrak{M}}} \cup B^{\mathfrak{M}} \in N(w)$. Therefore $\overline{A^{\mathfrak{M}}} \cup B^{\mathfrak{M}} = W$, since $N(g) = \{W\}$. So $A^{\mathfrak{M}} \subseteq B^{\mathfrak{M}}$. For the other direction, by assumption we have $\overline{A^{\mathfrak{M}}} \cup B^{\mathfrak{M}} = W$, so, for all $w \in \mathfrak{M}$, $\overline{A^{\mathfrak{M}}} \cup B^{\mathfrak{M}} \in N(w)$, i.e. $\mathfrak{M} \Vdash A \rightarrow B$.

(2) Follows immediately from (1). □

In [7] the question whether validity in NB-neighborhood frames and N-neighborhood frames is the same was left unanswered. In [4] it was discovered that the difference resides in the rule N. To the system WF we add this rule to obtain the logic WF$_N$:

$$\frac{A \to B \vee C \quad C \to A \vee D \quad A \wedge C \wedge D \to B \quad A \wedge C \wedge B \to D}{(A \to B) \leftrightarrow (C \to D)} \quad \text{(N)}$$

As usual a rule like N is considered to be valid in a frame $\mathfrak{F}$ if, on each $\mathfrak{M}$ in $\mathfrak{F}$ in which the premises of the rule are valid, the conclusion is valid as well. In the picture below the idea behind the rule is exhibited. The conclusion $(A \to B) \leftrightarrow (C \to D)$ of the rule is valid if $\overline{A} \cup B = \overline{C} \cup D$ or equivalently if $A \cap \overline{B} = C \cap \overline{D}$ or if $A \backslash B = C \backslash D$. In the picture the latter is the grey part. The four assumptions of the rule force the picture to be essentially as given (e.g. $A \to B \vee C$ means $A \subseteq B \cup C$) and to make sure that indeed $A \backslash B = C \backslash D$.

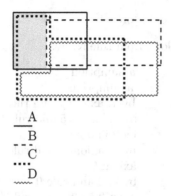

$$\underline{A}$$
$$\underline{B}$$
$$\underline{C}$$
$$\underline{D}$$

In deductions from assumptions the use of the rules shared with WF is similarly restricted. Also the new rule N can only be used if there are no assumptions.

Definition 8. *We define $\Gamma \vdash_{\text{WF}_N} A$ if there is a derivation of A from Γ and theorems of* WF$_N$ *using the rules $\frac{A \quad B}{A \wedge B}$ and $\frac{A \quad A \to B}{B}$, where in the latter case the restriction is that $A \to B$ has to be provable in* WF$_N$.

The proof of the following weak deduction theorem is easy and essentially the same as the one for WF ([7], Theorem 2.19).

Theorem 8 (Weak Deduction Theorem). $A \vdash_{\text{WF}_N} B$ *iff* $\vdash_{\text{WF}_N} A \to B$.

Theorem 9. *The logic* WF$_N$ *is sound with respect to the class of rooted N-neighborhood frames.*

Proof. Recall that, by Theorem 7(1), for all E, F, $\mathfrak{M} \Vdash E \to F$ iff $E^{\mathfrak{M}} \subseteq F^{\mathfrak{M}}$. We only check rule (N). Assume,

1. $\mathfrak{M} \Vdash A \to B \vee C$, i.e. $A^{\mathfrak{M}} \subseteq B^{\mathfrak{M}} \cup C^{\mathfrak{M}}$,
2. $\mathfrak{M} \Vdash C \to A \vee D$, i.e. $C^{\mathfrak{M}} \subseteq A^{\mathfrak{M}} \cup D^{\mathfrak{M}}$,

3. $\mathfrak{M} \Vdash A \wedge C \wedge D \rightarrow B$, i.e. $A^{\mathfrak{M}} \cap C^{\mathfrak{M}} \cap D^{\mathfrak{M}} \subseteq B^{\mathfrak{M}}$,

4. $\mathfrak{M} \Vdash A \wedge C \wedge B \rightarrow D$, i.e. $A^{\mathfrak{M}} \cap C^{\mathfrak{M}} \cap B^{\mathfrak{M}} \subseteq D^{\mathfrak{M}}$.

To get the conclusion $(A \rightarrow B) \leftrightarrow (C \rightarrow D)$ it is sufficient to prove that $\overline{A^{\mathfrak{M}}} \cup B^{\mathfrak{M}} = \overline{C^{\mathfrak{M}}} \cup D^{\mathfrak{M}}$ because then $w \Vdash A \rightarrow B$ iff $w \Vdash C \rightarrow D$ and hence $(A \rightarrow B)^{\mathfrak{M}} = (C \rightarrow D)^{\mathfrak{M}}$. By symmetry, it will suffice to show that $\overline{A^{\mathfrak{M}}} \cup B^{\mathfrak{M}} \subseteq \overline{C^{\mathfrak{M}}} \cup D^{\mathfrak{M}}$. Let $w \in \overline{A^{\mathfrak{M}}} \cup B^{\mathfrak{M}}$. Then $w \in \overline{A^{\mathfrak{M}}}$ or $(w \in A^{\mathfrak{M}}$ and $w \in B^{\mathfrak{M}})$. If $w \in \overline{A^{\mathfrak{M}}}$, we distinguish the cases $w \in D^{\mathfrak{M}}$ and $w \in \overline{D^{\mathfrak{M}}}$. In the first case we are done directly. In the second case, we can conclude from (2) that $w \in \overline{C^{\mathfrak{M}}}$ and we are done as well. If $w \in A^{\mathfrak{M}}$ and $w \in B^{\mathfrak{M}}$, we distinguish the cases $w \in \overline{C^{\mathfrak{M}}}$ and $w \in C^{\mathfrak{M}}$. In the first case we are done directly. In the second case, we can conclude from (4) that $w \in D^{\mathfrak{M}}$ and we are done as well. $\qquad\square$

Theorem 10. *The logic* $\mathsf{WF_N}$ *is sound with respect to the class of* N-*neighborhood frames.*

Proof. By Proposition 2 and Theorem 9. $\qquad\square$

Remark 1. The rule 11 follows from rule N, by:

1. $\vdash A \leftrightarrow C$	assumption
2. $\vdash B \leftrightarrow D$	assumption
3. $\vdash C \rightarrow A \vee D$	by 1, axiom 1 and rule 9
4. $\vdash A \wedge C \wedge B \rightarrow B$	by 2, axiom 5 and rule 9
5. $\vdash A \wedge C \wedge B \rightarrow D$	by 2 and 4
6. $\vdash A \rightarrow B \vee C$	by 1, axiom 2 and rule 9
7. $\vdash A \wedge C \wedge D \rightarrow D$	axiom 5
8. $\vdash A \wedge C \wedge D \rightarrow B$	by 7, 2 and rule 9
9. $\vdash (A \rightarrow B) \leftrightarrow (C \rightarrow D)$	from 3,5,6,8 using rule N

It is easy to prove the disjunction property for $\mathsf{WF_N}$. As in our previous papers this is simple by using Kleene's $|$ [5]; one needs to add a straightforward clause in the WF-proof for the use of the N-rule. The advantage of having it is that this shows the canonical model of $\mathsf{WF_N}$ to be rooted.

The definition of prime theories for $\mathsf{WF_N}$ is the same as the one for WF.

Theorem 11. *If* $\Sigma \nvdash_{\mathsf{WF_N}} D$ *then there is a prime theory* Δ *such that* $\Delta \supseteq \Sigma$ *and* $D \notin \Delta$.

Proof. The proof is similar to the WF case, Theorem 2.22 in [7]. $\qquad\square$

Definition 9. *Let* $W_{\mathsf{WF_N}}$ *be the set of all consistent prime theories of* $\mathsf{WF_N}$. *Given a formula* A, *we define* $[\![A]\!] = \{\Delta \mid \Delta \in W_{\mathsf{WF_N}}, A \in \Delta\}$. *The* **basic N-canonical model** $\mathfrak{M}_{\mathsf{WF_N}} = \langle W, N, \mathcal{X}, V \rangle$ *is defined by:*

1. $W = W_{\mathsf{WF_N}}$,

2. *For each* $\Gamma \in W$, $N(\Gamma) = \{\overline{[\![A]\!]} \cup [\![B]\!] \mid A \rightarrow B \in \Gamma\}$,

3. $\mathcal{X}$ *is the set of all* $[\![A]\!]$,

4. *If* $p \in At$, *then* $V(p) = [\![p]\!] = \{\Gamma \mid \Gamma \in W \text{ and } p \in \Gamma\}$.

To show that this definition is proper one W needs to be contained in $N(\Gamma)$ for each Γ. This is ensured by the following lemma.

Lemma 2. *In the basic N-canonical model* $\mathfrak{M}_{\mathsf{WF_N}} = \langle W, N, \mathcal{X}, V \rangle$, *for all* $\Gamma \in W$, $W \in N(\Gamma)$.

Proof. By definition of prime theories, for all $\Gamma \in W$ and for all formulas A, $A \to A \in N(\Gamma)$. Hence by definition of N in the basic N-canonical model we conclude that $\overline{[\![A]\!]} \cup [\![A]\!] \in N(\Gamma)$, that is $W \in N(\Gamma)$.   □

For a full N-model there are in this case no additional requirements. So, by the use of this basic canonical model we will immediately get completeness of $\mathsf{WF_N}$ with respect to full N-neighborhood frames.

Lemma 3. *For all formulas C and D, $[\![C]\!] \subseteq [\![D]\!]$ iff $\vdash_{\mathsf{WF_N}} C \to D$.*

Proof. Let $\nvdash_{\mathsf{WF_N}} C \to D$. Then by the Weak Deduction Theorem $C \nvdash D$. Let $\Sigma = \{C\}$, then by Theorem 11, there exists a prime theory Γ such that, $C \in \Gamma$ and $D \notin \Gamma$. That is $[\![C]\!] \nsubseteq [\![D]\!]$.

Now let $\Gamma \in W_{\mathsf{WF_N}}$, $C \in \Gamma$ and $\vdash_{\mathsf{WF_N}} C \to D$. Then by definition of theory $D \in \Gamma$.   □

Lemma 4. *Let $\overline{[\![A]\!]} \cup [\![B]\!] = \overline{[\![C]\!]} \cup [\![D]\!]$, then $\mathsf{WF_N} \vdash (A \to B) \leftrightarrow (C \to D)$.*

Proof. Assume $\overline{[\![A]\!]} \cup [\![B]\!] = \overline{[\![C]\!]} \cup [\![D]\!]$. By rule N it suffices to show

1. $\mathsf{WF_N} \vdash A \to B \vee C$,
2. $\mathsf{WF_N} \vdash A \wedge C \wedge D \to B$,
3. $\mathsf{WF_N} \vdash C \to A \vee D$,
4. $\mathsf{WF_N} \vdash A \wedge C \wedge B \to D$.

We will show 1 and 2; 3 and 4 are analogous.

1. From $\overline{[\![A]\!]} \cup [\![B]\!] = \overline{[\![C]\!]} \cup [\![D]\!]$ we get $[\![A]\!] \cap \overline{[\![B]\!]} = [\![C]\!] \cap \overline{[\![D]\!]}$. We have $[\![A]\!] \subseteq [\![B]\!] \cup [\![A]\!]$, so also, $[\![A]\!] \subseteq [\![B]\!] \cup ([\![A]\!] \cap \overline{[\![B]\!]})$, This means that $[\![A]\!] \subseteq [\![B]\!] \cup ([\![C]\!] \cap \overline{[\![D]\!]})$, so $[\![A]\!] \subseteq [\![B]\!] \cup [\![C]\!]$. By Lemma 3, this implies that $\mathsf{WF_N} \vdash A \to B \vee C$.
2. Again using $[\![A]\!] \cap \overline{[\![B]\!]} = [\![C]\!] \cap \overline{[\![D]\!]}$, we get $[\![A]\!] \cap [\![C]\!] \cap [\![D]\!] \cap \overline{[\![B]\!]} = [\![A]\!] \cap \overline{[\![B]\!]} \cap [\![C]\!] \cap [\![D]\!] = [\![C]\!] \cap \overline{[\![D]\!]} \cap [\![C]\!] \cap [\![D]\!] = \emptyset$. So, $[\![A]\!] \cap [\![C]\!] \cap [\![D]\!] \subseteq [\![B]\!]$, and, again by Lemma 3, $\mathsf{WF_N} \vdash A \wedge C \wedge D \to B$.   □

Lemma 5 (Truth Lemma). *In the N-canonical model* $\mathfrak{M}_{\mathsf{WF_N}}$,

$$E \in \Gamma \text{ iff } \Gamma \Vdash E.$$

Proof. By induction on E. The atomic case holds by definition of N-canonical model.

$(E := A \wedge B)$ Let $\Gamma \in W_{\mathsf{WF_N}}$ and $\Gamma \Vdash A \wedge B$ then $\Gamma \Vdash A$ and $\Gamma \Vdash B$. By the induction hypothesis $A \in \Gamma$ and $B \in \Gamma$. Γ is a theory so $A \wedge B \in \Gamma$.

Now let $A \wedge B \in \Gamma$. We have $\vdash A \wedge B \to A$ and $\vdash A \wedge B \to B$, hence by definition of theory we conclude that $A \in \Gamma$ and $B \in \Gamma$. By induction hypothesis $\Gamma \Vdash A$ and $\Gamma \Vdash B$ so $\Gamma \Vdash A \wedge B$.

$(E := A \vee B)$ Let $\Gamma \in W_{\mathsf{WF_N}}$ and $\Gamma \Vdash A \vee B$. Then $\Gamma \Vdash A$ or $\Gamma \Vdash B$. By the induction hypothesis $A \in \Gamma$ or $B \in \Gamma$. We have $\vdash A \to A \vee B$ and $\vdash B \to A \vee B$, so by definition of theory we conclude that $A \vee B \in \Gamma$.

Now let $A \vee B \in \Gamma$. Γ is a prime, so $A \in \Gamma$ or $B \in \Gamma$. By induction hypothesis we conclude that $\Gamma \Vdash A$ or $\Gamma \Vdash B$. That is, $\Gamma \Vdash A \vee B$.

$(E := A \to B)$ Let $\Gamma \in W_{\mathsf{WF_N}}$, then

$$\Gamma \Vdash A \to B \iff \overline{A^{\mathfrak{M}_{\mathsf{WF_N}}}} \cup B^{\mathfrak{M}_{\mathsf{WF_N}}} \in N(\Gamma)$$
$$\text{(by induction hypothesis)} \iff \overline{[\![A]\!]} \cup [\![B]\!] \in N(\Gamma)$$
$$\text{(by definition of } N(\Gamma)) \iff \overline{[\![C]\!]} \cup [\![D]\!] \in N(\Gamma) \text{ for some } C \to D \in \Gamma$$
$$\text{such that } \overline{[\![A]\!]} \cup [\![B]\!] = \overline{[\![C]\!]} \cup [\![D]\!]$$
$$\text{(by definition, Lemma 4)} \iff A \to B \in \Gamma. \qquad \square$$

Theorem 12 (Completeness of $\mathsf{WF_N}$). *The logic* $\mathsf{WF_N}$ *is sound and strongly complete with respect to the class of full N-neighborhood frames.*

Proof. Soundness already shown in earlier Theorem 9.

Let $\Sigma \nvdash A$, Then, by Theorem 11, there is a prime theory $\Delta \supseteq \Sigma$ such that $A \notin \Delta$. So, in the N-canonical model $\mathfrak{M}_{\mathsf{WF_N}}$ we will have $\mathfrak{M}_{\mathsf{WF_N}}, \Delta \Vdash \Sigma$ and $\mathfrak{M}_{\mathsf{WF_N}}, \Delta \nVdash A$. That is, $\Sigma \nVdash_{\mathsf{WF_N}} A$. $\qquad \square$

Definition 10. *Let* $\langle W, NB, \mathcal{X} \rangle$ *be an NB-neighborhood frame and* $\langle W, N, \mathcal{X} \rangle$ *be an N-neighborhood frame on the same set of worlds and* $\mathcal{X}$. *We say that* $\langle W, NB, \mathcal{X} \rangle$ *and* $\langle W, N, \mathcal{X} \rangle$ *are* **equivalent** *if for all* $X, Y \in \mathcal{X}$,

$$(X, Y) \in NB(w) \quad \text{iff} \quad \overline{X} \cup Y \in N(w).$$

Note that if $\langle W, NB, \mathcal{X} \rangle$ and $\langle W, N, \mathcal{X} \rangle$ are equivalent, then the two models that are based on those frames are pointwise equivalent.

Definition 11. *We define* (X, Y) *to be* **N-equivalent** *to* $(X', Y'), (X, Y) \equiv (X', Y')$, *if* $\overline{X} \cup Y = \overline{X'} \cup Y'$.

Lemma 6. *Let* $\langle W, N, \mathcal{X} \rangle$ *be an N-neighborhood frame. Then there exists an equivalent NB-neighborhood frame* $\langle W, NB, \mathcal{X} \rangle$. *This NB-frame is closed under N-equivalence, i.e., if* $(X, Y) \in NB(w)$ *and* $(X, Y) \equiv (X', Y')$, *then* $(X', Y') \in NB(w)$. *In addition, for all* X, Y, w, *if* $X \subseteq Y$, *then* $(X, Y) \in NB(w)$.

Proof. The proof is straightforward by considering, for each $w \in W$,

$$NB(w) = \left\{ (X, Y) \mid \overline{X} \cup Y \in N(w) \right\}. \qquad \square$$

Lemma 7. *Let* $\langle W, NB, \mathcal{X} \rangle$ *be an* N-*neighborhood frame closed under* N-*equivalence. Then there exists an equivalent* N-*neighborhood frame* $\langle W, N, \mathcal{X} \rangle$.

Proof. The proof is straightforward by considering, for each $w \in W$,

$$N(w) = \left\{ \overline{X} \cup Y \mid (X, Y) \in NB(w) \right\}. \qquad \square$$

Theorem 13. *The logic* WF$_N$ *is sound and strongly complete with respect to the class of* NB-*neighborhood frames that are closed under* N-*equivalence.*

Proof. By Theorem 12 and Lemma 6 applied with $\mathcal{X}$ is $\mathcal{P}(W)$. $\qquad \square$

4 Differences Between N-Validity and NB-Validity

In this section we will be interested in the following axiom schemas which make the difference between NB-frames and N-frames more concrete.

(N$_a$) $(A \rightarrow B) \leftrightarrow (A \vee B \rightarrow B)$
(N$_b$) $(A \rightarrow B) \leftrightarrow (A \rightarrow A \wedge B)$
(N$_c$) $(A \wedge B \rightarrow C) \leftrightarrow (A \wedge B \rightarrow A \wedge C)$
(N$_d$) $(A \rightarrow B \vee C) \leftrightarrow (B \vee A \rightarrow B \vee C)$

Lemma 8. *1.* WF$_N$ $\vdash$ N$_a$.
2. WF$_N$ $\vdash$ N$_b$.
3. WF$_N$ $\vdash$ N$_c$.
4. WF$_N$ $\vdash$ N$_d$.

Proof. The proofs are easy. We only prove 1:

1. $\vdash A \vee B \rightarrow A \vee B$ by axiom 3
2. $\vdash A \wedge (A \vee B) \wedge B \rightarrow B$ by axiom 4
3. $\vdash A \rightarrow B \vee (A \vee B)$ by axiom 1 and rule 9
4. $\vdash (A \rightarrow B) \leftrightarrow (A \vee B \rightarrow B)$ from 1,2 and 3 using rule N,
                                     by substitution of
                                       $A, B, A \vee B, B$ for A, B, C, D $\qquad \square$

Lemma 9. *1.* WF $\nvdash$ N$_a$.
2. WF $\nvdash$ N$_b$.
3. WF $\nvdash$ N$_c$.
4. WF $\nvdash$ N$_d$.

Proof. We will show 1 and 2; 3 and 4 are analogous.
1. We show that both directions of N$_a$ fail in WF.
$\nRightarrow$: Consider the rooted NB-neighborhood frame $\mathfrak{F} = \langle W, NB, \mathcal{P}(W) \rangle$ with

$$W = \{w, g\}, NB(w) = \{(\{g\}, \{w\})\} \cup \left\{ (X, Y) \in (\mathcal{P}(W))^2 \mid X \subseteq Y \right\}.$$

Also consider the valuation $p^{\mathfrak{M}} = \{g\}$, $q^{\mathfrak{M}} = \{w\}$. With this valuation we can conclude $(p \to q)^{\mathfrak{M}} = \{w\}$ and $(p \vee q \to q)^{\mathfrak{M}} = \varnothing$, so $(p \to q)^{\mathfrak{M}} \not\subseteq (p \vee q \to q)^{\mathfrak{M}}$, and hence $g \not\Vdash (p \to q) \to (p \vee q \to q)$.

$\Leftarrow$: Consider the rooted NB-neighborhood frame $\mathfrak{F} = \langle W, NB, \mathcal{P}(W) \rangle$ with

$$W = \{w, g\}, NB(w) = \{(\{g, w\}, \{w\})\} \cup \{(X, Y) \in (\mathcal{P}(W))^2 \mid X \subseteq Y\}.$$

Also consider the valuation $p^{\mathfrak{M}} = \{g\}$, $q^{\mathfrak{M}} = \{w\}$. With this valuation we can conclude $g \not\Vdash (p \vee q \to q) \to (p \to q)$.

2. $\not\Rightarrow$: Consider the rooted NB-neighborhood frame $\mathfrak{F} = \langle W, NB, \mathcal{P}(W) \rangle$ with

$$W = \{w, g\}, NB(w) = \{(\{g\}, \{w\})\} \cup \{(X, Y) \in (\mathcal{P}(W))^2 \mid X \subseteq Y\}.$$

Also consider the valuation $p^{\mathfrak{M}} = \{g\}$ and $q^{\mathfrak{M}} = \{w\}$. With this valuation we can conclude $g \not\Vdash (p \to q) \to (p \to p \wedge q)$.

$\Leftarrow$: Consider the rooted NB-neighborhood frame $\mathfrak{F} = \langle W, NB, \mathcal{P}(W) \rangle$ with

$$W = \{w, g\}, NB(w) = \{(\{g\}, \varnothing)\} \cup \{(X, Y) \in (\mathcal{P}(W))^2 \mid X \subseteq Y\}.$$

Also consider the valuation $p^{\mathfrak{M}} = \{g\}$ and $q^{\mathfrak{M}} = \{w\}$. With this valuation we can conclude $g \not\Vdash (p \to p \wedge q) \wedge (p \to q)$. $\square$

5 Completeness for Some Logics Above WF with NB-Neighborhood Frames

In this section we will be interested in the axiom schemes of the previous section and the following schemes and rule.

(C) $(A \to B) \wedge (A \to C) \to (A \to B \wedge C)$
(D) $(A \to B) \wedge (C \to B) \to (A \vee C \to B)$
($\widehat{\text{C}}$) $(A \to B \wedge C) \to (A \to B) \wedge (A \to C)$
($\widehat{\text{D}}$) $(A \vee B \to C) \to (A \to C) \wedge (B \to C)$
(C_{W}) $(A \to B) \to (C \wedge A \to C \wedge B)$
(D_{W}) $(A \to B) \to (C \vee A \to C \vee B)$
(N_2) $\dfrac{C \to A \vee D \quad A \wedge C \wedge B \to D}{(A \to B) \to (C \to D)}$

Actually, C and D are axiom schemes of Corsi's system F and $\widehat{\text{C}}$ and $\widehat{\text{D}}$ are their converses, which are not provable in WF either, see [4], Lemma 4.11, cf. The rule N_2 will be shown to axiomatize the logic of which basic monotonic modal logic with necessitation is a modal companion. The schemes C_{W} and D_{W} are typical examples of formulas provable using N_2 but not by means of N only.

Lemma 10. 1. $(p \to q) \leftrightarrow (p \vee q \to q)$ characterizes the class of rooted NB-neighborhood frames $\mathfrak{F}$ satisfying the N_a-condition: for all $w \in W$ and $X, Y \in \mathcal{X}$,

$$(X, Y) \in NB(w) \Leftrightarrow (X \cup Y, Y) \in NB(w).$$

2. $(p \to q) \leftrightarrow (p \to p \wedge q)$ *characterizes the class of rooted* NB-*neighborhood frames* $\mathfrak{F}$ *satisfying the* $\mathsf{N_b}$-*condition: for all* $w \in W$ *and* $X, Y \in \mathcal{X}$,

$$(X, Y) \in NB(w) \Leftrightarrow (X, X \cap Y) \in NB(w).$$

3. $(p \wedge q \to r) \leftrightarrow (p \wedge q \to p \wedge r)$ *characterizes the class of rooted* NB-*neighborhood frames* $\mathfrak{F}$ *satisfying the* $\mathsf{N_c}$-*condition: for all* $w \in W$ *and* $X, Y, Z \in \mathcal{X}$,

$$(X \cap Y, Z) \in NB(w) \Leftrightarrow (X \cap Y, X \cap Z) \in NB(w).$$

4. $(p \to q \vee r) \leftrightarrow (q \vee p \to q \vee r)$ *characterizes the class of rooted* NB-*neighborhood frames* $\mathfrak{F}$ *satisfying the* $\mathsf{N_d}$-*condition: for all* $w \in W$ *and* $X, Y, Z \in \mathcal{X}$,

$$(X, Y \cup Z) \in NB(w) \Leftrightarrow (Y \cup X, Y \cup Z) \in NB(w).$$

5. $(p \to q) \to (r \wedge p \to r \wedge q)$ *characterizes the class of rooted* NB-*neighborhood frames* $\mathfrak{F}$ *satisfying closure under* **weak intersection**: *for all* $w \in W$ *and* $X, Y \in \mathcal{X}$, *If* $(X, Y) \in NB(w)$, *then for all* $Z \in \mathcal{X}$, $(X, Y \cup Z) \in NB(w)$.

6. $(p \to q) \to (p \to r \vee q)$ *characterizes the class of rooted* NB-*neighborhood frames* $\mathfrak{F}$ *satisfying closure under* **weak union**: *for all* $w \in W$ *and* $X, Y \in \mathcal{X}$, *If* $(X, Y) \in NB(w)$, *then for all* $Z \in \mathcal{X}$, $(X, Y \cup Z) \in NB(w)$.

Proof. The proofs are easy. We only prove 6:

6. Let $\mathfrak{F} = \langle W, NB, \mathcal{X} \rangle$ be closed under weak union and $\mathfrak{M}$ be any model based on $\mathfrak{F}$. We have to prove for all $w \in W$

$$((p \to q)^{\mathfrak{M}}, (p \to r \vee q)^{\mathfrak{M}}) \in NB(w).$$

For this purpose it is sufficient to show that $(p \to q)^{\mathfrak{M}} \subseteq (p \to r \vee q)^{\mathfrak{M}}$. Let $w \in W$, $w \Vdash p \to q$. Then, $(p^{\mathfrak{M}}, q^{\mathfrak{M}}) \in NB(w)$. The frame is closed under weak union so, $(p^{\mathfrak{M}}, r^{\mathfrak{M}} \cup q^{\mathfrak{M}}) \in NB(w)$. That is, $w \Vdash p \to r \vee q$. Indeed, $(p \to q)^{\mathfrak{M}} \subseteq (p \to r \vee q)^{\mathfrak{M}}$.

For the other direction we use contraposition. Suppose that frame $\mathfrak{F}$ is not closed under weak union. Then there is a $w \in \mathfrak{F}$ such that $(X, Y) \in NB(w)$ and $Z \in \mathcal{X}$, but $(X, Y \cup Z) \notin NB(w)$. Consider the valuation such that $p^{\mathfrak{M}} = X$, $q^{\mathfrak{M}} = Y$ and $r^{\mathfrak{M}} = Z$. Then we will have

$$(p^{\mathfrak{M}}, q^{\mathfrak{M}}) \in N(w)$$

$$(p^{\mathfrak{M}}, (r \vee q)^{\mathfrak{M}}) \notin N(w).$$

So $(p \to q)^{\mathfrak{M}} \nsubseteq (p \to r \vee q)^{\mathfrak{M}}$. Then by the definition of rooted NB-neighborhood frames, $g \nVdash (p \to q) \to (p \to r \vee q)$. Therefore, $\mathfrak{F} \nVdash (p \to q) \to (p \to r \vee q)$. □

Basic NB-canonical models for logics L extending WF are defined exactly as in Definition 6 except that derivability in L is used instead of in WF in the definition of the prime theories.

Lemma 11.(a) *If* $\mathsf{WFN_a} \subseteq L$, *then the basic NB-canonical model of logic* L *is closed under the* $\mathsf{N_a}$-*condition.*

(b) *If* $\mathsf{WFN_b} \subseteq L$, *then the basic* NB-*canonical model of logic* L *is closed under the* $\mathsf{N_b}$-*condition.*

(c) *If* $\mathsf{WFN_c} \subseteq L$, *then the basic* NB-*canonical model of logic* L *is closed under the* $\mathsf{N_c}$-*condition.*

(d) *If* $\mathsf{WFN_d} \subseteq L$, *then the basic* NB-*canonical model of logic* L *is closed under the* $\mathsf{N_d}$-*condition.*

(e) *If* $\mathsf{WFC_W} \subseteq L$, *then the basic* NB-*canonical model of logic* L *is closed under weak intersection.*

(f) *If* $\mathsf{WFD_W} \subseteq L$, *then the basic* NB-*canonical model of logic* L *is closed under weak union.*

Proof. We only prove (a). The other cases are similar.

(a) Left to right of the $\mathsf{N_a}$-condition: Suppose that in the NB-canonical model of logic L, $(X, Y) \in NB(\Gamma)$. By definition of NB in the NB-canonical model there exist formulas A and B such that $(X, Y) = (\llbracket A \rrbracket, \llbracket B \rrbracket)$, where $A \to B \in \Gamma$. So using $(\mathsf{N_a})$, $A \vee B \to B \in \Gamma$. Hence $(\llbracket A \vee B \rrbracket, \llbracket B \rrbracket) \in NB(\Gamma)$. Therefore since $\llbracket A \rrbracket \cup \llbracket B \rrbracket = \llbracket A \vee B \rrbracket$, $(X \cup Y, Y) \in NB(w)$.

Right to left of the $\mathsf{N_a}$-condition: Suppose that in the NB-canonical model of logic L, $(X \cup Y, Y) \in NB(\Gamma)$. By definition of NB in the NB-canonical model there exist formulas A and B such that $X = \llbracket A \rrbracket$, $Y = \llbracket B \rrbracket$ and so $(X \cup Y, Y) = (\llbracket A \vee B \rrbracket, \llbracket B \rrbracket)$, where $A \vee B \to B \in \Gamma$. So using $(\mathsf{N_a})$, $A \to B \in \Gamma$. Hence $(\llbracket A \rrbracket, \llbracket B \rrbracket) = (X, Y) \in NB(\Gamma)$. So NB is closed under the $\mathsf{N_a}$-condition. $\square$

Lemma 12. *The rule* N *characterizes the class of rooted* NB-*neighborhood frames* $\mathfrak{F} = \langle W, NB, \mathcal{X} \rangle$ *that are closed under equivalence.*

Proof. Validity of N on NB-neighborhood frames closed under equivalence is immediate from Lemma 7. Now suppose that $\mathfrak{F}$ is an NB-frame and $w \in \mathfrak{F}$ such that $(X, Y) \in NB(w)$ and $U, V \in \mathcal{X}$ and $\overline{X} \cup Y = \overline{U} \cup V$, but $(U, V) \notin NB(w)$. Since $p \vee q \leftrightarrow p \vee q$, $p \wedge (p \vee q) \wedge q \to q$ and $p \to q \vee (p \vee q)$ are provable (the last one by axiom 1 and rule 9), it suffices to falsify $(p \to q) \leftrightarrow (p \vee q \to q)$ on the frame. Consider the valuation such that $p^{\mathfrak{M}} = X$, $q^{\mathfrak{M}} = Y$ and assume $U = X \cup Y$ and $V = Y$. It is easy to show that $\overline{X} \cup Y = \overline{U} \cup V$. So $(p^{\mathfrak{M}}, q^{\mathfrak{M}}) \in NB(w)$ and $((p \vee q)^{\mathfrak{M}}, q^{\mathfrak{M}}) \notin NB(w)$, and consequently, $(p \to q)^{\mathfrak{M}} \not\subseteq (p \vee q \to q)^{\mathfrak{M}}$. So, $\mathfrak{M}, g \not\Vdash (p \to q) \leftrightarrow (p \vee q \to q)$. $\square$

Definition 12. NB-*neighborhood frame* $\mathfrak{F}$ *is closed under* **superset equivalence** *if and only if for all* $w \in W$ *and* $X, Y, X', Y' \in \mathcal{X}$, *if* $(X, Y) \in NB(w)$ *and* $\overline{X} \cup Y \subseteq \overline{X'} \cup Y'$ *then* $(X', Y') \in NB(w)$.

Lemma 13. *The rule* $\mathsf{N_2}$ *characterizes the class of rooted* NB-*neighborhood frames* $\mathfrak{F} = \langle W, NB, \mathcal{X} \rangle$ *that are closed under superset equivalence.*

Proof. The validity of the rule $\mathsf{N_2}$ on NB-frames closed under superset equivalence has a proof very similar to Theorem 9, we skip it.

Suppose that $\mathfrak{F}$ and $w \in \mathfrak{F}$ are such that $(X, Y) \in NB(w)$ and $U, V \in \mathcal{X}$ and $\overline{X} \cup Y \subseteq \overline{U} \cup V$, but $(U, V) \notin NB(w)$. Since $p \wedge q \to q \vee (p \wedge r)$ and $q \wedge (p \wedge q) \wedge r \to p \wedge r$

are provable it suffices to falsify $(q \to r) \to (p \wedge q \to p \wedge r)$ on the frame. Consider the valuation such that $q^{\mathfrak{M}} = X$, $r^{\mathfrak{M}} = Y$, $(p \wedge q)^{\mathfrak{M}} = U$ and $(p \wedge r)^{\mathfrak{M}} = V$. It is easy to show that $\overline{(p)^{\mathfrak{M}}} \cup \overline{(q)^{\mathfrak{M}}} \cup (r)^{\mathfrak{M}} = \overline{(p \wedge q)^{\mathfrak{M}}} \cup (p \wedge r)^{\mathfrak{M}}$. Hence $\overline{X} \cup Y \subseteq \overline{U} \cup V$. So $(q^{\mathfrak{M}}, r^{\mathfrak{M}}) \in NB(w)$ and $((p \wedge q)^{\mathfrak{M}}, (p \wedge r)^{\mathfrak{M}}) \notin NB(w)$, and consequently, $(q \to r)^{\mathfrak{M}} \nsubseteq (p \wedge q \to p \wedge r)^{\mathfrak{M}}$. So, $\mathfrak{M}, g \nVdash (q \to r) \to (p \wedge q \to p \wedge r)$.   □

Notation. The rule N can be derived from N_2; N_2 applied to the first and third assumption of N gives the left-to-right direction, the second and fourth assumption give the other direction. So, in the following we will write WF_{N_2} instead of $WF_N N_2$.

Lemma 14. *Let $\overline{[\![A]\!]} \cup [\![B]\!] \subseteq \overline{[\![C]\!]} \cup [\![D]\!]$ in a canonical model of WF_{N_2}, then $WF_{N_2} \vdash (A \to B) \to (C \to D)$.*

Proof. The proof is similar to Lemma 4.   □

Lemma 15. *If $WF_{N_2} \subseteq L$, then the basic NB-canonical model of logic L is closed under superset equivalence.*

Proof. Suppose that in the basic NB-canonical model of logic L, $(X, Y) \in NB(\Gamma)$ and $\overline{X} \cup Y \subseteq \overline{X'} \cup Y'$. By definition of NB in the basic NB-canonical model there exist formulas A, B, C and D such that $(X, Y) = ([\![A]\!], [\![B]\!])$ and $(X', Y') = ([\![C]\!], [\![D]\!])$, where $A \to B \in \Gamma$. Using Lemma 14, $\vdash (A \to B) \to (C \to D)$. Hence $C \to D \in \Gamma$ and $(X', Y') \in NB(\Gamma)$.   □

Definition 13. *The N-neighborhood frame $\mathfrak{F} = \langle W, N, \mathcal{X} \rangle$ is closed under superset if and only if for all $w \in W$, if $Y \in N(w)$ and $V \in \mathcal{X}$ and $Y \subseteq V$, then $V \in N(w)$.*

Lemma 16. *Let $\langle W, NB, \mathcal{X} \rangle$ be an NB-neighborhood frame closed under superset equivalence. Then the equivalent N-neighborhood frame $\langle W, N, \mathcal{X} \rangle$ constructed in Lemma 7 is closed under superset.*

Proof. Assume $Y \in N(w)$ and $Y \subseteq V$. By construction we know that $N(w) = \{\overline{X} \cup Y \mid (X, Y) \in NB(w)\}$. Since $Y = \overline{W} \cup Y$, $\overline{W} \cup Y \in N(w)$. Moreover, $V = \overline{W} \cup V$, so $\overline{W} \cup Y \subseteq \overline{W} \cup V$. So, by superset equivalence $V = \overline{W} \cup V \in N(w)$. So, $N(w)$ is closed under superset.   □

Theorem 14. *The logic WF_{N_2} is sound and strongly complete with respect to the class of N-neighborhood frames that are closed under superset.*

Proof. By Lemmas 13, 15 and 16.   □

Lemma 17. *1. $WF_{N_2} \vdash \widehat{C}$.*
2. $WF_{N_2} \vdash \widehat{D}$.
3. $WF_{N_2} \vdash C_w$.
4. $WF_{N_2} \vdash (A \to A \wedge B \wedge C) \to (A \to A \wedge B)$.
5. $WF_{N_2} \vdash (A \to A \wedge B) \to (C \wedge A \to C \wedge A \wedge B)$.

Proof. The proofs are easy.   □

Proposition 3. *1.* $\mathsf{WF\widehat{C}} \vdash (A \to A \wedge B \wedge C) \to (A \to A \wedge B)$.
2. $\mathsf{WF\widehat{D}} \vdash (A \vee B \vee C \to A) \to (A \vee B \to A)$.
3. $\mathsf{WFC_W} \vdash (A \to A \wedge B) \to (C \wedge A \to C \wedge A \wedge B)$.

Proof. The proofs are easy.   □

The logic $\mathsf{WF\widehat{CD}}$ is complete with respect to the class of NB-neighborhood frames that are closed under upset and downset [7], i.e. for all $w \in W$,

(Upset) if $(X, Y) \in NB(w)$ and $Y \subseteq Z$ then $(X, Z) \in NB(w)$,
(Downset) if $(X, Y) \in NB(w)$ and $Z \subseteq X$ then $(Z, Y) \in NB(w)$.

Lemma 18. $\mathsf{WF\widehat{CD}} \nvdash (p \to q) \to (r \wedge p \to r \wedge q)$.

Proof. Consider the rooted NB-neighborhood frame $\mathfrak{F} = \langle W, NB, \mathcal{P}(W) \rangle$ with

$$W = \{w, v, g\},$$
$$NB(v) = NB(g),$$
$$NB(w) = \{(\{w, v\}, \{v, g\}), (\{w\}, \{v, g\}), (\{w\}, \{v\}), (\{w, v\}, \{v\})\}$$
$$\cup \left\{ (X, Y) \in (\mathcal{P}(W))^2 \mid X \subseteq Y \right\}.$$

The frame $\mathfrak{F}$ is closed under upset and downset. Then consider the valuation $p^{\mathfrak{M}} = \{w, v\}$, $q^{\mathfrak{M}} = \{v\}$ and $r^{\mathfrak{M}} = \{w\}$. With this valuation we can conclude, $g \nVdash (p \to q) \to (r \wedge p \to r \wedge q)$, since $(p \to q)^{\mathfrak{M}} \nsubseteq (r \wedge p \to r \wedge q)^{\mathfrak{M}}$.   □

5.1 Completeness of $\mathsf{WF_{N_2}}$ with Respect to Full N-Neighborhood Frames

In this subsection we will just discuss the completeness with respect to full frames of $\mathsf{WF_{N_2}}$. For the other extensions of WF that we described we haven't yet succeeded in proving completeness with respect to full frames.

Definition 14. *Let* $\mathfrak{M} = \langle W, N, V \rangle$ *be an N-neighborhood model, the* ***supplementation*** *of* $\mathfrak{M}$, *denoted* $sup(\mathfrak{M})$, *is the N-neighborhood model* $sup(\mathfrak{M}) = \langle W, N^{sup}, V \rangle$, *such that for every* $w \in W$ *and* $Y, V \in \mathcal{P}(W)$:

$$N^{sup}(w) = \{V \mid \exists\, Y \in N(w) \text{ where } Y \subseteq V\}.$$

By the argument in Lemma 16, for every $X, Y, U, V \in \mathcal{P}(W)$:

$$N^{sup}(w) = \left\{ \overline{U} \cup V \mid \exists\, \overline{X} \cup Y \in N(w) \text{ where } \overline{X} \cup Y \subseteq \overline{U} \cup V \right\}.$$

Theorem 15. *Suppose that* $\mathfrak{M}$ *is the supplementation of the basic N-canonical model* $\mathsf{WF_{N_2}}$, $\mathfrak{M} = sup(\mathfrak{M}_{\mathsf{WF_{N_2}}})$. *Then* $\mathfrak{M}$ *is a full N-canonical model for* $\mathsf{WF_{N_2}}$.

Proof. Suppose that $\mathfrak{M}_{\mathsf{WF}_{N_2}} = \langle W, N, V \rangle$, then for each $\Gamma \in W$,

$$N(\Gamma) = \{ \overline{[\![A]\!]} \cup [\![B]\!] \mid A \to B \in \Gamma \}.$$

By definition of supplementation we conclude that $\mathfrak{M} = \langle W, N^{sup}, V \rangle$, such that

$$N^{sup}(\Gamma) = \{ \overline{X} \cup Y \mid \overline{[\![A]\!]} \cup [\![B]\!] \subseteq \overline{X} \cup Y \ for \ some \ A \to B \in \Gamma \}.$$

To show that $\mathfrak{M}$ is a full N-canonical model, it is enough to prove that for all formula A, B and for each $\Gamma \in W$,

$$A \to B \in \Gamma \ \text{if and only if} \ \overline{[\![A]\!]} \cup [\![B]\!] \in N^{sup}(\Gamma).$$

From left to right: Assume that $A \to B \in \Gamma$ then $\overline{[\![A]\!]} \cup [\![B]\!] \in N(\Gamma)$. But we know that $N(\Gamma) \subseteq N^{sup}(\Gamma)$, hence $\overline{[\![A]\!]} \cup [\![B]\!] \in N^{sup}(\Gamma)$.

From right to left: Assume that $\overline{[\![A]\!]} \cup [\![B]\!] \in N^{sup}(\Gamma)$, so that for some $C \to D \in \Gamma$,

$$\overline{[\![C]\!]} \cup [\![D]\!] \subseteq \overline{[\![A]\!]} \cup [\![B]\!].$$

By Lemma 14, we conclude that $\mathsf{WF}_{N_2} \vdash (C \to D) \to (A \to B)$, and so $A \to B \in \Gamma$. $\square$

Theorem 16. *The logic* WF_{N_2} *is sound and strongly complete with respect to the class of full N-neighborhood frames that are closed under superset.*

Proof. The proof is easy by Lemma 13 and Theorem 15. $\square$

6 Modal Companions

In this section we clarify the connection with modal logic, mainly with classical modal logic and monotonic modal logic, but also some stronger modal logics. We consider the translation $\square$ (see [2]) from $\mathcal{L}$, the language of propositional logic, to $\mathcal{L}_\square$, the language of modal propositional logic. It is given by:

1. $p^\square = p$;
2. $(A \wedge B)^\square = A^\square \wedge B^\square$;
3. $(A \vee B)^\square = A^\square \vee B^\square$;
4. $(A \to B)^\square = \square(A^\square \to B^\square)$.

Definition 15. *A logic* $L_\square$ *in* $\mathcal{L}_\square$ *is called a* **modal companion** *of logic* L *if, for all* A *in* $\mathcal{L}$, $\vdash_L A$ *iff* $\vdash_{L_\square} A^\square$.

Definition 16. *A system of modal logic is* **classical** *iff it is closed under RE* $(\frac{A \leftrightarrow B}{\square A \leftrightarrow \square B})$ [1].

E is the smallest classical modal logic. The logic $\mathsf{E}_{\mathsf{Nec}}$ extends E by adding the axiom $\square\top$, or equivalently, the necessitation rule. Completeness holds for $\mathsf{E}_{\mathsf{Nec}}$ with respect to (full) modal neighborhood frames that contain the unit, i.e. for all $w \in W$, $W \in N(w)$ [1]. The clause for $\square A$ in the neighborhood models is: $w \Vdash \square A$ iff $A^{\mathfrak{M}} \in N(w)$. We will write $\Vdash_N$ for truth in our N-neighborhood models (Definition 7) if we need to distinguish it from the modal $\Vdash$.

Lemma 19. *Let* $\mathfrak{M} = \langle W, N, V \rangle$ *be an* N-*neighborhood model. Then for all* $w \in W$,

$$\mathfrak{M}, w \Vdash_N A \quad \text{iff} \quad \mathfrak{M}, w \Vdash A^{\square}.$$

Proof. The proof is by induction on A. The atomic case holds by induction and the conjunction and disjunction cases are easy. We only check the implication case. So let $A = C \to D$, then

$$\mathfrak{M}, w \Vdash_N C \to D \quad \Longleftrightarrow \{v \mid v \nVdash_N C\} \cup \{v \mid v \Vdash_N D\} \in N(w)$$

$$\text{(by induction hypothesis)} \Longleftrightarrow \left\{v \mid v \nVdash C^{\square}\right\} \cup \left\{v \mid v \Vdash D^{\square}\right\} \in N(w)$$

$$\Longleftrightarrow \left\{v \mid v \Vdash \neg C^{\square}\right\} \cup \left\{v \mid v \Vdash D^{\square}\right\} \in N(w)$$

$$\Longleftrightarrow \left\{v \mid v \Vdash \neg C^{\square} \vee D^{\square}\right\} \in N(w)$$

$$\Longleftrightarrow \mathfrak{M}, w \Vdash \square(\neg C^{\square} \vee D^{\square})$$

$$\Longleftrightarrow \mathfrak{M}, w \Vdash (C \to D)^{\square}. \qquad \square$$

Our completeness Theorem 12 then enables us to prove:

Theorem 17. ([4], Theorem 5.17) *For all formulas A,*

$$\vdash_{\mathsf{WF_N}} A \quad \text{iff} \quad \vdash_{\mathsf{E_{Nec}}} A^{\square}.$$

Definition 17. *A system of modal logic is* **monotonic** *iff it is closed under* RM $\left(\frac{A \to B}{\square A \to \square B}\right)$ [3].

EM (M) is the smallest monotonic modal logic. Completeness holds for M with respect to (full) monotonic modal neighborhood frames, i.e. in $\mathfrak{F} = \langle W, N \rangle$ in which for all $w \in W$, $N(w)$ is closed under superset [3]. The modal logic corresponding to $\mathsf{WF_{N_2}}$ will be $\mathsf{M_{Nec}}$, the extension of M by the neccesitation rule. $\mathsf{M_{Nec}}$ is complete for monotonic neighborhood frames containing the unit.

An N-neighborhood model $\langle W, N, V \rangle$ is closed under superset if and only if the N-neighborhood frame $\langle W, N \rangle$ is closed under superset.

Theorem 18. *For all formulas A,*

$$\vdash_{\mathsf{WF_{N_2}}} A \quad \text{iff} \quad \vdash_{\mathsf{M_{Nec}}} A^{\square}.$$

Proof. By Theorem 14 and Lemma 19. $\qquad \square$

The classical modal logic $\mathsf{EC_{\square}}$ extends E by adding the axiom scheme $\mathsf{C_{\square}}$, $(\square A \wedge \square B) \to \square(A \wedge B)$. Completeness holds for $\mathsf{EC_{\square}}$ with respect to the class of neighborhood frames that are closed under intersection, i.e., if $X, Y \in N(w)$, then $X \cap Y \in N(w)$ [6].

Definition 18. N-*neighborhood frame* $\mathfrak{F}$ *is closed under* **N-*intersection*** *if and only if for all* $w \in W$, *if* $\overline{X} \cup Y \in N(w)$ *and* $\overline{X} \cup Z \in N(w)$ *then* $\overline{X} \cup (Y \cap Z) \in N(w)$.

Lemma 20. *The formula* $(p \to q) \wedge (p \to r) \to (p \to q \wedge r)$ *characterizes the class of rooted* N-*neighborhood frames* $\mathfrak{F}$ *satisfying closure under* N-*intersection.*

Proof. Let $\mathfrak{F} = \langle W, N, \mathcal{X} \rangle$ be closed under N-intersection and $\mathfrak{M}$ be any model based on $\mathfrak{F}$. We have to prove for all $w \in W$,

$$\{v \mid v \Vdash (p \to q) \wedge (p \to r) \Rightarrow v \Vdash p \to q \wedge r\} \in N(w).$$

For this purpose it is sufficient to show that

$$K = \{v \mid v \Vdash (p \to q) \wedge (p \to r) \Rightarrow v \Vdash p \to q \wedge r\} = W.$$

Let $w \in W$, $w \Vdash p \to q$ and $w \Vdash p \to r$ then

$$\overline{p^{\mathfrak{M}}} \cup q^{\mathfrak{M}} \in N(w)$$

$$\overline{p^{\mathfrak{M}}} \cup r^{\mathfrak{M}} \in N(w).$$

The frame is closed under N-intersection, so $\overline{p^{\mathfrak{M}}} \cup (q \wedge r)^{\mathfrak{M}} \in N(w)$, that is $w \Vdash p \to q \wedge r$. Hence $W = K$ and so for all $w \in W$, $K \in N(w)$, since by the definition of N-neighborhood frames for all $w \in W$, $W \in N(w)$.

Let N-frame $\mathfrak{F}$ and $w \in \mathfrak{F}$ be such that $\overline{X} \cup Y \in N(w)$ and $\overline{X} \cup Z \in N(w)$, but $\overline{X} \cup (Y \cap Z) \notin N(w)$. To falsify $(p \to q) \wedge (p \to r) \to (p \to q \wedge r)$ in the N-frame $\mathfrak{F}$ we should find $u \in W$ such that $u \nVdash (p \to q) \wedge (p \to r) \to (p \to q \wedge r)$.

For this purpose consider the valuation such that, $p^{\mathfrak{M}} = X$, $r^{\mathfrak{M}} = Z$ and $q^{\mathfrak{M}} = Y$. Then we will have

$$\overline{p^{\mathfrak{M}}} \cup q^{\mathfrak{M}} \in N(w)$$

$$\overline{p^{\mathfrak{M}}} \cup r^{\mathfrak{M}} \in N(w)$$

$$\overline{p^{\mathfrak{M}}} \cup (q \wedge r)^{\mathfrak{M}} \notin N(w).$$

So, $w \Vdash (p \to q) \wedge (p \to r)$ and $w \nVdash p \to q \wedge r$. That is $w \in (p \to q)^{\mathfrak{M}}$, $w \in (p \to r)^{\mathfrak{M}}$ and $w \notin (p \to q \wedge r)$. Therefore,

$$w \notin \overline{(p \to q)^{\mathfrak{M}}} \cup \overline{(p \to r)^{\mathfrak{M}}} \cup (p \to q \wedge r)^{\mathfrak{M}}.$$

So we have

$$\overline{(p \to q)^{\mathfrak{M}}} \cup \overline{(p \to r)^{\mathfrak{M}}} \cup (p \to q \wedge r)^{\mathfrak{M}} \neq W.$$

Hence,

$$\mathfrak{M}, g \nVdash (p \to q) \wedge (p \to r) \to (p \to q \wedge r),$$

and $\mathfrak{F} \nVdash (p \to q) \wedge (p \to r) \to (p \to q \wedge r)$. $\qquad\square$

Lemma 21. *If* $\mathsf{WF_NC} \subseteq L$, *then the basic* N-*canonical model of logic* L *is closed under* N-*intersection.*

Proof. Suppose that in the basic N-canonical model of logic L, $\overline{X} \cup Y \in N(\Gamma)$ and $\overline{X} \cup Z \in N(\Gamma)$. By definition of N in that model there exist formulas A, B and C such that $[\![A]\!] = X$, $[\![B]\!] = Y$ and $[\![C]\!] = Z$, where $A \to B \in \Gamma$ and $A \to C \in \Gamma$. Hence $(A \to B) \wedge (A \to C) \in \Gamma$ and so using (C), $A \to B \wedge C \in \Gamma$. Hence $\overline{[\![A]\!]} \cup [\![B \wedge C]\!] = \overline{X} \cup (Y \cap Z) \in N(\Gamma)$. So, N is closed under N-intersection. $\square$

Remark 2. On N-neighborhood frames, N-intersection and intersection are the same thing. Assume the N-neighborhood frame $\mathfrak{F}$ is closed under N-intersection and $X, Y \in N(w)$. We know that $X = \overline{W} \cup X$ and $Y = \overline{W} \cup Y$, so, by the N-intersection condition we conclude that $\overline{W} \cup (X \cap Y) = X \cap Y \in N(w)$. The other way around is immediate by the intersection condition.

Theorem 19. *The logic* $\mathsf{WF_NC}$ *is sound and strongly complete with respect to the class of full* N-*neighborhood frames that are closed under intersection.*

Proof. By Lemma 21. The basic N-canonical model is on a full frame. $\square$

Theorem 20. *For all formulas* A,

$$\vdash_{\mathsf{WF_NC}} A \quad \text{iff} \quad \vdash_{\mathsf{E_{Nec}C_\square}} A^\square.$$

Proof. By Theorem 19 and Lemma 19. $\square$

The logic K is the smallest normal, or Kripkean, modal logic. Corsi in [2] showed that the modal companion of subintuitionistic logic F is the logic K. Completeness holds for K with respect to the class of augmented neighborhood frames, i.e. closed under superset and containing for all $w \in W$, its core $\bigcap_{X \in N(w)} X \in N(w)$. Already in [1] it was shown that the logic $\mathsf{M_{Nec}C_\square}$ equals the logic K.

Theorem 21. *The logic* $\mathsf{WF_{N_2}C}$ *is sound and strongly complete with respect to the class of full* N-*neighborhood frames that are closed under superset and intersection.*

Proof. By Theorems 16 and 19. $\square$

Theorem 22. *For all formulas* A,

$$\vdash_{\mathsf{WF_{N_2}C}} A \quad \text{iff} \quad \vdash_{\mathsf{K}} A^\square.$$

Proof. By Theorem 21 and Lemma 19. $\square$

Because the modal companions of $\mathsf{WF_{N_2}C}$ and F are the same we then immediately have

Corollary 3. *The logic* $\mathsf{WF_{N_2}C}$ *equals the logic* F.

7 Conclusion

In the above the similarities and differences between N-frames and NB-frames were fully clarified. An interesting new logic WF_{N_2} was axiomatized and it was established that the monotonic logic M_{Nec} is a modal companion of this logic. It will be worthwhile to further study this logic, which takes a central place amid subintuitionistic logics. An open question is to find a modal comapanion for the basic logic WF itself. A simple answer seems very difficult, because one would need a modal logic in which the rule E is restricted, since the stronger logic WF_N is tied up with the rule E. A restriction of E is not easy to imagine. We are looking for less orthodox ways to solve the problem as a modal logic with a binary modality.

Acknowledgement. This paper greatly profited from the remarks of an anonymous referee who read the paper very carefully. Many points were clarified and readability has improved strongly. The second author expresses thanks for the financial support of the project number 95012990 of the Iran National Science Foundation (INSF) as a post-doc researcher at school of mathematics, statistics and computer science of university of Tehran.

References

1. Chellas, B.: Modal Logic: An Introduction. Cambridge University Press, Cambridge (1980)
2. Corsi, G.: Weak logics with strict implication. Zeitschrift für Mathematische Logik und Grundlagen der Mathematik **33**, 389–406 (1987)
3. Hansen, H.H.: Monotonic modal logics. Master thesis, University of Amsterdam (2003)
4. de Jongh, D., Shirmohammadzadeh Maleki, F.: Subintuitionistic logics and the implications they prove. Indagationes Mathematicae. https://doi.org/10.1016/j.indag.2018.01.013
5. Kleene, S.C.: Disjunction and existence under implication in elementary intuitionistic formalisms. J. Symb. Log. **27**, 11–18 (1962)
6. Pacuit, E.: Neighborhood Semantics for Modal Logic. Springer, Cham (2017). https://doi.org/10.1007/978-3-319-67149-9
7. Shirmohammadzadeh Maleki, F., de Jongh, D.: Weak subintuitionistic logics. Log. J. IGPL **25**(2), 214–231 (2017)

Bare Nouns and the Hungarian Mass/Count Distinction

Kurt Erbach[⊠], Peter R. Sutton, and Hana Filip

Heinrich Heine Universität, Düsseldorf, Germany
erbach@uni-duesseldorf.de, peter.r.sutton@icloud.com, hana.filip@gmail.com

Abstract. We argue that in Hungarian notionally count, singular nouns like *könyv* ('book'), *toll* ('pen'), and *ház* ('house') are semantically number-neutral (see also Farkas and de Swart (2010)). This departs from the view that such nouns are dual-life with respect to being count or mass, such as *brick* or *stone* in English, as recently argued by Rothstein (2017) and Schvarcz and Rothstein (2017), who rely on two assumptions: that pseudo-partitive (measure) NPs require mass predicates denoting measured entities (Rothstein 2011); and that classifiers modify mass nouns (Chierchia 1998, 2010). We provide evidence against these two assumptions and argue that, together with (i) the commonly accepted analysis of measure DPs on which they require *cumulative* predicates to denote what is measured (i.a, Krifka 1989; Filip 1992, 2005; Nakanishi 2003; Schwarzschild 2006; and (ii) for an analysis of classifiers (Krifka 1995; Sudo 2017) in which they combine with numerical expressions rather than nouns, a number neutral analysis of Hungarian notionally count, singular nouns covers a wider range of data than a dual-life analysis does. We build on the use of context to specify what counts as one (Landman 2011; Rothstein 2010; Sutton and Filip 2016) and the analyses of counting and measuring in Filip and Sutton (2017) yielding a novel analysis in which Hungarian has many count nouns and many mass nouns, rather than many dual-life and mass nouns, but few count nouns.

Keywords: Count/mass · Number-neutral · Classifiers · Context-sensitivity

---

First, we would like to thank the participants and organizers of the *Twelfth International Tbilisi Symposium on Language, Logic and Computation*. We especially enjoyed the inspirational environment and welcoming hospitality provided by our Georgian friends and colleagues. We are very grateful to Zsofia Gyarmathy and Karoly Varasdi for their help with Hungarian data during the preparation of this manuscript. Finally, we thank our colleagues and collaborators in the Department of Linguistics and the affiliated *Collaborative Research Center 991* at Heinrich Heine University, and in particular the participants of the *Semantics and Pragmatics Exchange (SemPrE)* colloquium. This research was funded by the German Research Foundation (DFG) CRC 991, Project C09.

A. Silva et al. (Eds.): TbiLLC 2018, LNCS 11456, pp. 86–107, 2019.
https://doi.org/10.1007/978-3-662-59565-7_5

1   Introduction

This paper offers a new perspective on semantics of counting and measuring in Hungarian. We argue that Hungarian notionally count, singular nouns like *könyv* ('book') are semantically number-neutral, denoting countable individuals and sums thereof (see also Farkas and de Swart 2010).[1] This analysis better captures the available interpretations of such nouns, namely that they can refer to one or more individuals in certain contexts. Our analysis also better reflects the empirical facts of Hungarian than the recent analyses by Rothstein (2017) and Schvarcz and Rothstein (2017) in which such nouns are claimed to be dual-life—i.e. can occur straightforwardly felicitously in count and mass syntax—similar to *brick* or *stone* in English.[2] In particular, we account for the fact that most Hungarian notionally count, singular nouns occur bare with quantifiers and in counting constructions, and lack a mass reading in argument position.

Rothstein's (2017) and Schvarcz and Rothstein's (2017) dual-life analysis of Hungarian notionally count, singular nouns rests on the claim that these nouns have a count denotation, because they are directly countable, i.e., they freely occur in counting constructions with count cardinal quantifiers, and they also have a mass denotation because they occur in two environments which Rothstein (2017) and Schvarcz and Rothstein (2017) claim to be mass syntax: pseudo-partitive (measure) DPs and classifier constructions. The latter claim presupposes two (not uncontroversial) assumptions: (i) pseudo-partitive (measure) DPs require their constituting nominal predicates (which denote what is measured) to have a mass interpretation (Rothstein 2011, 2017), and (ii) nouns in bona fide classifier constructions are mass (Chierchia 1998, 2010; Rothstein 2017).

We argue that if Hungarian count nouns are semantically number-neutral, we can cover more data than a dual-life based analysis. It not only explains why such nouns are predicted to be felicitous in measure DPs (under a widespread analysis of measure DPs), but can also explain why they can occur bare with

---

[1] A note on terminology. We use *count, mass* and sometimes *grammatically count/mass* as grammatical categories. For example, the English noun *chair* is count, since it is straightforwardly felicitous in syntatctic environments diagnostic of count nouns (such as direct numerical modification). The English nouns *mud* and *furniture* are mass, since they are straightforwardly felicitous in syntatctic environments diagnostic of mass nouns (such as occurring as bare singulars in the argument position). We use the terms *count denotation* and *mass denotation* in a theory dependent way. Most semantic analyses of the count/mass distinction differentiate count nouns from mass nouns in terms of some property of their denotation, be it *semantic atomicity* (Rothstein 2010) or *disjoint counting base set* (Landman 2016). The distinction between a *count denotation* and a *mass denotation* is just whatever the relevant semantic distinction is in the theory in question.

[2] Typically, when the syntactic environment is ambiguous, dual life nouns can have both a count reading and a mass reading. For example, *Alex's stone is in the yard* is ambiguous between the count reading in which one single stone is referred to, and a mass reading in which some portion of stone-stuff is referred to.

quantifiers and in counting constructions, and why they have no mass reading in argument position. We reject the claim that nouns in bona fide classifier constructions must be mass, because cross-linguistic evidence shows that there are classifier languages that have a grammaticized lexical mass/count distinction, and in which classifiers modify numerical expressions rather than modifying mass nouns. With these counter-arguments and novel data, we conclude that Hungarian singular count nouns are best interpreted as number-neutral predicates, as was done on independent grounds by Farkas and de Swart (2010).

The outline of the paper is as follows. First, we summarize the relevant details of the Hungarian mass/count distinction. Section 3 summarizes the main arguments Rothstein (2017) and Schvarcz and Rothstein (2017) give for their claim, one based on measure (pseudo-partitive) DPs, the other on classifiers. Section 4, provides reasons to doubt some of Rothstein's (2017) and Schvarcz and Rothstein's (2017) assumptions. We argue that a number-neutral analysis of notionally count, singular nouns in Hungarian covers a wider range of data in Sect. 6, and we give a formal account of these data based on the context sensitive analysis of counting and measuring in Filip and Sutton (2017).

2   The Hungarian Mass/Count Distinction

2.1   Morphosyntactic Tests for the Hungarian Mass/Count Distinction

Rothstein (2017) and Schvarcz and Rothstein (2017) argue that Hungarian has a grammaticized lexical mass/count distinction, which is evident in three morphosyntactic tests: (i) number marking, (ii) counting constructions, and (iii) WH-quantification.

Using plural morphology as a litmus test for the mass/count distinction in Hungarian, Rothstein (2017) and Schvarcz and Rothstein (2017) distinguish nouns that occur in plural as count, and nouns that do not as mass:

(1)   rózsa / rózsá-k
       rose / rose-PL
       'rose/roses'   (Schvarcz and Rothstein 2017, p. 185)

(2)   *kosz-ok
       dirt-PL
       'dirts'   (Schvarcz and Rothstein 2017, p. 193)

Second, Hungarian nouns which are straightforwardly compatible with numerical expressions are count, while incompatible nouns are mass. Note that plural morphology is never used on nouns in Hungarian counting constructions.

(3)   három könyv(*-ek)
       three book(*-PL)
       'three books'   (Schvarcz and Rothstein 2017, p. 185)

(4) *három kosz
     three dirt
     'three dirts'   (Schvarcz and Rothstein 2017, p. 193)

Third, nouns that can straightforwardly occur with *hány* ('how many') are count, and nouns that cannot are mass. As in counting constructions, nouns do not take plural morphology when composed with WH-quantifiers:

(5) hány   könyv(*-ek)?
     how.many book(*-PL)
     'How many books?'

(6) *hány   szemét/ sár/ kosz?
     how.many trash   mud dirt
     'How many trash/mud/dirt?'   (Schvarcz and Rothstein 2017, p. 195)

Schvarcz and Rothstein (2017) use these data and more to argue that Hungarian has a true mass/count distinction and that it is not a genuine classifier language, contrary to the claim made in Csirmaz and Dékány (2014). Count nouns are nouns that occur with plural morphology, are directly compatible with numerical expressions—i.e. can be counted—and are compatible with the WH-quantifier *hány* ('how many'). Mass nouns are nouns that do not occur with plural morphology, and are not directly compatible with numerical expressions or *hány* ('how many'). While Hungarian does in fact use classifiers in counting constructions (see below), they are not mandatory, unlike in true classifier languages, such as Mandarin and Japanese. These characteristics make Hungarian a sort of 'mixed' language with respect to counting given it has a straightforward mass/count distinction and optional classifiers.

2.2 Hungarian Measure DPs, and Classifiers

In addition to claiming that Hungarian has a mass/count distinction, Rothstein (2017) and Schvarcz and Rothstein (2017) point out three syntactic environments in which Hungarian differs from other number marking languages with a mass/count distinction. First, measure DPs in Hungarian are only felicitous with bare singular (count or mass) nouns e.g. (7).

(7) Ki   cipelte fel a   harminc kg könyvet?
     who hauled up the thirty   kg book.SG.ACC
     'Who hauled the thirty kilos of books upstairs?'   (Elicited data)[3]

The Hungarian WH-quantifier *mennyi* ('what quantity of') is likewise only felicitous with bare singular nouns be they count or mass. Generally, when mass nouns occur with the WH-quantifier *mennyi* ('what quantity of'), the question

---

[3] The novel Hungarian data and readings thereof were elicited in correspondence with native speakers including Zsofia Gyarmathy and Károly Varasdi.

can only be felicitously answered with measure˙ of weight, but not cardinality. Hence *három kiló-t* ('three kilos') is a felicitous answer to (8), but *hármat* ('three') is not.

(8)   mennyi   szemét/ sár/ kosz?
       what.quantity.of trash   mud dirt
       'What quantity of trash/mud/dirt?' (Schvarcz and Rothstein 2017, p. 195)

Nouns like *könyv* ('book'), which are felicitous in count syntax, also occur with *mennyi* ('what quantity of'). However, the relevant questions with *mennyi* ('what quantity of') can be felicitously answered in terms of weight or cardinality. Both *három kiló-t* ('three kilos') and *hármat* ('three') are felicitous answers to (9) (Schvarcz and Rothstein 2017, p. 200, example (42)):

(9)   Mennyi   könyvet   tudsz   cipelni?
       what.quantity.of book.SG.ACC able.you to.carry
       'What quantity of books can you carry?'

Lastly, as Dékány (2011) and Csirmaz and Dékány (2014) have shown, most notionally count, singular nouns in Hungarian can occur with optional classifiers:

(10)   a.   három *(darab) sár
            three CL$_{general}$ mud
            'three pieces of mud'   (Schvarcz and Rothstein 2017, p. 194, ex. 27a)
       b.   három (szál)   rózsa
            three CL$_{thread}$ rose.SG
            'three roses'   (Schvarcz and Rothstein 2017, p. 185, ex. 3a)
       c.   három (darab)   könyv
            three CL$_{general}$ book.SG
            'three books'   (Schvarcz and Rothstein 2017, p. 185, ex. 3b)

Analyses of number-marking languages like English cannot straightforwardly be applied to Hungarian, because data like (7)–(10) are not found in most number marking languages. While measure DPs in Hungarian take bare singular nouns, measure DPs in English are felicitous with plural and mass terms (e.g. *thirty kilos of books/mud*). Also, while Hungarian has optional classifiers for counting objects and can also use classifiers for counting portions of substances (while classifiers are required for counting anything in classifier-languages like Mandarin), English only has classifier-like constructions for mass nouns (e.g. *three pieces of mud/kitchenware*). How one analyses singular nouns, measure DPs, and classifier(-like) constructions, will therefore shape how such nouns are characterized in respect to the mass/count distinction.

3   The Dual-Life Analysis of Hungarian Nouns

Rothstein (2017) and Schvarcz and Rothstein (2017) claim that, unlike many number-marking languages, Hungarian has few nouns that only have a count denotation and it has many nouns with both count and mass denotations—i.e. dual-life nouns like *stone* in English, which can occur in either count or mass syntax. This claim that most notionally count, singular nouns in Hungarian, like *könyv* ('book') are dual-life rests on the (i) occurrence of these nouns in measure DPs and classifier constructions, and (ii) the controversial assumption that nouns that occur in these environments are mass. Rothstein (2017) and Schvarcz and Rothstein (2017) build their analysis of Hungarian on the semantic theory of the mass/count distinction in Rothstein (2010) also taking inspiration from the analysis of measure DPs in Rothstein (2011) and the analysis of classifier constructions in Chierchia (1998, 2010).

In respect to measure DPs, Rothstein (2017) and Schvarcz and Rothstein (2017) rely on the assumption that nouns in measure DPs are mass. Nouns like *könyv* ('book'), they argue, must therefore have a mass denotation in addition to having a count denotation (the latter being shown with data such as (5)). Additionally, Rothstein (2017) and Schvarcz and Rothstein (2017) take the fact that questions like (9) can be felicitously answered in terms of measure (e.g., *három kiló-t* 'three kilos') as evidence that most Hungarian notionally count, singular nouns have bona fide mass denotations, and so are best viewed as dual-life. As they argue, given that most of these nouns can both be straightforwardly individuated in terms of cardinality, and directly measured (by e.g., weight), they must have both count and mass denotations, which means that they are dual-life nouns.

With respect to their mass denotations, Hungarian dual-life nouns are interpreted as object mass nouns, such as *furniture* in English (Schvarcz and Rothstein 2017), rather than as shifted by a universal grinder-like mechanism into a substance interpretation. Object mass nouns, like *furniture* in English, denote discrete objects, as opposed to substance denoting mass nouns like *water*, which do not. *Objects* and *substances* in the sense of Soja et al. (1991), respectively refer to concrete solids like *knives* that hold their shape across the space-time continuum, and non-solids like *mud*. Object mass nouns are of particular importance because they show a mismatch between grammatical mass behavior and conceptual individuation. However, Gyarmathy (2016) has suggested that Hungarian has no object mass nouns, though this claim has not been the thoroughly investigated.

The empirical basis for Rothstein's (2017) and Schvarcz and Rothstein's (2017) claim that nouns in measure DPs are mass is examples like (11), taken from Rothstein (2011) [p. 24, example (45b)].

(11)   #Twenty kilos of books are lying on top of each other on the floor.

(Rothstein 2011, p. 24)

According to Rothstein (2011, 2017), (11) is infelicitous, and the individual books are not semantically accessible by the reciprocal operator *on top of each other*, so *each* has no grammatical antecedent. This is precisely because the plural count noun *books* must first shift into a mass interpretation in order to intersectively combine with the measure phrase *twenty kilos (of)*. On this intersective analysis, the whole pseudo-partitive (measure) DP is mass, which is also supported by data like (12), according to Rothstein (2011).

(12)   a. #I have read many of the twenty kilos of books that we sent.
         b. I have(n't) read much of the twenty boxes/kilos of books in our
            house.   (Rothstein 2011, p. 23)

In summary, according to Schvarcz and Rothstein (2017), measure readings of *mennyi* N ('what quantity of N') questions are aligned with mass interpretations of N, and cardinality interpretations of *mennyi* N ('what quantity of N') questions are aligned with count interpretations of N, hence, given that ('what quantity of N') questions formed with singular nouns like *könyv* ('book') admit both interpretations, singular nouns like *könyv* ('book') have both a count and a mass interpretation (are dual life).

In respect to classifiers, Rothstein (2017) and Schvarcz and Rothstein (2017) assume that any noun in a classifier construction must have a mass denotation, as argued by Chierchia (1998, 2010). For Chierchia (2010), all nouns in classifier languages are kind denoting predicates of type k. Given the assumption that numerical expressions are universally of the adjectival type $\langle\langle e, t\rangle, \langle e, t\rangle\rangle$, Chierchia (2010) argues that classifiers are of type $\langle k, \langle e, t\rangle\rangle$, meaning they combine with mass nouns to form a countable NP. Following this line of thought, and given the fact that most notionally count, singular nouns in Hungarian can occur with classifiers, Rothstein (2017) and Schvarcz and Rothstein (2017) argue that most Hungarian notionally count, singular nouns must (also) have a mass denotation.

The formal representation of the claim in Rothstein (2017) and Schvarcz and Rothstein (2017) that Hungarian notionally count, singular nouns like *könyv* ('book') are simultaneously mass and count is as follows. On its mass reading, *könyv* ('book') denotes a root noun, a plural subset of the domain M equal to the upward closure of a set of atoms ($N_{root} = $ *A where *X $= m \in M: \exists\, Y \subseteq X: m = \sqcup_M Y$). Count nouns are derived from the root via the COUNT$_k$ operation, which picks out a set of *semantic atoms*. Semantic atoms ($\langle d,k\rangle$: $d \in k$) are atoms relative to the context $k : k \subseteq M$ (Rothstein 2011). On its count reading, *könyv* ('book') denotes a set of countable *semantic atoms*. Each dual-life noun in Hungarian, therefore, has two denotations, one mass and one count.

4   Counterarguments to the Dual-Life Analysis

As we have seen above, the claim that notionally count nouns in Hungarian are dual-life relies on two assumptions: (i) all nouns in measure DPs are mass, and (ii) all nouns in classifier constructions are mass. It also rests on the assumption that notionally count, singular nouns in count syntax denote only single entities. In what follows we will provide four main arguments, including novel data from native speakers, against this claim and the assumptions it relies on.

First, we provide data against the claim that pseudo-partitive (measure) DPs require only mass predicates. The key evidence comes from the observation that the individuals denoted by a plural count noun are accessible to semantic operations, even if that noun is in a pseudo-partitive (measure) DP (see also Landman 2016). We first note that native English speakers are divided on the acceptability of sentences such as (13).

(13)   Twenty kilos of books are lying on top of each other on the floor.

Some native speakers, including the native English speaking authors of this paper and six consultants we have asked, straightforwardly interpret (13) as meaning that the books are stacked one on top of the other—i.e. the individual books are accessible by *on top of each other*, and their cumulative weight is twenty kilos. We, therefore, think that placing any theoretical burden on the felicity or infelicity of such constructions should be at least questioned until further empirical work has been done to clarify matters.

More compellingly, perhaps, continuations of measure DPs show that plural count nouns retain their atomic denotations, but atoms in the denotation of object mass nouns are not straightforwardly accessible to semantic operations. This can be shown by the observation that it is possible to anaphorically refer to the countable individuals in the plural count denotation of the dual-life noun *chocolate* in a measure DP (14). Such an anaphoric reference is impossible with a substance denoting mass noun like *hummus* (15), without a shift to a portion reading, and anaphoric reference is also excluded with an object mass noun in the same context (16), despite its denoting perceptually and conceptually salient objects (e.g. individual pieces of furniture).

(14)   I bought 200 g of chocolates, each of which was filled with a different kind of ganache.

(15)   #I made 1.5 kg of hummus, each of which was eaten at the party.

(16)   #I shipped 200 kg of furniture, each of which went to a different address.

The view that nouns in measure DPs uniformly have an (object) mass interpretation (Rothstein 2011, 2017)—i.e. lack denotations with an accessible atomic structure—cannot straightforwardly explain the differences in (14)–(16). Further complicating the picture are nouns like *livestock* and *cattle* which seem more acceptable than those like *furniture* in such a construction (17). Allan (1980) has shown that such nouns like *cattle* belong in a class of their own, separate from other nouns given they do not pattern with object mass nouns like *furniture* or substance denoting mass nouns like *water* (18).

(17)   ?I sold 50 tonnes of livestock, each of which went to a different farmer.

(18)   a.   Quite a few livestock/cattle have disappeared today.
          b. #Quite a few furniture/water have disappeared today.

In sum, our first argument against the assumption that nouns in measure DPs are mass denoting is that plural nouns retain their atomicity when used in measure DPs, and the objects they denote are semantically accessible by reciprocal operators. Based on this evidence, we conclude that measure DPs do not require nouns to be mass, rather they also sanction plural count predicates (denoting what is measured), contrary to Rothstein (2011). Given the claim of Rothstein (2011) is fundamental to the arguments of Rothstein (2017) and Schvarcz and Rothstein (2017), their claims are weakened as well.

Our second counterargument is that Hungarian notionally count, singular nouns do not have available mass interpretations when used in a (full) argumental position. If these nouns were truly dual-life, then they could have either a count or a mass reading—i.e. it should be able to refer to either one book or a collection of books—but this prediction is not borne out. In Hungarian, an object mass reading is not available, at least with a definite determiner, and cannot be enforced with context.

(19) A könyv 2 kg-ot nyom.
      the book 2 kg-ot weigh
      'The book weighs 2 kg.' (only refers to one book)   (Elicited data)

While an anonymous reviewer points out that *a könyv* ('the book') might only have a singular count reading and not a mass reading in (19) because of competition with the definite plural, which would be used to refer to individuals and sums thereof. However, for dual-life nouns in other languages, we do not see any blocking of the mass reading for sentences in which the dual-life noun is in a definite DP in the object position. For example, in English, the dual life noun used in a definite DP, has either the count or mass definite reading straightforwardly available in (20), depending on its context.

(20)   The seed in the shed was damaged by the cold and dampness.

Context A: Alex had two sunflower seeds. One single seed was stored in the shed over the winter, the other was stored indoors. In this context, (20) refers to a single seed.
Context B: Alex, a farmer, had several sacks of seed. Some sacks were stored in the shed over the winter, the others were stored in an indoor storage room. In this context, (20) refers to a collection of seeds (all those in the relevant sacks).

Furthermore, in Hungarian, a small number of object denoting nouns can get an object mass reading in such a context. *Löszer* ('ammunition'), *felszerelés* ('equipment'), and *csomagolás* ('packaging') can equally felicitously refer to one or more than one object depending on the situation as in (21).

(21) A löszer   2 kg-ot nyom.
      the ammunition 2 kg-ot weigh
      'The ammunition weighs 2 kg.' (one or several pieces)

                                                              (Elicited data)

True dual-life nouns, therefore, have two readings when singular and definite, though this is not seen with Rothstein's (2017) and Schvarcz and Rothstein's Schvarcz and Rothstein (2017) proposed dual-life nouns. Also, certain object denoting singular nouns can have mass interpretations when definite while most do not. These characteristics of singular count nouns like *könyv* ('book') is not addressed by Rothstein (2017) or Schvarcz and Rothstein (2017), though we take it as evidence that most Hungarian notionally count, singular nouns do not have a mass denotation and therefore are not dual-life.

Turning from measure phrases to classifier phrases, we now present two counterarguments against the claim that Hungarian nouns like *könyv* are dual-life is that nouns need not be mass in classifier constructions. Recall that Rothstein (2017) and Schvarcz and Rothstein (2017) argue that the co-occurrence of nouns with classifiers indicates the availability of a mass reading. This adheres to one of two main approaches to the analysis of classifiers, namely those like Chierchia (1998, 2010) in which classifiers combine with nouns thereby making the nouns countable. In the other approach, e.g. Krifka (1995), classifiers combine with numerals and form a numerical determiner. Here, we collate two arguments from the body of research which favor an approach to classifiers like that in Krifka (1995). We use the data therein to argue against the claim made in Rothstein (2017) and Schvarcz and Rothstein (2017).

Our first argument against the view that all nouns have uniform denotation in classifier languages is that classifier languages like Japanese have at least some reflexes of a grammaticized lexical mass/count distinction (Bale and Barner 2012; Sudo 2016, 2017). Rothstein's (2017) and Schvarcz and Rothstein's (2017) analyses of Hungarian classifiers emulate Chierchia's (2010) analysis of classifiers in classifier languages like Mandarin, which also presupposes Chierchia (1998). However, there are reasons to doubt this is the right analysis for classifiers in all languages, given that it fails to yield the right predictions for Japanese, as Sudo (2016, 2017) shows. He suggests that there are nouns in Japanese, namely those for objects like houses, books and the like, that can be straightforwardly used in certain quantifying constructions without any classifier, while nouns for substances, liquids, and gases are infelicitous in such constructions. Good examples are the quantifiers *nan-byaku-to-iu* ('hundreds of') or *dono mo* ('every'), as in (22) below:

(22)   a.   dono-ie-mo   totemo furui
            which-house-MO very   old
            'Every house is very old.'
       b.   #dono-ase-mo   arainagashita
            which-sweat-MO washed.off
            Intended: '(I) washed off all the sweat.'   (Sudo 2017,p. 6, ex. 12)

While the existence of examples like (22) in classifier languages was not noticed by Chierchia (2010, 1998), Sudo (2017), based on such data, argues for a novel analysis of Japanese classifiers, which is similar to Krifka's (1995) analysis of classifiers in Mandarin. Sudo (2017) suggests that Japanese classifiers are required by numerical expressions, rather than by nouns: Numericals are of type n, classifiers are of type $\langle n, \langle e, t \rangle \rangle$, and together they form a predicate of type $\langle e, t \rangle$. Sudo (2016, 2017) concludes that there are nouns in Japanese with countable denotations, and that nominal denotations in Japanese, a classifier language, are not so different from those in non-classifier languages like English.

Our second argument that nouns need not be mass with classifiers is from analyses of languages like Chol (Mayan) that have an idiomatic use of classifiers that is insensitive to the mass/count distinction (Bale and Coon 2014). On their view, the idiosyncratic classifier use in languages like Chol (Mayan) speaks for Krifka's (1995) analysis of classifiers, and against Chierchia's (1998; 2010) analysis; i.e. classifiers combine with numerals to form numerical determiners rather than combining with nouns to form countable nouns.

The idiosyncratic use of classifiers in Chol results from the contact of the Chol counting system with Spanish. Chol numerical expressions for numbers 1–6, 10, 20, 40, 60, 80, 100, and 400 require the use of classifiers (23-a), and Spanish-based numerical expressions cannot be used with classifiers (23-b).

(23)   a.   ux-*(p'ej) tyumuty
              three-CL   egg
              'three eggs'
        b.   nuebe-(*p'ej) tyumuty
              nine-CL   egg
              'nine eggs'   (Bale and Coon 2014,p. 701)

Given that languages like Chol use classifiers only with certain numbers, but never with others, Chierchia's (1998) analysis, as Bale and Coon (2014) argue, is implausible. Moreover, it would require the ad hoc assumption that all countable nouns in Chol are mass when used with Chol numerical expressions, but count when used with Spanish-based numerical expressions. This would mean that all countable nouns in Chol are dual-life and that there would have to be rules specifying which of the noun's denotations is to be used with each number expression. Such an ad hoc assumption is avoided on Krifka's (1995) analysis. Chol numerical expressions would denote numbers, e.g. $[\![ux]\!] = 3$, and, therefore, require a classifier in order to combine with a noun. Spanish-based numbers in Chol would have a built in cardinality function and, therefore, straightforwardly occur with a noun but not be able to occur with a classifier.

We take the evidence from Japanese and Chol to mean that classifier languages may have a grammaticized lexical mass/count distinction in which classifiers combine with numericals, contrary to the common view, and also capitalize on the arguments made in the studies cited above that the most straightforward analysis of the relevant data in Japanese and Chol is one in which the classifier combines with numericals rather than nouns. The presence of classifier constructions in Hungarian, therefore, does not require that notionally count, singular nouns like *könyv* ('book') have a mass denotation, *pace* Rothstein (2017) and Schvarcz and Rothstein (2017).

To summarize our counterarguments, we have given four negative arguments against Rothstein's (2017) and Schvarcz and Rothstein's (2017) dual-life analysis. We have provided reasons to doubt that all nouns in pseudo-partitive (measure) DPs are mass (their atoms can be accessed with distributive determiners like *each* in some contexts), and we have provided data which casts doubt on the claim that notionally count Hungarian nouns in the singular are in fact dual-life (they) lack mass readings in some grammatical contexts). Lastly, we provided arguments that the Hungarian classifiers need not be analyzed as requiring mass nouns, rather a growing body of evidence supports an analysis in which classifier languages have a grammaticized lexical mass/count distinction in which classifiers combine with numericals. Each of these counterarguments miltate against the assumptions of Rothstein (2017) and Schvarcz and Rothstein (2017), and they point to an analysis in which notionally count, singular nouns like *könyv* are number neutral rather than dual-life.

5 Evidence for Number Neutrality

If most notionally count, singular nouns in Hungarian are not dual life, then an alternative is that they are number neutral (denote countable entities and sums thereof). Indeed, there is direct evidence for number neutrality: Hungarian notionally count, singular nouns in count syntax cannot denote only single entities, because they often refer to sums of individuals as well. Farkas and de Swart (2010) have shown that, in addition to denoting individuals, notionally count, singular nouns also denote sums of individuals with the data in (24-a).

(24)   a.   sok   / több / mindenféle gyerek / *gyerekek
           many / more / all.kind   child   / child.pl
           'many/more/all kinds of children'
     b.   egy pár   gyerek / *gyerekek
           a   couple child   / child.pl
           'a couple of/some children'   (Farkas and de Swart 2010)

Most notionally count, singular nouns denote both singularities and pluralities in yet more syntactic environments. For example, one can use the bare singular as in (25) to announce that one or more books have arrived, and then follow up with a specific number, for instance, four. The plural *könyvek érkeztek* ('Books arrived') could also be used, but would entail exclusive plural reference (only to sums), while the singular makes no commitment to the reference of sums.

(25)   könyv   érkezett.   Négy.
       book.SG.NOM arrived.3SG four
       'Books arrived. Four.'   (Elicited data)

Hungarian notionally count, singular nouns in the translative case can also denote individuals and sums thereof. In (26), the use of the singular *pillangó* ('butterfly') in the translative case entails that one or more butterflies have undergone a complete transformation from caterpillars.

(26)   Láttuk,   ahogy a   hernyók   pillangó-vá
       see.1PL.PST as   the caterpillar.PL butterfly.SG-TRANS
       váltak.
       become.3PL.PST
       'We saw the caterpillars become butterflies.'   (Elicited data)

The examples above provide evidence that notionally count, singular nouns denote individuals and sums thereof. In other words, notionally count, singular nouns are number neutral. The fact that singular nouns receive number-neutral interpretation in Hungarian has also been discussed as the result of pseudo-incorporation in Farkas and De Swart (2003), though the examples discussed therein involve nouns in the accusative case. The examples we provide and those from Farkas and de Swart (2010) show that bare singular nouns can also receive number-neutral interpretation when outside of a pseudo-incorporation environment.

6 Analysis

Based on the evidence from the previous section, our analysis starts out with the assumption that Hungarian singular count nouns like *könyv* ('book') denote number-neutral predicates. As we argue below, this assumption explains their straightforward acceptability in constructions in which they occur bare and can be interpreted as singular or plural, in counting constructions, measure DPs, and quantifed DPs. At the same time, it also prompts the following question: What purpose do classifiers serve in Hungarian? We answer this question by proposing that sortal classifiers, such as *darab*, restrict what counts as 'one' individual in counting constructions.

On the widespread view of the semantics of measure phrases (e.g Krifka 1989; Filip 1992, 2005; Nakanishi 2003; Schwarzschild 2006, i.a.), measure phrases like *twenty kilos (of)* select for cumulative predicates, which in English are expressed either by mass (e.g. *flour*) or plural count nouns (e.g. *books, apples*), and are built with extensive measure functions (e.g., KILO) which can only apply to cumulative Ps (27) (or non-quantized predicates, see Krifka 1989) to yield quantized predicates (e.g. *twenty kilos of flour/books*), modified from (28) (Krifka 1989).

(27) $\forall P[\text{CUM}(P) \leftrightarrow \forall x\, \forall y[P(x) \land P(y) \to P(x \sqcup y)]]$
(28) $\forall P[\text{QUA}(P) \leftrightarrow \forall x\, \forall y[P(x) \land P(y) \to \neg y \sqsubset x)]]$

Crucially, English measure phrases (e.g. *twenty kilos (of)*) cannot apply to singular count nouns like *book*, because they are already quantized[4]. In other words, since the evidence above supports an analysis in which Hungarian notionally singular nouns are number-neutral, they have cumulative (and thus not quantized) reference, and can straightforwardly occur in measure DPs. We, therefore, have an alternative answer for why Hungarian sentences like (9): *Mennyi könyvet tudsz cipelni?* ('What quantity of books can you carry?'), can be answered with both *hármat* ('three') and *három kiló-t* ('three kilos'): Number-neutral count nouns are atomic, and so suitable for counting, and cumulative, and so suitable for measuring. Such an analysis better accounts for the data above and is

---

[4] Not all singular count nouns like *fence* and *wall* are quantized, as originally observed by Zucchi and White (1996). Notably, however, such nouns are also felicitous in measure phrases, e.g., *Alex put up 400 m of fence* (Filip and Sutton 2017).

compatible with a widely accepted account of measure DPs on which measure phrases sanction cumulative predicates (Krifka 1989; Filip 1992, 2005; Nakanish 2003; Schwarzschild 2006, i.a.).

An analysis in which Hungarian singular count nouns like *könyv* ('book') denote number-neutral predicates is also compatible with an analysis of classifier constructions like, that in Krifka (1995), in which classifiers modify numericals rather than nouns. While it is the case that classifiers are required for counting portions of mass nouns in Hungarian, this does not force us to conclude that nouns must be mass in order to occur in a counting construction with a classifier. We propose a novel analysis of classifiers in Hungarian on which they introduce selectional restrictions to counting DPs. We propose that Hungarian classifiers combine with numerical expressions to restrict what counts as 'one' individual in counting constructions. In addition to the arguments by Sudo (2016, 2017) and Bale and Coon (2014) for an analysis like Krifka's (1995), in which classifiers combine with numeral interpretations of numerical expressions, we provide independent observations that classifiers like *darab*, *szál*, and *szem* restrict what counts as 'one': while a 'numerical + noun' combination can be used to count individuals or kinds (29), a 'numerical + $darab_{CL}$ + noun' combination, for instance, only sanctions counting of individuals (30).

(29)   három sütemény
        three   cookie
        'three (individual/kinds of) cakes/cookies.'   (Elicited data)

(30)   három darab   sütemény
        three   CL$_{piece}$ cookie
        'three (individual/*kinds of) cakes/cookies.'   (Elicited data)

We assume that numerical expressions in Hungarian (*egy* 'one', *kettő* 'two', *három* 'three') denote numerals—e.g. $[\![három]\!] = 3$. Similar to Krifka's (1995) analysis, classifiers such as *darab* shift numeral denoting numerical expressions into numerical determiners (i.e. function from predicates, P, to the set of individuals that have the relevant cardinality with respect to that predicate). However, our novel proposal is that the semantics of classifiers like *darab* also introduces a selectional restriction on P, namely, that the atoms of P are individuals (as opposed to (sub)kinds).

6.1   Counting and Measuring Singular NPs

Recent analyses of the mass/count distinction all incorporate some notion of context in order to account for interactions between nominal denotation and countability. For instance, object(s) in the denotation of a given noun can be counted in more than one way, depending on our counting perspective or context (Filip and Sutton 2017; Landman 2011, 2016; Rothstein 2010, 2017; Sutton and Filip 2016). Sutton and Filip (2016) synthesize the two distinct, but related, notions of context formalized in Rothstein (2010) and Landman (2011) in order to account for variations in countability within a particular language and across languages. For Rothstein (2010), count nouns denote *semantic atoms* (formally, entity-context pairs, see Sect. 3). This context-indexing, crucially, allows her to capture how non-quantized count nouns like *fence*, are nonetheless countable: they denote (possibly different) sets of entities in different contexts. In Landman (2011), sets of entities that count as one are called *generator sets*. Count nouns have disjoint generator sets and mass nouns have overlapping generator sets. Object mass nouns such as *kitchenware* and *furniture* have overlapping generator sets because these sets admit of different *variants*, namely, maximally disjoint subsets of the generator sets, that also contain entities that count as 'one'. For example, a teacup and saucer sum would be a different variant of the generator set for *kitchenware* than the individual teacup and individual saucer. *Kitchenware* is thus mass on Landman's account because, for example, "the cup, the saucer, the cup and saucer all count as kitchenware and can all count as one simultaneously in the same context" (Landman 2011, pp. 34–35).

Sutton and Filip (2016) argue that crosslinguistic patterns in the encoding of countability are the result of two functions on predicates, $P_{\langle e,t \rangle}$. The function $\mathbf{IND}_{\langle \langle e,t \rangle, \langle e,t \rangle \rangle}$ identifies a, possibly overlapping, set of individuals in the denotation of a noun, and the function $c_{\langle \langle e,t \rangle, \langle e,t \rangle \rangle}$ identifies what individuals are counted in a specific context, i.e., subjected to grammatical counting operation. Similar to Landman's (2011) variants, the c function can have different results for certain sets of objects. For example, take the set of things that count as one for *kitchenware* in c: $c(\mathbf{IND}(\text{KITCHENWARE}))$ (the counting base set for *kitchenware* at c). A mortar and pestle could be counting base set when $c = c_1$, and their sum could be in the counting base set when $c = c_2$. With a specific counting schema, c_i, applied to X, the denotation will be maximally disjoint subset of X. However, with the null counting scheme, c_0 applied to X, the denotation is a set of all, possibly overlapping partitions in X (which comes out as equivalent to X). So at the null counting schema (when $c = c_0$), the pestle and mortar sum, the individual pestle, and the individual mortar are all members of the counting base set which means that $c_0(\mathbf{IND}(\text{KITCHENWARE}))$ has members which overlap.

Grammatical counting is possible when the counting base is a disjoint set, c_i, but counting goes wrong when the counting base is an overlapping set, c_0. The possibility of resolving an overlap at specific counting schemas explains variation in mass/count lexicalization patterns for collective artifact nouns like *furniture* and *meubel* ('(piece of) furniture', Dutch). Collective artifact nouns interpreted at c_0 are mass, e.g. *furniture*, and collective artifact nouns interpreted at a specific counting schema, c_i, are count, e.g. *meubel* ('(piece of) furniture', Dutch).

Sutton and Filip (2016) also argue that predicates for substances and objects are semantically distinguished in their lexical entries. This is supported by the ability of pre-linguistic infants to distinguish substances from objects (Soja et al. 1991). Formally, only object denoting nouns have the **IND** function in their lexical entries. Sutton & Filip therefore account for the fact that the notional distinction between substances and objects does not perfectly mirror the grammatical mass/count distinction in that the interaction of **IND** and c gives rise to the misalignment of these categories.

Filip and Sutton (2017) build on the analysis of the mass/count distinction in Sutton and Filip (2016), and accounts for the fact that nouns like *fence, wall,* and *twig* can be counted, thereby displaying the characteristic property of count nouns, but can also occur in the singular in a pseudo-partitive measure DP, thereby behaving like a mass noun. Filip and Sutton (2017) argue that English count nouns like *fence* can be straightforwardly counted and admitted to pseudo-partitive (measure) NPs because they are quantized at specific counting schemas, which is required for counting, and they are non-quantized at the null counting schema, which is required for admittance to pseudo-partitive NPs. The same is true of Hungarian notionally count, singular nouns like *könyv* ('book') under our number neutral analysis. Furthermore, for Filip and Sutton (2017), counting in English resembles counting in Mandarin under Krifka's (1995) analysis in that numericals denote numerals and therefore require a modifier to compose with nouns. Erbach et al. (2017) make use of this resemblance to account for object mass nouns in Japanese are infelicitous with certain count quantifiers but can nevertheless be counted with classifiers.

Because the analysis in Filip and Sutton (2017) can account for non-quantized singular count NPs in measure DPs and classifier constructions, it is uniquely situated in that it can straightforwardly be applied to the nominal semantics in Hungarian. The empirical facts about Hungarian singular nouns that we have discussed in this paper require an analysis of counting and measuring that can accommodate non-quantized, singular count NPs in measure DPs, straightforward counting, and classifier constructions. Filip and Sutton (2017) provide such an analysis of such measure constructions and counting in English, which has already been extended to capture classifier constructions in Japanese (Erbach et al. 2017). We therefore build our analysis of Hungarian on Filip and Sutton (2017) rather than adopting other analyses like Krifka (1995) or Landman (2016), which fit some of the data but would require further adaptation to capture the Hungarian data.

6.2   Nominal Semantics in Hungarian

We interpret concrete nouns as tuples of the kind ⟨*extension, counting_base, pre-condition*⟩, following Filip and Sutton (2017). The first projection is the extension of the predicate, and the second projection is the set of entities that count as one relative to a counting schema. The third projection lists the preconditions and/or presuppositions relating to, for example, the selectional restrictions of classifiers. Count nouns like *könyv* ('book') are interpreted at a specific counting schema that specifies disjoint counting base (31). Mass nouns like *kosz* ('dirt', as in (32)) are substance denoting and lack the **IND**-function in their denotation.

This analysis is in agreement with Farkas and de Swart (2010) in that singular notionally count nouns in Hungarian are number-neutral.[5]

(31) $[\![k\ddot{o}nyv]\!]^{c_i} = \lambda x.\langle {}^*c_i(\mathbf{IND}(\mathrm{BOOK}))(x),\ \lambda y.c_i(\mathbf{IND}(\mathrm{BOOK}))(y),\ \varnothing\rangle$

(32) $[\![kosz]\!]^{c_i} = \lambda x.\langle {}^*c_0(\mathrm{DIRT})(x),\ \lambda y.c_0(\mathrm{DIRT})(y),\ \varnothing\rangle$

We use the projection functions π_1, π_2, and π_3 in order to modify the projections of the lexical entries in composition with other expressions.

If $X = \langle \phi, \psi, \chi\rangle_{\langle a\times b\times c\rangle}$, then: $\pi_1(X) = \phi_a$, $\pi_2(X) = \psi_b$ and, $\pi_3(X) = \chi_c$

Numerical expressions in Hungarian denote numerals: e.g., *egy* 'one', *kettő* 'two', *három* 'three' denote 1, 2, 3, respectively. Numeral denoting numerical expressions can be type-shifted with a modifying operation that allows the numerical expressions to combine with nouns and count individuals or kinds.

(33) $\mathrm{MOD} = \lambda n.\lambda P.\begin{cases} \lambda x.\left\langle \begin{array}{l} \pi_1(P(x)), \\ \mu_{card}(x, \lambda y.\pi_2(P(y))) = n, \\ \mathrm{QUA}(\lambda y.\pi_2(P(y))) \end{array}\right\rangle, \\[2em] \lambda c.\lambda k.\left\langle \begin{array}{l} c(\mathbf{SK}(^{\cap}P))(k), \\ \mu_{card}(k, \lambda k'.c(\mathbf{SK}(^{\cap}P))(k')) = n, \\ \mathrm{QUA}_k(\lambda k'.c(\mathbf{SK}(^{\cap}P))(k')) \end{array}\right\rangle \end{cases}$

The function takes, as an argument, a numeral n, and the interpretation of a common noun P and returns either a set of Ps that have a cardinality n with respect to a quantized counting base, or a context indexed set of subkinds of P that have a cardinality n with respect to a quantized counting base ($^{\cap}$ is a kind forming operator, $\mathbf{SK}$ applies to a kind and returns a set of subkinds, k is a variable over (sub)kinds). (See Sutton and Filip (2017) for a notion of disjointness for subkinds (and a mereology for subkinds) that could be extended to a notion of a quantized set of subkinds.) When combined with a numeral, the MOD function selects for count nouns (since mass nouns do not have quantized counting bases). It entails that, for example, *három könyv* 'three books', is ambiguous between the set of (sums of) entities that are books and that number three relative to a quantized counting base in counting schema c_i and the set of (sums of) subkinds of books that number three with respect to a quantized counting base in counting schema c_i.[6] We use **book** to refer to the *book* kind.

---

[5] Farkas and de Swart (2010) also argue that plural morphology includes explicit reference to pluralities and whether singular individuals are also specifically referred to is determined by the strongest meaning hypothesis for plurals. However, we remain agnostic as to whether Hungarian plurals are inclusive or exclusive.

[6] This enrichment of Filip and Sutton (2017) requires that counting schemas be made polymorphic with respect to applying to sets of individuals or sets of subkinds.

(34) $$[\![\textit{három}]\!]^{c_i} = 3$$

(35)
$$[\![\textit{három könyv}]\!]^{c_i} = \text{MOD}([\![\textit{három}]\!]^{c_i})([\![\textit{könyv}]\!]^{c_i}) =$$
$$\begin{cases} \lambda x. \left\langle \begin{array}{l} {}^{*}c_i(\mathbf{IND}(\mathbf{BOOK}))(x), \\ \mu_{card}(x, \lambda y.c_i(\mathbf{IND}(\mathbf{BOOK})(y)) = 3, \\ \text{QUA}(\lambda y.c_i(\mathbf{IND}(\mathbf{BOOK}))(y)) \end{array} \right\rangle, \\[2em] \lambda k. \left\langle \begin{array}{l} {}^{*}c_i(\mathbf{SK}(\mathbf{book})(k), \\ \mu_{card}(k, \lambda k'.c_i(\mathbf{SK}(\mathbf{book}) = 3, \\ \text{QUA}_k(\lambda k'.c_i(\mathbf{SK}(\mathbf{book})(k')) \end{array} \right\rangle \end{cases}$$

Classifiers like *darab* combine with numerical expressions in much the same way as the type shifting counting modification, though they restrict counting schemas to those that count individuals. They also introduce the counting schema of utterance into the argument noun interpretation so allowing mass nouns to be modified by numerical-classifier combinations. When a numerical is combined with *darab*, the result is a numerical determiner that has a precondition that the extension of the argument noun is a solid object which matches the intuitions of Hungarian speakers that *darab* denotes, for example dried up clumps of mud when it composes directly with *sár* ('mud')[7].

(36)
$$[\![\textit{darab}]\!]^{c_i} = \lambda n.\lambda c.\lambda P.\lambda x.$$
$$\left\langle \begin{array}{l} \pi_1(P(x)), \\ \mu_{card}(x, \lambda y.\pi_2(P(y)) = n, \\ \text{QUA}(\lambda y.\pi_2(P(y))) \wedge \forall z.(\lambda y.\pi_2(P(y)))(z) \wedge z \sqsubseteq x \to \text{SOLID}(z) \end{array} \right\rangle$$

(37)
$$[\![\textit{három darab könyv}]\!]^{c_i} =$$
$$\lambda x. \left\langle \begin{array}{l} {}^{*}c_i(\mathbf{IND}(\mathbf{BOOK}))(x), \\ \mu_{card}(x, \lambda y.c_i(\mathbf{IND}(\mathbf{BOOK})(y)) = 3, \\ \text{QUA}(\lambda y.c_i(\mathbf{IND}(\mathbf{BOOK}))(y)) \wedge \\ \forall z.c_i(\mathbf{IND}(\mathbf{BOOK}))(z) \wedge z \sqsubseteq x \to \text{SOLID}(z) \end{array} \right\rangle$$

Measure NPs (pseudo-partitives) are also represented as tuples with three projections. For instance, *három kiló könyvet* ('three kilos of books') has an extension consisting of sums of whole, quantized, countable books. The second projection is a set of books that measures three kilos in weight. The third projection contains the precondition that the extension of the predicate is not quantized.

(38) $$[\![\text{kiló}]\!] = \lambda n.\lambda P.\lambda x.\langle \pi_1(P(c_0)(x)), \ \mu_{kg}(x) = n, \ \neg\text{QUA}(\lambda y.\pi_1(P(c_0)(y)))\rangle$$

(39)
$$[\![\textit{három kiló könyvet}]\!]^{c_i} = \lambda x.\langle {}^{*}c_0(\mathbf{IND}(\mathbf{BOOK}))(x), \mu_{kilo}(x) = 3,$$
$$\neg\text{QUA}({}^{*}\lambda y.c_0(\mathbf{IND}(\mathbf{BOOK}))(y))\rangle$$

---

[7] Gyarmathy, PC.

This representation allows us to capture the insight that bare plural count nouns (which are semantically cumulative) retain their atomicity when used in pseudo-partitive (measure) DPs (Krifka 1989; Landman 2016), *pace* Rothstein (2011) (see e.g. (14)), on the assumption that measure phrases select for cumulative predicates. Most importantly, analyzed in this way, pseudo-partitive (measure) DPs straightforwardly can admit Hungarian notionally count, singular nouns like *könyv* ('book') as long as they are cumulative predicates, i.e., denoting a whole semi-lattice from which either mass or plural count nouns take their denotation. This reflects our conclusion that such Hungarian notionally count, singular nouns are semantically number-neutral. Moreover, it can be shown that while such nouns are straightforwardly acceptable in the pseudo-partitive (measure) DP and in this respect pattern with mass nouns, they fail to behave like mass nouns in a number of other syntactic environments, contrary to Rothstein (2017) and Schvarcz and Rothstein (2017).

While our analysis based on Filip and Sutton (2017) is not alone in its ability to capture these insights, it has provided us the means to formally distinguish the Hungarian and English nominal systems, and to account for classifiers and kinds in a way that would not be possible without enriching the systems of Krifka (1989) and Landman (2016). Our system, unlike that of Krifka (1989) places the context sensitivity of individuation at its core, thus accounting for counterexamples raised against this account such as non-quantized count nouns like *twig* (see e.g. Zucchi and White 1996; Rothstein 2010, and references therein). Landman's (2016) treatment of pseudo-partitive measure phrases makes no use of the property of *not-quantized* which we have exploited here. It is possible that an account in the spirit of the one we have provided here could be given within Landman's (2016) system, but we leave that as a matter for further research.

Taken all around, we converge on the same insight independently reached in Farkas and de Swart (2010) in proposing that the meaning of notionally count, singular nouns like *könyv* ('book') in Hungarian, taken as lexical items, corresponds to the number-neutral property whose denotation is built from the set of entities that count as single books closed under sum. From this, we can conclude that the denotation of *könyv*-like ('book') nouns in Hungarian can be assimilated to that of count nouns and therefore that balance of the distribution of mass, count and dual-life nouns in Hungarian is closer to English, *pace* the analysis of Rothstein (2017) and Schvarcz and Rothstein (2017).

7   Conclusion

Hungarian is a number marking language that allows the use of classifiers in counting constructions with both mass nouns and count nouns. Furthermore, singular nouns are straightforwardly used in counting constructions, pseudo-partitive (measure) DPs, and classifier constructions. In order to account for this puzzling property, Rothstein (2017) and Schvarcz and Rothstein (2017) argue that notionally count, singular nouns like *könyv* ('book') are dual-life, presupposing both the mass/count theory of Rothstein (2011), who argues that nouns must be mass to be used in pseudo-partitive (measure) DPs, and the analysis of classifiers by Chierchia (2010) who assumes that classifiers combine with nouns in order to make them countable.

We provided several arguments based on novel data against such analyses. We first argued against the analysis of pseudo-partitive measure DPs in Rothstein (2011) by showing that count nouns do not behave like object mass nouns when in a pseudo-partitive measure DP—i.e. atoms denoted by count nouns are accessible by reciprocal operators while those of object mass nouns are not—and therefore count nouns are not mass in such an environment. Second, we showed there is reason to doubt that notionally count, singular nouns like *könyv* ('book') are dual-life: they do not get an object-mass interpretation when in the argument position, which should not be the case if they are dual-life. Lastly, drawing from work on other classifier languages (Japanese and Chol) we showed that there are classifier constructions in natural languages which do not require mass nouns.

We then provided data in support of the view that notionally count, singular nouns in Hungarian are compatible with both singular and plural interpretations in many morphosyntactic environments, which strongly suggests that they are number-neutral. This formed the basis for our formal analysis. An advantage of analyzing notionally count, singular nouns like *könyv* ('book') as semantically number-neutral is that such an analysis is compatible with the standard analysis of pseudo-partitive (measure) DPs (Krifka 1989; Filip 2005, 1992; Nakanishi 2003; Schwarzschild 2006), and also under the sort of analysis of the mass/count distinction proposed by Krifka (1995), thereby providing a sound alternative to previous analyses of Hungarian.

If our proposal that Hungarian notionally count, singular nouns are number-neutral, is correct, then this has the following major implication. Not only is it compatible with a widespread view of pseudo-partitive (measure) DPs, but also Hungarian turns out to pattern with English, rather than with Brazilian Portuguese (as analyzed by Pires de Oliveira and Rothstein (2011)), when it comes to the distribution of nouns across countability classes—namely, a substantial number of mass and count nouns, but few dual life nouns—and therefore shifts the typological classification of Hungarian.

References

Allan, K.: Nouns and countability. Language **56**, 541–567 (1980)

Bale, A., Barner, D.: Semantic triggers, linguistic variation and the mass-count distinction. In: Count and Mass Across Languages, pp. 238–260. Oxford University Press (2012)

Bale, A., Coon, J.: Classifiers are for numerals, not for nouns: consequences for the mass/count distinction. Linguist. Inq. **45**, 695–707 (2014)

Chierchia, G.: Plurality of mass nouns and the notion of "semantic parameter". In: Rothstein, S. (ed.) Events and Grammar: Studies in Linguistics and Philosophy, vol. 7, pp. 53–103. Kluwer, Dordrecht (1998)

Chierchia, G.: Mass nouns, vagueness and semantic variation. Synthese **174**(1), 99–149 (2010)

Csirmaz, A., Dékány, É.: Hungarian is a classifier language. In: Simone, R., Masini, F. (eds.) Word Classes: Nature, Typology and Representations, pp. 141–160. John Benjamins Publishing Company, Amsterdam (2014)

Dékány, É.: A profile of the Hungarian DP. Ph.D. thesis. Center for Advanced Study in Theoretical Linguistics, University of Tromsø (2011)

Erbach, K., Sutton, P., Filip, H., Byrdeck, K.: Object mass nouns in Japanese. In: Cremers, A., van Gessel, T., Roelofsen, F., (eds.) Proceedings of the 21st Amsterdam Colloquium, pp. 235–244. Institute for Logic, Language, and Computation at the University of Amsterdam (2017)

Farkas, D., De Swart, H.: The Semantics of Incorporation. Stanford Monographs in Linguistics. CSLI Publications, Stanford (2003)

Farkas, D., de Swart, H.: The semantics and pragmatics of plurals. Semant. Pragmat. **3**, 1–54 (2010)

Filip, H.: Aspect and interpretation of nominal arguments. In: Proceedings of the Twenty-Eighth Meeting of the Chicago Linguistic Society, pp. 139–158. University of Chicago, Chicago (1992)

Filip, H.: Measures and indefinites. Reference and quantification: the Partee effect, pp. 229–289 (2005)

Filip, H., Sutton, P.: Singular count NPs in measure constructions. Semant. Linguist. Theory **27**, 340–357 (2017)

Gyarmathy, Z.: Technical report on the count/mass distinction in hungarian from the perspective of rothstein's theory. Heinrich Heine University (2016)

Krifka, M.: Nominal reference, temporal constitution and quantification in event semantics. In: Bartsch, R., van Benthem, J.F.A.K., van Emde Boas, P. (eds.) Semantics and Contextual Expression, pp. 75–115. Foris Publications, Dordrecht (1989)

Krifka, M.: Common nouns: a contrastive analysis of English and Chinese. In: Carlson, G., Pelletier, F.J. (eds.) The Generic Book, pp. 398–411. Chicago University Press, Chicago (1995)

Landman, F.: Count nouns-mass nouns, neat nouns-mess nouns. Balt. Int. Yearb. Cogn. Log. Commun. **6**(1), 12 (2011)

Landman, F.: Iceberg semantics for count nouns and mass nouns: the evidence from portions. Baltic Int. Yearbook Cogn. Logic Commun. **11**(1), 6 (2016)

Nakanishi, K.: The semantics of measure phrases. In: Proceedings-NELS, vol. 33, pp. 225–244 (2003)

Pires de Oliveira, R., Rothstein, S.: Bare singular noun phrases are mass in Brazilian Portugese. Lingua **121**, 2153–2175 (2011)

Rothstein, S.: Counting and the mass/count distinction. J. Semant. **27**(3), 343–397 (2010)

Rothstein, S.: Counting, measuring and the semantics of classifiers. Balt. Int. Yearb. Cogn. Log. Commun. **6**(1), 15 (2011)

Rothstein, S.: Semantics for Counting and Measuring. Cambridge University Press, Cambridge (2017)

Schvarcz, B.R., Rothstein, S.: Hungarian classifier constructions, plurality and the mass-count distinction. In: Approaches to Hungarian, Papers from the 2015 Leiden Conference, vol. 15, pp. 183. John Benjamins Publishing Company (2017)

Schwarzschild, R.: The role of dimensions in the syntax of noun phrases. Syntax **9**(1), 67–110 (2006)

Soja, N.N., Carey, S., Spelke, E.S.: Ontological categories guide young children's inductions of word meaning: object terms and substance terms. Cognition **38**(2), 179–211 (1991)

Sudo, Y.: The semantic role of classifiers in japanese. Balt. Int. Yearb. Cogn. Log. Commun. **11**(1), 10 (2016)

Sudo, Y.: Countable nouns in Japanese. Proc. WAFL **11**(1), 11 (2017)

Sutton, P., Filip, H.: Counting in context: count/mass variation and restrictions on coercion in collective artifact nouns. Semant. Linguist. Theory **26**, 350–370 (2016)

Sutton, P.R., Filip, H.: Restrictions on subkind coercion in object mass nouns. In: Proceedings of Sinn und Bedeutung, vol. 21 (2017, to appear)

Zucchi, S., White, M.: Twigs, sequences and the temporal constitution of predicates. In: Proceedings of SALT, vol. 6 (1996)

Why Aktionsart-Based Event Structure Templates Are not Enough – A Frame Account of Leaking and Droning

Jens Fleischhauer[✉], Thomas Gamerschlag, and Wiebke Petersen

Heinrich-Heine-Universität, Düsseldorf, Germany
fleischhauer@phil.uni-duesseldorf.de

Abstract. In the paper, we present a frame approach to emission verbs and demonstrate how this framework enables us to account for their different uses and the constructions they can occur in. The frame model we apply is based on Barsalou's ideas about frames as the fundamental structures of cognitive representation (Barsalou 1992). More precisely, frames are conceived as recursive attribute-value structures that allow one to zoom into conceptual structures to any desired degree and to access meaning components by attribute paths (cf. Petersen 2007/2015). We argue that such a formal frame-based account of meaning is highly suited for capturing the way particular uses of emission verbs are constrained by the interaction of grammar and cognition. The focus of the analysis is on degree gradation of substance emission verbs such as in *sehr lecken* 'leak a lot' as well as sound emission verbs as in *sehr dröhnen* 'drone a lot'. We show that a proper treatment of both of these phenomena requires lexical decomposition that goes beyond the traditional event structural templates as applied by Rappaport Hovav and Levin (1998) among others.

Keywords: Emission verbs · Verb gradation · Frame analysis

1 Introduction: Aktionsart and Event Structure Templates

Starting, at least, with the seminal analysis of Vendler (1957), it has been recognized that verbs can be grouped into different aktionsart classes based on inherent temporal characteristics such as dynamicity, telicity and durativity. Dynamicity separates stative predicates from dynamic predicates which denote a 'happening' in the world. Telicity refers to the sub-classification of dynamic predicates into atelic and telic ones. Telic predicates entail that a specific endpoint is reached within the event denoted by the predicate, whereas atelic predicates do not.[1] Durativity, finally, distinguishes durative from punctual predicates. The eventuality denoted by a predicate can either happen instantaneously – in case of a punctual predicate – or be temporally extended (i.e., durative).

---

[1] We refer the reader to Borik (2006) for a comparison of different theoretical explications of the notion of telicity.

© Springer-Verlag GmbH Germany, part of Springer Nature 2019
A. Silva et al. (Eds.): TbiLLC 2018, LNCS 11456, pp. 108–127, 2019.
https://doi.org/10.1007/978-3-662-59565-7_6

Dowty (1979) has proposed a sub-lexical representation of verb meaning which directly reflects aktionsart classification. This approach, well known under the notion of 'lexical decomposition', is at the heart of most current approaches to verb semantics. Representations of this type, also called 'event structure representations', consist of structural and idiosyncratic meaning components. The structural meaning components represent the different aktionsart classes and are used for formulating 'event structure templates'. A particular account on representing 'event structure templates' has been developed by Levin and Rappaport Hovav (e.g. Levin and Rappaport Hovav 1995, 2005; Rappaport Hovav and Levin 1998). In (1), the event structure templates applied by Levin and Rappaport Hovav (2005) are shown. Structural meaning components are written in bold face, the idiosyncratic meaning components are written in angled brackets. ACT, BECOME and CAUSE are predicates used to define aktionsart classes. The predicate ACT represents activity predicates which are dynamic and atelic while BECOME indicates change of state predications, which are often but not necessarily telic. So-called 'degree achievement predicates' – a notion going back to Dowty (1979) – show variable telicity, with some of them (e.g. *to grow*) being only atelic. Durativity is not reflected in the event structure templates, thus the contrast between punctual and durative changes is neglected. Instead causation – represented by CAUSE – is taken up as a defining event structural property. We ignore this aspect for the current discussion but see Van Valin (2005: 38) for an argumentation against treating causation as a property relevant for aktionsart classification and Croft (1991) for an approach to verb classification essentially based on the notion of 'causal chains'.

(1)   a.   State predicate   [x ⟨State⟩]
      b.   Activity predicate   [x **ACT**⟨Manner⟩]
      c.   Achievement predicate   [**BECOME** [x ⟨RESULT⟩]]
      d.   Accomplishment predicate [x **CAUSE** [**BECOME** [y ⟨RESULT⟩]]]

The structural components are class-building predicates, which means – for example – that every activity predicate has the event structure shown in (1b). Although *hit* and *kiss* are both activity predicates and therefore show the same event structural representation, they differ with respect to the idiosyncratic meaning component as shown in (2). The idiosyncratic meaning component is called 'root' by Rappaport Hovav and Levin (1998). The root acts as a (manner) modifier in case of activity predicates – modifying the ACT predicate – but serves as an argument of the BECOME predicate.

(2)   a.   *hit:* [x **ACT**⟨hit⟩ y]
      b.   *kiss:* [x **ACT**⟨kiss⟩ y]

The aim of decompositional approaches is to represent "components of meaning that recur across significant sets of verbs" (Levin and Rappaport Hovav 2005: 69). An intention is capturing those meaning components which are grammatically relevant for classes of verbs. An aktionsart-based decompositional representation accounts for combinatory restrictions of time adverbials but also for argument linking (e.g. Van Valin 2005). Certain aspects of grammatical behavior depend on the structural meaning components of event structure templates. The root, on the other hand, "is [...] opaque to grammar" (Rappaport Hovav and Levin 1998: 254). This means that grammatical differences should not depend on the root element.

In the remainder of the paper, we will present a case study of gradation of emission verbs which shows that there are phenomena not sufficiently captured by classical template representations. In Sect. 3, we focus on templatic accounts since there exists some previous work (e.g. Tsujimura 2001) dealing with degree modification within such a type of approach. As an alternative to such an account, we present a sketch of how verb gradation can be dealt with by a frame account. Given the space confinements of the paper, we concentrate on a single phenomenon and will not discuss other aspects of emission verbs which call for a frame approach such as the extended use as verbs of directed motion (see Fleischhauer et al. 2017 for a frame account of this use).

2 · Emission Verbs and the Limits of Event Structure

For illustrating the limits of approaches relying on aktionsart-based event structure templates, we focus on a single class of verbs (verbs of emission) and show that grammatical differences within this semantic class of verbs are caused by the respective root element. Verbs of emission are basically intransitive verbs denoting the emission of a stimulus like smell, sound, light or substance. Depending on the type of stimulus, four subclasses of emission verbs are distinguished (Levin 1993: 233ff.):[2]

(3) a. Verbs of smell emission: *smell, stink*
     b. Verbs of light emission: *light, shine, glitter, sparkle*
     c. Verbs of sound emission: *drone, bark, clapper*
     d. Verbs of substance emission: *leak, bleed, fester*

Rappaport Hovav and Levin (2000: 283) state that "verbs of emission fall along a continuum of stativity, with verbs of smell emission being the most stative, verbs of light emission slightly less stative, followed by verbs of sound emission and substance emission, which are the most process-like." Under this analysis, verbs of sound emission and verbs of substance emission receive the same event structural representation, as indicated for the examples in (4).

(4) a. *drone:* [x ACT$_{\langle drone \rangle}$ y]
     b. *leak:*   [x ACT$_{\langle leak \rangle}$ y]

Verbs of emission typically allow for verb gradation in English, German and other languages like Japanese (Tsujimura 2001). Gradation in general is the linguistic process

---

[2] There exists a bunch of literature (e.g. Perlmutter 1978; Gerling and Orthen 1979; Atkins et al. 1988; Atkins and Levin 1991; Levin 1991; Potashnik 2012) discussing various aspects of verbs of emission – e.g. argument realization patterns of verbs of sound emission – which are not relevant for the current discussion. The reader is referred to the mentioned literature.

of comparing two (or possibly more) degrees (Fleischhauer 2016a: 16). Prototypically, gradation is associated with adjectives and degree constructions such as the comparative or superlative construction. In the comparative construction in (5), Peter's degree of height is compared to the one of his brother and it is expressed that Peter exceeds his brother on the height scale.

(5)   *Peter is taller than his brother.*

Scales figure crucially in the analysis of gradable expressions. A scale is conceived as a linearly ordered set of degrees in a certain measurement dimension (Kennedy 1999a; Kennedy and McNally 2005). A gradable adjective like *tall* is then analyzed as a measure function mapping the referent of its argument on a height scale (e.g. Kennedy 1999b, 2007).

   Verb gradation turns out to be a bit more complex than gradation in the adjectival domain. The reason is that verb gradation comes in two subtypes, called 'extent' and 'degree gradation' (Bolinger 1972; Löbner 2012; Fleischhauer 2016a). (6a) is an example of 'extent gradation', the degree expression *viel* 'much' specifies the frequency of leaking events. A suitable paraphrase for (6a) is 'the pipe leaked often'. Extent gradation either consists in such a frequentative reading or in a specification of an event's temporal duration. An example of degree gradation is shown in (6b). *Sehr* 'very' specifies the quantity of the emitted liquid and the example can be paraphrased as 'the pipe emitted a large quantity of liquid'.[3]

(6)   a.   *Das Rohr leckte viel.*
               the   pipe leaked much
               'The pipe leaked a lot.'
        b.   *Das Rohr leckte sehr.*
               the   pipe leaked very
               'The pipe leaked a lot.'

Extent gradation is largely regular and results in the same interpretation for all dynamic predicates (see Doetjes 1997, 2007 on this issue). By contrast, the interpretational patterns of verbal degree gradation depend on the semantic class of the verb (Löbner 2012; Fleischhauer 2016a). This can easily be seen by comparing (6b) – a graded verb of substance emission – with (7): *sehr* specifies the quantity of emitted substance in (6b) but the loudness of the emitted sound in (7).

---

[3] Whereas German uses different degree expressions for extent and degree gradation, English applies the same expression for both. See Doetjes (2008) and Fleischhauer (2016a, 2016b) for the cross-categorical distribution of degree expressions and also for a cross-linguistic comparison.

(7)   *Der Motor dröhnt sehr.*
        the   engine drones very
        'The engine is droning a lot.'

The verbal root is relevant for determining the interpretation of verbal degree gradation. Moreover, a second interpretational contrast depends on the root element. The perfect construction in (8a) licenses a perfective interpretation (see Löbner 2002 on the aspectual interpretation of the German perfect construction). To emphasize the perfective interpretation, we provide a context which requires a sequential interpretation of the two events. The first event – the leaking of the pipe – has to be finished before the second event starts. Outside of such a context, the German perfect also allows for an imperfective reading, in which case (8a) is interpreted like (8b). In the perfective interpretation of (8a), *sehr* specifies the total quantity of liquid emitted in the event. The sentence can be paraphrased as 'The pipe emitted a large amount of water'. (8b) is an instance of the periphrastic *am*-progressive (e.g. Andersson 1989; Ebert 2000).[4] The progressive construction focusses on a stage of the event and *sehr* specifies the quantity of liquid emitted at that event stage. The progressive sentence requires a different paraphrase than the perfective one because 'The pipe emitted a large amount of water' does not paraphrase its meaning. A suitable paraphrase for (8b) is 'The pipe was leaking heavily/badly'.

(8)   a.   *Bevor der Klempner das Rohr reparierte, hat es*
               before the plumber   the pipe   repaired   has it
               sehr geleckt
               very leaked
               'Before the plumber repaired it, the pipe leaked a lot.'
        b.   *Das Rohr war sehr am   Lecken.*
               the   pipe was very at.the leaking
               'The pipe was leaking a lot.'

The interpretations of the two sentences in (8) are clearly related, nevertheless the sentences do not necessarily entail each other. If a lot of liquid has been emitted in a single event, it does not necessarily follow that at any single stage of the event, a large quantity of liquid has been emitted. Rather, a pipe can be slightly leaking, but the overall quantity of emitted liquid can sum up as large. On the other hand, if a lot of liquid has been emitted at a single stage of the event, this does not necessarily result in a large quantity at the event's end. The pipe can be leaking for a while, only emitting a single drop but at one stage of the event, it emits a larger amount of water for a very short while. Thus, in the overall event the quantity of liquid emitted by the pipe may still count as 'small'.

---

[4] Various varieties of German make use of a periphrastic construction for the expression of progressive aspect. As the construction is still on its way of getting grammaticalized, native speakers vary with respect to its acceptability.

Similar examples are found in other languages as well. In (9), a perfective use of the verb *saigner* 'to bleed' is contrasted with a progressive one. The periphrastic *passé composé* used in (9a) receives a perfective reading, the sentence is interpreted as 'The subject referent emitted a large quantity of blood'. The interpretation of the periphrastic progressive construction (9b) is – like in German – that a large quantity of blood is emitted at a certain stage of the event. Nothing is said about the quantity of blood emitted prior or later to that stage.

(9)   a.   *Il   a   beaucoup   saigné.*
               he   has   a lot   bled
               'He bled a lot.'

        b.   *Il   est   en train de   saigner   beaucoup.*
               he   is   PROG   bleed.INF   a lot
               'He is bleeding a lot.' (Fleischhauer 2016b: 228)

As the German and French data show, the interpretation of degree gradation is dependent on grammatical aspect. The reason why aspect can affect the interpretation of degree gradation is that degree gradation of verbs of substance emission is event-dependent (Fleischhauer 2013, 2016a). The quantity of the emitted substance increases, as long as the event progresses. The interpretational difference between the perfective and the imperfective aspect results from the fact that the perfective aspect denotes complete events (e.g. Comrie 1976), whereas the progressive aspect does not but restricts denotations to a proper (not initial and not final) part of the event.

Verbs of sound emission represent event-independent degree gradation. This is illustrated by the examples in (10). The perfect construction in (10a) again licenses a perfective interpretation. *Sehr*, as already discussed above, specifies the loudness of the emitted sound. The degree of loudness is indicated as being 'high'. The progressive construction in (10b) results in the same interpretation: the emitted sound is very loud. Both sentences receive the same interpretation, irrespective of grammatical aspect.

(10)   a.   *Der Motor   hat   sehr   gedröhnt.*
                the   engine   has   very   droned
                'The engine droned a lot.'
         b.   *Der Motor   war   sehr   am   Dröhnen.*
                the   engine   was   very   at.the   droning
                'The engine was droning a lot.'

The crucial difference between verbs of substance emission and verbs of sound emission is that the quantity of substance emitted in the event increases when the event unfolds. The loudness of the sound emitted in the event does not (necessarily) increase when the event progresses. Thus, there is a homomorphic relationship between the progression of the event and the quantity of substance emitted, while there is no such relationship between the event's progression and the loudness of the emitted sound.

Accounting for the contrast between event-dependent and event-independent degree gradation within the class of emission verbs requires lexical decomposition of the root element since the gradation scales are differently related to the emission process. In the next section, we present a further need for deeper lexical decomposition – meaning decomposition beyond event templates – based on a distinction between lexically scalar and lexically non-scalar verbs.

3 Lexically Scalar Verbs

One particular question with respect to verbal degree gradation is: Which verbs license a particular modifier like German *sehr* or English (*very*) *much; a lot*? The question has not been discussed in much detail in the previous literature, two notable exceptions are a paper by Tenny (2000) and one by Tsujimura (2001). Tenny (2000) argues that measure adverbs like *completely* and *partly* only combine with verbs having a core event in their event structural representation. She associates core events – also called inner events – with the expression of changes and the attainment of a final result state. Thus, only verbs having a BECOME predicate in their event structural representation have a core event. Verbs having a core events also have a scale – which Tenny calls measure or path – as part of their lexical meaning. Tenny (2000: 296) states that in such cases the final state of the core event is a gradable predicate, which admits degree modification.

In her analysis, Tenny makes a statement about the use of degree expressions like *completely*. Tenny states that such adverbs combine with verbs having a core event, which are change of state verbs (e.g. *break*), verbs of directed motion (e.g. *run to NP*) and incremental theme verbs (e.g. *eat*). Ernst (2002) – building on Tenny's analysis – adds degree adverbs like (*very*) *much* to the expressions which require a core event in the event structural representation of the verb they modify.

In the last section, we already saw examples of graded verbs of sound/substance emission. These verbs neither qualify as result verbs – see below – nor as verbs having a core event (they do not express a change of state). It can also easily be shown that (*very*) *much* modifies verbs which neither qualify as result verbs nor verbs having a core event. One particular example at hand is the verb *to love* in (11). (For a more detailed criticism of Tenny's analysis see Fleischhauer 2016a: 93ff.).

(11)   *The boy hates his teacher very much.*

Tsujimura (2001) provides an analysis of Japanese gradable verbs within a templatic approach. She aims at showing that a deep connection between a verb's event structure and the licensing of the degree modifier *totemo* 'very' exists. She basically states (p. 47) that a verb admits degree modification by *totemo* if: (i) the verb has a state component in its event structural representation, (ii) the state component refers to a gradable property, and (iii) the gradable property must have a nontrivial standard (i.e., non-maximal/minimal degree).[5] One predication of this analysis is that verbs, which do not have a state component in their event structure representation, do not license *totemo*. Contrary to fact, however, activity predicates license *totemo*, as the example in (12) shows.

(12)   *Taroo-wa totemo waratta.*
        Taro-TOP   very   laughed
        'Taro laughed very much.' (Fleischhauer 2016a: 100)

The same is true for German and English as verbs of substance emission are clearly activity predicates. This is seen for English by the fact that – as a contrast to stative verbs – verbs like *to bleed* and *to drone* receive a habitual interpretation if used in the simple present.

In the limited number of studies devoted to degree modification of verbs, gradability has usually been related to event structural properties. But, as the discussion above has shown, gradability is independent from a verb's event structure. Nevertheless, we like taking up the previous studies relating event structure and scalarity as they show, in our view, one crucial difference between two classes of gradable verbs. We follow Rappaport Hovav and Levin's (2010) analysis of results verbs as denoting scalar changes. The authors basically argue that verbs like *to break* and *to crack* express a directed change within a single scalar dimension. Stating the truth conditions of such verbs always requires a comparison of degrees. The clearest instances are degree achievement predicates such as *to widen, to darken* and *to lengthen*. To evaluate whether the sentence in (13) is true, one has to compare the crack's width at the beginning of the event and its width at the event's end. Only if the crack's width increased, the sentence in (13) can be true.

(13)   *The crack widened.*

In the analysis of Rappaport Hovav and Levin (2010), manner verbs do not denote a directed change along a single scalar dimension but can either denote undirected changes or changes in multiple dimensions.[6] Manner verbs, as for example the sound emission verb *drone*, do not require the comparison of (at least) two degrees to evaluate whether a sentence like the one in (14) is true. The sentence in (14) neither means 'the engine droned more than some other mechanical device/the engine droned more than usual' nor 'the engine droned with a specific loudness'.

(14)   *The engine droned.*

---

[5] The notion of a 'trivial/non-trivial standard' goes back to Kennedy and McNally (1999). A trivial standard defaults with an endpoint of a scale, whereas a nontrivial standard does not.

[6] Manner/result complementarity, as explicated by Rappaport Hovav and Levin (2010), only applies to dynamic predicates since the discussion crucially relies on the notion of 'change'.

The same holds for the substance emission verb *to leak*. *The pipe leaked* does not mean 'the pipe leaked more than something else/than it did at some other occasion' and it also does not mean 'the pipe emitted a specific quantity of liquid'. The meaning of *leak* is simply that some unspecified amount of liquid is emitted out of a container. Stating the truth conditions of the predicate does not require a comparison of degrees and it is not obvious what would be compared. However, *leak* requires that at least some liquid is emitted and *drone* requires that the emitted sound exceeds some volume, otherwise the predicates could not truthfully apply. But this only means that there is some quantity of emitted liquid, resp. some loudness of the emitted sound. But it does not follow that the verbs make a predication about the quantity degree, resp. loudness degree. The quantity degree/loudness degree is only relevant within a degree context. Thus, only a construction like *to drone a lot* requires a comparison of degrees – the actual loudness degree has to exceed the standard of comparison introduced by the degree expression – but not the ungraded verb *to drone*.

To capture the intuitive difference between verbs like *to widen* on the one hand and verbs like *to leak* and *to drone* on the other hand, Fleischhauer (2015, 2016a, 2018) introduces the notion of a 'lexically scalar predicate', which is defined as in (15). The crucial notion of the definition of lexically scalar predicate is 'scalar predication'.

(15)   (a) Scalar predication: A predication is taken to be scalar iff it expresses a
        comparison of degrees on a scale.
        (b) Lexically scalar predicate: A predicate is lexically scalar iff it expresses
        a scalar predication in every context of use.
        (Fleischhauer 2015:58; 2016a:174, 2018: 240)

Given the definition in (15), gradable adjectives qualify as scalar. An adjective such as *tall* compares the degree of the referent of its individual argument with a comparison degree. *The boy is tall* can mean something like 'the boy is tall for a boy of his age/for a basketball player' (e.g. Kennedy 1999b). Result verbs, in the sense of Rappaport Hovav and Levin, also qualify as lexically scalar. This is obvious for degree achievement predicates as explicating the truth conditions of a verb like *to widen* requires comparing degrees.

Adopting the view that emission verbs are lexically non-scalar results in the need to explain where the gradation scale comes from. It is implausible to assume that the scale is introduced by the degree expression. As discussed in the last section, *sehr* – and similarly *a lot* in English – applies to different scales, e.g. quantity scales (verbs of substance emission), loudness scales (verbs of sound emission) and further scales in the context of other gradable verbs (e.g. intensity scales in the context of psych verbs like *to hate* and *to frighten*). The degree expression requires a gradation scale which is somehow contributed by the graded predicate and which is activated in a degree context.

A principle account of the activation of gradation scales in lexically non-scalar verbs is required which needs to answer the following questions:

(i)   How are gradation scales licensed?

What licenses the activation of a quantity scale in verbs of substance emission and the activation of a loudness scale in verbs of sound emission? In both cases, the scale represents a property of the emittee, which is an implicit event participant (Goldberg

2005: 20f. speaks of an implicit theme argument). As we are dealing with properties of implicit event participants, the implicit argument has to be part of the semantic representation too.

(ii) How are gradation scales constrained?

Why does degree gradation of verbs of substance emission exactly apply to the quantity scale of the emittee argument and not to a different scale? Irrespective of the type of emitted substance, it is always only the quantity scale the degree expression applies too. The examples in (16) exemplify different types of emitted substances – *hair* (a), *fester* (b), *dust* (c) – which does not affect degree gradation (e.g., it is not the case that 'sehr' specifies that the lost hair in (16a) is very long).

(16) a. *Die Katze hat sehr gehaart.*
          The cat   has very lost.hair
          'The cat lost a lot of hair.'
      b. *Die Wunde hat sehr geeitert.*
          The wound has very festered
          'The wound festered a lot.'
      c. *Die Bücher haben sehr gestaubt.*
          The books   have   very raised.dust
          'The books raised a lot of dust.'

An answer to the second question also requires answering why degree gradation singles out a unique scale within a certain semantic class (e.g. always loudness in case of verbs of sound emission) but different scales for different semantic verb classes (e.g. a loudness scale in the case of sound emission verbs but a quantity scale in the case of substance emission verbs). We are not addressing this question in the current paper but see Fleischhauer (2015, 2016a) for a discussion of this question.

4 Frame Analysis

4.1 Dynamic Frames

In the previous sections it has turned out that traditional aktionsart-based event structural approaches are not fine-grained enough in order to account for the different distributional patterns and interpretational restrictions coming with the semantic class of emission verbs. The examples we have discussed show that the following information must be part of a semantic representation of emission verbs:

1. The relations between an event and the objects participating in it, which may be either explicitly stated (e.g., *the pipe* in *the pipe leaks*) or not (e.g., *the liquid* in *the pipe leaks*), must be represented.
2. The internal structures of the objects involved in an event must be represented by specifying their properties and mereological structures (e.g., it is the engine of a motorbike that drones and that means that the sound emitted by the engine is of a specific quality).

3. Events may change some properties of objects involved in the event (e.g., the quantity of the emitted liquid in *the pipe leaks*). Thus, a semantic representation has to capture the temporal evolution of the event and the changing of object properties in order to account for the aspectual and entailment restrictions described in the previous sections.

To account for these requirements and to overcome the coarse-graininess of aktionsart-based event templates, we will use a dynamic frame approach instead. Static frames are recursive attribute-value structures in which attributes act as functions that assign unique values to entities. Naumann (2013) points out that simple attribute-value structures are static descriptions that are not sufficient to represent the dynamic nature of events. Therefore, Naumann (2013) develops a dynamic theory of frames that provides the desired representational levels needed to capture the dynamic evolution of an event. An illustration of his account is given in the following figure representing the example *The boy has grown 5 cm*.

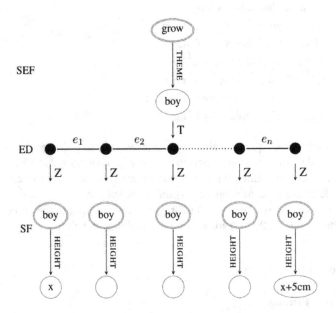

Fig. 1. Dynamic frame for 'The boy has grown 5 cm'

The relations between event participants (here *the boy*) and the event (here *grow*) are captured in a 'static event frame' (SEF, top of the figure).[7] In such a SEF the objects undergoing a change are represented as atoms that are statically linked to the event node. On this level it is not possible to specify the properties of the objects as they are not static but undergo a change. That is why objects at the level of SEFs are atoms. The properties of the objects are represented at the level of 'situational frames' (SFs, bottom

---

[7] In the frame graphs, we mark the central node of a frame that specifies what the frame is about by a double line. For a graph-based definition of static frames see Petersen (2007/2015).

of the figure),which are originally termed 'temporalized SEFs' (t-SEFs) in Naumann (2013). At this level, the object representations are temporalized, that is participating objects like motorbikes, boys, engines or liquids are at any time point described by the properties that hold for the object. The properties and relations to other objects are represented by a static attribute-value structure. The SEF- and the SF-level are connected by the 'event decomposition level' (ED, middle of the figure) which represents the temporal evolution of the event and that links the temporalized representations of the objects participating in the event to the stages of the event. On the event decomposition level, the event is decomposed into a temporal sequence of subevents. From the boundaries of the subevents one can zoom into the object representations at the SF-level. The formal details of the temporalization and the zooming relation are given in Naumann (2013). Often it is not necessary to temporally decompose an event into more than one subevent in order to grasp its main properties. However, sometimes it is not sufficient to only represent the changed values at the beginning and the end of the full event and it is necessary to represent the value changes while the event evolves; this can be captured by a decomposition into subevents.

An alternative way of capturing the changing of a value while an event evolves has been proposed by Gamerschlag et al. (2014) who combine Naumann's account on dynamic frames with the path semantics approach of Zwarts (2005). The key idea is that the values of the attributes that change while the event progresses leave a trace path in the value space. The following figure shows again the frame for 'The boy has grown 5 cm' but this time, the change of the value of the attribute HEIGHT is represented as a path in the height space:

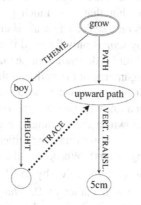

Fig. 2. Condensed dynamic frame for 'The boy has grown 5 cm'

The trace path of the HEIGHT value has to be a path that is continuously going upwards and for which the vertical distance between its initial and end point is 5 cm. Such frames are termed 'condensed dynamic frames'. Note that the formation of the trace path in the course of the event is technically captured by the 'dynamic attribute' TRACE. This special type of attribute, which is indicated by a broken line, is projected into this frame from the event decomposition frame in the preceding figure. The function of the TRACE-attribute is to map the changing values of the HEIGHT-attribute to the record of its trace in the time span of the event. Thus the value of the TRACE attribute is a path object with start and end point that is 1-dimensional and corresponds to an interval of the height scale. As usual we think of 1-dimensional scales as being vertically oriented. An increasing value thus leaves an upward trace path on such a scale (see Fig. 2). The uncondensed dynamic frame in Fig. 1 and the condensed one in Fig. 2 are directly translatable into each other. The trace path captures the change of the object properties at the 'situational frames' level in the uncondensed dynamic frame. The two dynamic frame versions differ in the perspective they take on the represented event: While the uncondensed frame emphasizes the different status of the relations involved in an event by capturing them at different levels (static relations versus dynamic relations), the condensed frame represents the change itself as a single object (namely a path object) and thereby enables one to express direct properties of this change.

4.2   Verbal Degree Gradation and Dynamic Frames

In Sect. 2 we have seen that while the interpretation of extent gradation is largely regular, the interpretational patterns of verbal degree gradation depend on the semantic class of the verb. In the following we will analyze verbal degree gradation with *sehr* in our dynamic frame model. Before we focus on the verbs of emission in the examples given in (6b) and (7) in Sect. 2, we illustrate our approach with *grow* as an example of a typical gradable change of state verb. In *Der Junge ist sehr gewachsen* 'The boy has grown a lot', the change of state verb *wachsen* 'grow' refers to a change in height. The degree expression *sehr* expresses that the height of the referent of the subject argument is not only greater than before but that it is much greater. In general, the degree expression *sehr* always operates on a scale. The scale is either a value scale or a degree of change scale. Modification by *sehr* is only licensed if a threshold value is exceeded. This threshold value is first and for all dependent on the modified verb and it is further contextually restricted by the arguments of the verb and the time span of the event. For example, a boy growing 5 cm in a month can be described as *Der Junge ist sehr gewachsen* 'The boy has grown a lot' while a water melon growing 5 cm in half a year cannot. This contrast is well-known in the literature on comparison and usually analyzed by making reference to some contextually specified comparison class (cf. Kennedy and McNally 2005 among others). We denote the contextually given threshold value for *sehr*-gradation of *wachsen* 'grow' by $\Delta_{c,t}^{\text{sehr}}(\text{grow})$. Figure 3 shows the uncondensed and condensed dynamic frames of *Der Junge ist sehr gewachsen*.

The only difference between these frames and the ones for 'grown 5 cm' in Figs. 1 and 2 is the way how the amount of growth is specified. In the 'grown 5 cm' example the amount is fixed, while in the graded example it is restricted by the contextually given lower limit $\Delta_{c,t}^{\text{sehr}}(\text{grow})$. In the uncondensed frame in Fig. 3 this is depicted by the additional constraint that the difference between the height at the beginning of the event and at the end of the event has to be at least $\Delta_{c,t}^{\text{sehr}}(\text{grow})(x_n - x_0 > \Delta_{c,t}^{\text{sehr}}(\text{grow}))$.

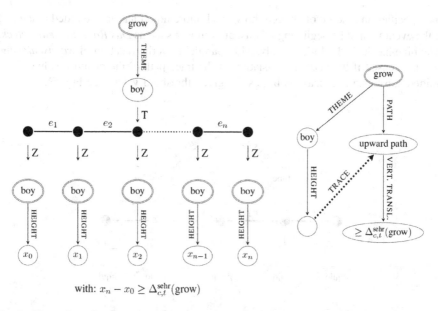

with: $x_n - x_0 \geq \Delta_{c,t}^{\text{sehr}}(\text{grow})$

Fig. 3. Uncondensed and condensed dynamic frame for *Der Junge ist sehr gewachsen* 'The boy has grown a lot'

Next, we turn to verb gradation of verbs of sound emission. Although the example in (7), *Der Motor dröhnt sehr* ('The engine is droning a lot'), involves a threshold value as well, the example differs fundamentally from the *grow*-example. The degree expression *sehr* is licensed for a growing event only if the HEIGHT values at the beginning and at the end differ by at least the given threshold value. Thus, it addresses the vertical translation of the full trace path of the HEIGHT value and not the HEIGHT value as such. In contrast, in a droning event *sehr* is licensed if the loudness of the droning sound is at least of the degree of a given threshold value. Hence, *sehr* is not determining the minimal difference between two values but setting an absolute minimal value. Therefore, the trace path of the loudness value is of no interest and it is sufficient to model the expression as a static event frame.

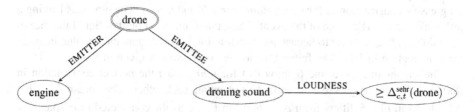

Fig. 4. Static event frame for *Der Motor dröhnt sehr* 'The engine is droning a lot'

By contrast, with verbs of substance emission such as *leak*, the degree expression *sehr* targets the quantity of the substance which is emitted in the course of the event and thus specifies an amount of change (how much more liquid has been emitted at the end of the event than at its beginning). Thus, an example such as *Das Rohr hat sehr geleckt* 'The pipe has leaked a lot', is analyzed in parallel to verb gradation of *wachsen/grow*: *sehr* is licensed if the vertical translation of the trace path of the quantity value of the emitted substance is at least as big as the given threshold value (see Fig. 5).

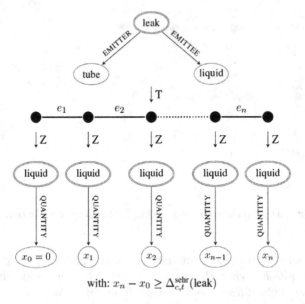

with: $x_n - x_0 \geq \Delta^{\mathrm{sehr}}_{c,t}(\mathrm{leak})$

Fig. 5. Dynamic frame for *Das Rohr hat sehr geleckt* 'The pipe has leaked a lot'

The contrast between event-dependent and event-independent degree gradation as described in Sect. 2 shows up immediately in the frames for *sehr dröhnen* (Fig. 4) and *sehr lecken* (Fig. 5). While the *sehr dröhnen*-frame does not refer to any global property of the trace path as such, the *sehr lecken*-frame specifies a minimal vertical translation of the full trace path. Remember that the minimal threshold value is among others contextually determined by the time span of an event. In a perfective interpretation of (8a) the time span of the described event is the entire leaking event. In contrast, the *am*-progressive construction in (8b), *Das Rohr war sehr am lecken* 'The pipe was leaking a lot', picks out a single stage of the event. Thus, the minimal amount of liquid that has to be emitted in (8a) in order to license the construction is bigger than the one that licenses the construction in (8b). The frame for the *am*-progressive is given in Fig. 6.

The frames in Figs. 5 and 6 show that for *sehr lecken* the perfect construction in (8a) and the *am*-progressive in (8b) do not entail each other. The threshold value $\Delta^{\mathrm{sehr}}_{c,t}(\mathrm{leak})$ in Fig. 5 differs from $\Delta^{\mathrm{sehr}}_{c,t'}(\mathrm{leak})$ in Fig. 6 as the considered time span t' in the *am*-progressive case is shorter than the considered time span t in the perfect case. It is possible to reach the threshold value $\Delta^{\mathrm{sehr}}_{c,t}(\mathrm{leak})$ for the whole leaking event, without reaching $\Delta^{\mathrm{sehr}}_{c,t'}(\mathrm{leak})$ in any single subevent. At the same time, reaching the threshold value $\Delta^{\mathrm{sehr}}_{c,t'}(\mathrm{leak})$ in one subevent does not entail that $\Delta^{\mathrm{sehr}}_{c,t}(\mathrm{leak})$ is reached for the

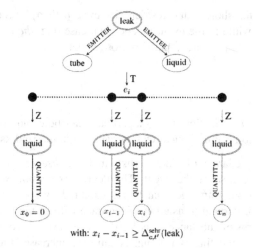

with: $x_i - x_{i-1} \geq \Delta_{c,t'}^{\text{sehr}}(\text{leak})$

Fig. 6. Dynamic frame for *Das Rohr war sehr am lecken* 'The pipe was leaking a lot'

whole leaking event. In contrast, in the case of *sehr dröhnen*, the vertical translation of the trace path is not restricted and thus the threshold value is not time dependent. Hence, both, the perfect and the progressive of *sehr dröhnen* can be represented by the frame in Fig. 4. It follows that both propositions necessarily entail each other. By means of the event decomposition level, a dynamic frame analysis is able to account for the contrast between event-dependent and event-independent degree gradation.

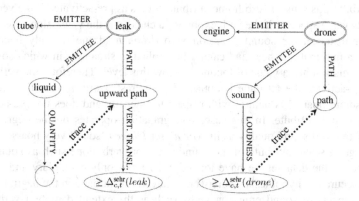

Fig. 7. Condensed dynamic frames for 'The pipe leaked a lot' (left) and 'The engine droned a lot'

Finally, Fig. 7 shows the condensed dynamic frames for *the pipe leaked a lot* and *the engine droned a lot*. In the condensed frames, the difference between event dependent and event independent degree gradation is expressed by whether the threshold value introduced by the intensifier *sehr* restricts the spread of the trace path or not. In the case of an event dependent gradation (e.g. *the pipe leaked a lot*) the contextually given threshold value restricts the trace path by specifying its minimal vertical translation, while in the event independent case (e.g. *the engine droned a lot*) the spatial dimensions of the trace path are not restricted.

5   Conclusion

The way in which *sehr*-gradation targets a particular attribute and its scale in the frame of verbs belonging to specific lexical classes can be summarized as follows:

- *wachsen* 'grow' (change of state verbs): in a *wachsen*-event the HEIGHT value increases in the course of the event. The degree expression *sehr* intensifies this value change. In our frame account this is captured by specifying a contextually specified minimal vertical translation of the trace path of the HEIGHT value.
- *lecken* 'leak' (verbs of substance emission): a *lecken*-event involves a change of the QUANTITY of the liquid that is leaking. Again *sehr* increases the minimal vertical translation of the trace path of the changing QUANTITY value.
- *dröhnen* 'drone' (verbs of sound emission): verbs of this class express the emission of a lexically specified sound. Degree gradation increases the value of the LOUDNESS attribute.

In Sect. 3 we have raised the two major questions how a particular gradation scale is selected by degree gradation and how its values are constrained. In this regard the verb class specific generalizations formulated above are only a first approximation. Our preliminary hypothesis towards the formulation of a universal rule is that the gradation scale targeted by *sehr* has to be introduced by an attribute that is already 'pre-activated' by the verb. In this sense a verb frame attribute counts as pre-activated if the verb either restricts its value space or if the verb expresses a change of the value along the attribute scale. For instance, the sound emission verb *dröhnen* 'drone' is only licensed for sounds of a minimal loudness and cannot be applied to situations in which some kind of sound emission has a value of loudness below this level. Therefore, the verb can be said to pre-activate the attribute LOUDNESS. By contrast, *klingen* 'sound' as in *heiser klingen* 'sound hoarse' is neutral with respect to intensity and does not pre-activate a particular sound attribute. In this case, *sehr*-gradation operates on the degree of the externally realized sound quality as in *sehr heiser klingen* 'sound very hoarse' while it cannot target a scalar attribute of the sound emission verb alone such as LOUDNESS.

Of course, the data set we have looked at so far is only exemplary and does not suffice to secure our hypothesis. Moreover, as already mentioned in the beginning there are other aspects of sound emission verbs such as the extended motion verb use as discussed by Fleischhauer et al. (2017) which can be treated in a particularly fruitful way in a frame account giving access to the necessary aspects of verb meaning as well as to fine-grained semantic properties of the verb's arguments. Therefore, further research has to be done with regard to emission verbs to arrive at a full-fledged phenomenology of this verb class and the way it can be treated within a frame account.

Acknowledgements. The research presented in this paper was funded by the Deutsche Forschungsgemeinschaft (DFG) with a grant to the Collaborative Research Centre (SFB) 991 "The Structure of Representations in Language, Cognition, and Science". We are grateful to the two reviewers of this paper for many valuable comments. We would also like to thank the audiences of TbiLLC 2017 for their feedback on an earlier version.

References

Andersson, S.-G.: On the Generalization of Progressive Constructions. Ich bin das Buch am Lesen – status und usage in three Varieties of German. In: Larsson, L.-G. (ed.) Proceedings of the Second Scandinavian Symposium on Aspectology, pp. 95–106. Almqvist & Winkel, Uppsala (1989)

Atkins, B.T., Kegl, J., Levin, B.: Anatomy of a verb entry: from linguistic theory to lexicographic practice. Int. J. Lexicogr. **1**(2), 84–126 (1988)

Atkins, B.T., Levin, B.: Admitting impediments. In: Uri, Z. (ed.) Lexical Acquisition: Exploiting On-line Resources to Build a Lexicon, pp. 233–262. Lawrence Erlbaum, Hillsdale (1991)

Barsalou, L.W.: Frames, concepts, and conceptual fields. In: Lehrer, A., Kittay, E.F. (eds.) Frames, Fields, and Contrasts. New Essays in Semantic and Lexical Organization, Chapter 1, pp. 21–74. Lawrence Erlbaum Associates, Hillsdale (1992)

Bolinger, D.: Degree Words. Mouton, The Hague (1972)

Borik, O.: Aspect and Reference Time. Oxford University Press, Oxford (2006)

Comrie, B.: Aspect. Cambridge University Press, Cambridge (1976)

Croft, W.: Syntactic Categories and Grammatical Relations. University of Chicago Press, Chicago (1991)

Doetjes, J.: Quantifiers and Selection. Holland Institute of Generative Linguistics, Dordrecht (1997)

Doetjes, J.: Adverbs and quantification: degree versus frequency. Lingua **117**, 685–720 (2007)

Doetjes, J.: Adjectives and degree modification. In: McNally, L., Kennedy, C. (eds.) Adjectives and Adverbs - Syntax, Semantics and Discourse, pp. 123–155. Oxford University Press, Oxford (2008)

Dowty, D.: Word Meaning and Montague Grammar. Reidel, Dordrecht (1979)

Ebert, K.: Progressive markers in Germanic languages. In: Dahl, Ö. (ed.) Tense and Aspect in the Languages of Europe, pp. 605–653. Mouton de Gruyter, Berlin (2000)

Ernst, T.: The Syntax of Adjuncts. Cambridge University Press, Cambridge (2002)

Fleischhauer, J.: Interaction of telicity and degree gradation in change of state verbs. In: Arsenijevic, B., Gehrke, B., Marin, R. (eds.) Studies in Composition and Decomposition of Event Predicates. Studies in Linguistics and Philosophy, vol. 93, pp. 125–152. Springer, Dordrecht (2013). https://doi.org/10.1007/978-94-007-5983-1_6

Fleischhauer, J.: Activation of attributes in frames. In: Pirrelli, V., Marzi, C., Ferro, M. (eds.) Word Structure and Word Usage, pp. 58–62 (2015). http://ceur-ws.org

Fleischhauer, J.: Degree Gradation of Verbs. Düsseldorf University Press, Düsseldorf (2016a)

Fleischauer, J.: Degree expressions at the syntax-semantics interface. In: Fleischhauer, J, Latrouite, A., Osswald, R. (eds.) Explorations of the Syntax-Semantics-Pragmatics-Interface, pp. 209–246. Düsseldorf University Press, Düsseldorf (2016b)

Fleischhauer, J.: Graduierung nicht skalarer Verben. Zeitschrift für germanistische Linguistik **46** (2), 221–247 (2018)

Fleischhauer, J., Gamerschlag, T., Petersen, W.: A frame-analysis of the interplay of grammar and cognition in emission verbs. In: Hartmann, S. (ed.) Yearbook of the German Cognitive Linguistics Association, vol. 5, pp. 177–194. Mouton de Gruyter, Berlin (2017)

Gamerschlag, T., Geuder, W., Petersen, W.: Glück auf der Steiger kommt – a frame account of extensional and intensional 'steigen'. In: Gerland, D., Horn, Ch., Latrouite, A., Ortmann, A. (eds.) Meaning and Grammar of Nouns and Verbs, pp. 115–144. Düsseldorf University Press, Düsseldorf (2014)

Gerling, M., Orthen, N.: Deutsche Zustands- und Bewegungsverben. Eine Untersuchung zu ihrer semantischen Struktur und Valenz. Narr, Tübingen (1979)

Goldberg, A.: Argument realization: the role of constructions, lexical semantics and discourse factors. In: Ostman, J., Fried, M. (eds.) Construction Grammars: Cognitive Grounding and Theoretical Extensions, pp. 17–43. John Benjamins, Amsterdam (2005)

Kennedy, C.: Gradable adjectives denote measure functions, not partial functions. Stud. Linguist. Sci. 29(1), 65–80 (1999a)

Kennedy, C.: Projecting the Adjective – The Syntax and Semantics of Gradability and Comparison. Garland, New York (1999b)

Kennedy, C.: Vagueness and grammar: the semantics of relative and absolute gradable adjectives. Linguist. Philos. 30(1), 1–45 (2007)

Kennedy, C., McNally, L.: From event structure to scale structure: degree modification in deverbal adjectives. In: Matthews, T., Strolovitch, D. (eds.) Proceedings of Semantics and Linguistics Theory, vol. 9, pp. 163–180. CLC Publications, Ithaca (1999)

Kennedy, C., McNally, L.: Scale structure, degree modification, and the semantics of gradable predicates. Language 81(2), 345–381 (2005)

Levin, B.: Building a Lexicon: the contribution of linguistics. Int. J. Lexicogr. 4(3), 205–226 (1991)

Levin, B.: English Verb Classes and Alternations. Chicago University Press, Chicago (1993)

Levin, B., Rappaport Hovav, M.: Unaccusativity: At the Syntax-Lexical Semantics Interface. MIT Press, Cambridge (1995)

Levin, B., Rappaport Hovav, M.: Argument Realization. Cambridge University Press, Cambridge (2005)

Löbner, S.: Is the German Perfekt a perfect Perfect? In: Kaufmann, I., Stiebels, B. (eds.) More than Words, pp. 369–391. Akademie-Verlag, Berlin (2002)

Löbner, S.: Sub-compositionality. In: Werning, M., Hinzen, W., Machery, E. (eds.) The Oxford Handbook of Compositionality, pp. 220–241. Oxford University Press, Oxford (2012)

Naumann, R.: An outline of a dynamic theory of frames. In: Bezhanishvili, G., Löbner, S., Marra, V., Richter, F. (eds.) TbiLLC 2011. LNCS, vol. 7758, pp. 115–137. Springer, Heidelberg (2013). https://doi.org/10.1007/978-3-642-36976-6_9

Perlmutter, D.M.: Impersonal passives and the unaccusativity hypothesis. In: Proceedings of the 4th Annual Meeting of the Berkeley Linguistics Society, pp. 157–190 (1978)

Petersen, W.: Decomposing concepts with frames. Balt. Int. Yearb. Cogn. Log. Commun. 2, 151–170 (2007/2015). Reprint 2015 in Gamerschlag, T., Gerland, D., Osswald, R., Petersen, W. (eds.) Meaning, Frames, and Conceptual Representation, pp. 43–67. Düsseldorf University Press, Düsseldorf

Potashnik, J.: Emission verbs. In: Everaert, M., Marelj, M., Siloni, T. (eds.) The Theta System, pp. 251–278. Oxford University Press, Oxford (2012)

Rappaport Hovav, M., Levin, B.: Building verb meanings. In: Butt, M., Geuder, W. (eds.) The Projection of Arguments: Lexical and Syntactic Constraints, pp. 97–134. CSLI Publications, Stanford (1998)

Rappaport Hovav, M., Levin, B.: Classifying single argument verbs. In: Coopmans, P., Everaert, M., Grimshaw, J. (eds.) Lexical Specification and Insertion, pp. 269–304. John Benjamins, Amsterdam (2000)

Rappaport Hovav, M., Levin, B.: Reflections on Manner/Result Complementarity. In: Rappaport Hovav, M., Doron, E., Sichel, I. (eds.) Lexical Semantics, Syntax and Event Structure, pp. 21–38. Oxford University Press, Oxford (2010)

Tenny, C.: Core events and adverbial modification. In: Tenny, C., Pustejovsky, J. (eds.) Events as Grammatical Objects, pp. 148–185. CSLI Publications, Stanford (2000)

Tsujimura, N.: Degree words and scalar structure in Japanese. Lingua **111**, 29–52 (2001)

Valin, V., Robert Jr., D.: Exploring the Syntax-Semantics Interface. Cambridge University Press, Cambridge (2005)

Vendler, Z.: Verbs and times. Philos. Rev. **56**, 143–160 (1957)

Zwarts, J.: Prepositional aspect and the algebra of paths. Linguist. Philos. **28**(6), 739–779 (2005)

The Athlete Tore a Muscle: English Locative Subjects in the Extra Argument Construction

Katherine Fraser[✉]

University of the Basque Country (UPV/EHU), Vitoria-Gasteiz, Spain
katherineelizabeth.fraser@ehu.eus

Abstract. This paper investigates a special class of English change-of-state verbs exhibiting unexpected argument structure when the change-of-state is interpreted as unintentional and the subject is a LOCATION: e.g., *the skier tore a muscle* or *the boat broke a rudder*. This construction provides evidence for a change-of-state construction beyond the uniform class of anti-causatives (reflexively marked or not) and their inchoative alternants. I argue that there is a necessary part-whole relation between the extra argument subject and the object. This paper describes the semantic constraints of the construction, both how the subject's unexpected semantic role restricts the interpretation, and how changing the part-whole relationship can have implications for the event structure. Finally, I argue that in this configuration, the extra argument subject is the Perspectival Center, like in the Genitive of Negation in Russian.

1 Introduction

This paper describes an unexpected English structure with an unaccusative verb[1] and a non-THEME, non-AGENT subject. The sentences in (1) are examples of the construction; the subject DPs are boldfaced.

(1)   {Hurricane Irma suddenly switched paths, and residents of Southwest Florida are rushing to prepare. Among the damage from the hurricane}

For helpful comments and discussion, I would like to thank Elena Castroviejo, Berit Gehrke, Daniel Hole, Henk Zeevat, two anonymous reviewers, and the audiences at TbiLLC in Lagodekhi and the Event Semantics Workshop in Köln. The author gratefully acknowledges the predoctoral grant BES-2016-076783 (Spanish Ministry of Economy, Industry, and Competitiveness, MINECO), project FFI2015-66732-P, (MINECO and FEDER), the IT769-13 Research Group (Basque Government), and UFI11/14 (UPV/EHU). Part of this research has also been supported by SFB Project B8: "Alternations and Binding" at Universität Stuttgart.

[1] Here, "unaccusative" is taken to refer to cases where there is no external argument in the structure, following, e.g., Embick (2004). One additional terminological note: I use "construction" informally; I am not assuming this structure is a construction in the technical sense.

A. Silva et al. (Eds.): TbiLLC 2018, LNCS 11456, pp. 128–146, 2019.
https://doi.org/10.1007/978-3-662-59565-7_7

    a.   **Kevin's car** burst a tire.
    b.   **The school window** cracked a pane.
    c.   **The bucket by the pool** spilled water.

In each sentence, a change-of-state (CoS) occurred. What is unusual, particularly for the typically inflexible English word order, is that the sentences in (1) include an inanimate subject. This excludes the possibility that the subject is the AGENT, and as I will show below, it is also not a "non-agentive" causer. Instead, an external AGENT caused the CoS. Here, the causer would be the hurricane.

The literature on argument structure, including unaccusatives and reflexives, mostly focuses on accounting for decreased valencies of the root verb. Here we have the opposite situation, as there is an additional argument in the event structure. Similar to the external possession structures in Romance (2), the English construction adds an additional argument (see Deal 2017 for a cross-linguistic overview of external possession structures and the various accounts). Unlike the Romance construction, the English version is possible with inanimates.

(2)   A *Juan se le rompió una pierna.*
       to Juan REFL 3SG broke a leg
       'Juan broke his leg.'   SPANISH

Following Hole (2006), I call the unusual subjects "extra" arguments and claim that the semantic role of these subjects is, in fact, LOCATION.[2] As LOCATIONs are typically considered adjuncts, not arguments, it is interesting to have a LOCATION subject. This paper intends to join the discussion on UTAH (Uniformity of Theta Assignment Hypothesis)-violating English subjects, in that the thematic relationships are not corresponding to identical structural relationships (Baker 1988). Another construction which violates this principle is in (3).[3]

(3)   ACCOMMODATION construction
    a.   This bed sleeps five people.
    b.   The game plays up to 7 players.   (Sailor and Ahn 2012)

Similar to the extra argument construction of (1), the subjects of (3) are not AGENTS or causers. If one were to interpret these constructions as if the subject showed agentive properties, the resulting interpretation would be comical: a bed causing a certain number of people to sleep or a game causing a certain number of players to be able to participate. Instead, the interpretation is as the respective names suggest; namely, (3) hosts constructions where the subject has the capacity for a certain number of entities, hence the name ACCOMMODATION construction. Although similar in appearance, the main difference between the ACCOMMODATION construction and the extra argument one is that the former expresses a possibility, whereas the latter describes the actuality of a CoS.

---

[2] Hole uses the term LANDMARK (cf. Langacker 1987), instead of LOCATION, as LANDMARK signifies an intermediate Ground. For the purposes of this short paper, such a specificity is not needed.

[3] Thanks to Marcel Pitterof for bringing this article to my attention.

A more specific motivation for investigating the extra argument phenomenon is that it is understudied. Besides being mentioned in passing in some descriptive grammars, the only literature of which I am aware is Rohdenburg's (1974) dissertation, which describes Germanic secondary subjects but lacks a formal analysis, and Hole's (2006) cross-linguistic examination of extra arguments, which examines affecteehood extra arguments in Mandarin and German, but also discusses inanimate, or landmark, extra arguments in English. The binding analysis that he further develops in later work (cf. especially Hole 2014) is concerned with free datives in German, a kind of extra argument which is omissible without semantic or syntactic "residue". The English construction is very similar to the free dative one, and I adopt many of Hole's assumptions about extra arguments. However, the English construction includes only two arguments, neither of which are free datives, so it needs further examination. In this paper, I build on Hole's cross-linguistic descriptive generalisations by detailing the criteria for the English extra argument. For the analysis, I focus on the information structure interface in the English version.

The claims for this paper are the following. As mentioned above, the first, "extra", argument in the construction is a semantic LOCATION, not an AGENT. This extra argument can be either animate or inanimate. With this semantic role, there is an entailment that the CoS happened AT the subject. Not only is the subject a LOCATION, but there is a necessary part-whole relation between the subject and object. Contrary to Hole, I will argue that this relation is a part of the relational noun's lexical entry, not a presupposition, and as the extra argument subject c-commands the THEME, the possessor argument in the lexical entry can be filled. This part-whole relation is a necessary, but not sufficient, condition for the extra argument construction. This part-whole relation must be intact: when the relational THEME is elsewhere in the structure or there is no part-whole relation, an instrumental causer reading arises, which is often infelicitous in these configurations, especially with inanimate subjects.

Section 2 begins by outlining the descriptive generalisations, or the prerequisites for my claims. I will describe the type of predicates which can participate in the construction, and the constraints within this subclass. The section continues by motivating LOCATION as a thematic role. Section 3 discusses the background of possession constructions and argues for the necessity of an intact part-whole relation. Section 4 introduces the Perspectival Center Principle to account for the construction. Section 5 concludes.

2   The Extra Argument Construction

This section describes the extra argument construction in English. To begin, the eligible predicate must be a change of state (CoS), but without an agent. Within the class of CoS predicates, there are four subclasses that I have delineated (following Levin 1993). The subsection continues by illustrating that there is no constraint on gradability of the CoS for this construction. In the next subsection, I argue for the LOCATION role of the extra argument.

2.1 Eligible Predicates

As already noted in Rohdenburg's (1974) dissertation, the participating predicates, of varying verb classes, are *ungesteuerte Zustandsveränderungen* 'uncontrolled'/'unintentional CoS'. That is, even with an animate entity in the sentence, the interpretation is that the CoS occurred because of something uncontrolled; in other words, there is no entailment that the volition of some agent was involved. Below is an eligible example with an animate extra argument in (4).

(4)   Johnny broke a leg. $\nrightarrow$ 'Johnny caused the leg to break.'

In (4), it is not necessarily entailed that Johnny caused the relevant leg to be broken, even if it might be salient in the interpretation that Johnny was involved in the action; that is, it is also not necessary that he was unconscious or lying somewhere when somebody/-thing else broke his leg. Levin (1993: 226) notes that "these verbs relate to the occurrence of damage to the body through *a process that is not under control of the person that suffers the damage*. Although some of these verbs may take agentive subjects, *the subject does not receive an agentive interpretation on the use considered here*" (my emphasis, KF). The relevant interpretation is in (5).

(5)   'Johnny's leg broke; causer undefined.'

Not only is there no entailment of intentionality, there is also no implicit agent in the structure. Some constructions with non-causer subjects, such as passives, do have an implicit agent; middles and unaccusatives do not. Examples in (6) illustrate this difference with an embedded subject control predicate. If there is an implicit agent in the structure, it controls PRO. If not, the extra material results in an infelicitous or ungrammatical sentence.

(6)   **Implicit agents of passives (vs. middles and unaccusatives)**
    a.   This ship was sunk [PRO to collect the insurance].   *passive*
    b.   #This ship sank [PRO to collect the insurance].   *unaccusative*
    c.   *This ship sinks easily [PRO to collect the insurance].   *middle*

(Bhatt and Pancheva 2006: 254)

In (6), only the passive has an implicit agent; it is possible in (a) to attribute the ship-sinking action to an external agent only. Unaccusatives and middles, on the other hand, cannot (easily) control PRO (though, as an anonymous reviewer pointed out, (c) would be felicitous if the embedded phrase was something like *to escape capture*).

This diagnostic is applied to extra arguments in (7).

(7)   a.   #The window cracked a pane [PRO to collect the insurance].
    b.   #The bucket spilled water [PRO to clean the terrace].

Patterning like an unaccusative, the extra argument construction precludes an implicit agent construal so that combination with control constructions results in pragmatic incongruity. For the sentences in (7) to be felicitous, the subject would have to be the one collecting insurance. In a non-cartoon world, these utterances are odd: a window is inanimate and therefore incapable of intentionally causing its glass pane to break, with the motive of collecting insurance; similarly with the bucket. Therefore, the extra argument construction has no implicit agent.[4] Next, we will discuss which verbs can participate in the construction.

In addition to the BODILY HARM class seen in (5), I have identified three verb classes (of the classes in Levin 1993) which can be CoS unaccusatives and can have extra argument subjects: DESTRUCTION, APPEARANCE, SUBSTANCE EMISSION; with this, my proposal is a more systematic definition of the participating predicates than Rohdenburg's. Table 1 provides examples for each.[5]

Table 1. Verb classes participating in the extra argument construction

DESTRUCTION	APPEARANCE
(a) The car **burst** a tire	(g) James **developed** a tumour
(b) The sofa **splintered** a leg	(h) The lizard **grew** a tail
(c) The cellos **popped** strings	
BODILY HARM	SUBSTANCE EMISSION
(d) The climber **broke** a rib	(i) The tunnel **seeped** water
(e) The skater **chipped** a tooth	(j) The bucket **spilled** water
(f) Duncan **sprained** a finger	(k) The brownies **oozed** caramel

As can be seen in the table, singular and plural subjects, as well as count and mass objects, are possible; if there are any constraints along these lines, they

---

[4] The ACCOMMODATION construction, mentioned in the introduction, patterns like the extra argument construction, as seen in (i).

(i)   a. #This bed sleeps tall guests [PRO to please them].
       b. #This game plays up to 7 players [PRO to please the kids].

Semantically, however there are differences. Namely, a CoS is entailed in the extra argument construction (depending on tense/mood, of course, but in the simple past this is true), whereas the ACCOMMODATION one entails no event—it only expresses the potential of the subject being able to accommodate the object (regardless of tense); this potential can be translated as modality, which has implications for the structure. And, even though (i) shows there is no implicit agent in the structure, the active variant's external argument is present in the non-active variant (*this bed sleeps five people*), unlike for the extra argument construction, where no external argument is present at all. For these reasons, I will no longer make comparisons with the ACCOMMODATION construction, but see Ahn and Sailor (2012) for an analysis.

[5] There are only two sentences for the APPEARANCE class because *grow* and *develop* are the sole predicates in this class that allow an extra argument subject.

are due to the lexical semantics of the predicates themselves. People, animals, artifacts, and other concrete entities all are possible—as long as they fulfill the semantic constraints. For example, the lexical semantics of *splinter* dictates that the THEME be composed of wood. With respect to the semantic type, the extra argument subject is free from, e.g., an animacy restriction. Instead, the entity need only be coherent with the entire predicate's semantics and world knowledge; e.g., *?the bucket spilled people* or *??Duncan sprained water/brownies* would be incoherent. With this in mind, let us consider types of scalar change.

SCALAR CHANGE involves predicates with a CoS that is measurable along a scale (cf. Kennedy and Levin 2008; Kennedy and McNally 2005; Beavers 2008, a.m.o.). That is, there is a SCALE with a set of degrees on a particular dimension with an ordering relation. Each of the three types of scalar change predicates are concerned with a different scale: INCREMENTAL THEMES measure spatial extent, DIRECTED MOTION measures the path of an argument's motion, and CoS measures the value of a gradable property. Often, the latter two types of predicates are differentiated from the case of incremental themes where the scale of spatial extent is introduced by the argument. In contrast, the other two types concern lexicalised scales (see, e.g., Rappaport Hovav 2008). Here, only the lexicalised property scale is relevant, as the eligible predicates are CoS predicates.

The CoS must be unintentional or uncontrolled, as noted above. Besides this requirement, the CoS can be either non-gradable•or gradable (8), depending on the complexity of the predicate, in terms of event structure (Hay et al. 1999; Beavers 2008, 2013; a.o.). Gradability is dependent on the individual predicate's (degree) semantics, so not a relevant constraint.

(8)   a.   The car (#slowly) burst a tire.   *non-gradable change*
      b.   The lizard (slowly) grew a tail.   *gradable change*

Similarly, concreteness is not a constraint of the extra argument construction.[6] This is seen in (9); agentive sentences with *spoil* are provided in (10).

(9)   a.   Society spoils people.   *extra argument*
      b.   19th Century English society spoiled the Victorians.

(10) a.   The mother spoils her son.   *transitive*
      b.   The king spoils his subjects [in 19th Century English society].

In the abstract sentences of (9), there is arguably still an uncontrolled (gradable) CoS, of the DESTRUCTION type. The subject, *society*, is not an AGENT. Instead, I would argue that the subject, *society*, is where the object, *people/Victorians* are located. Although not as common as the concrete examples, it is possible to have an abstract CoS with an extra argument.

This subsection has argued that the CoS described by the extra argument construction must be unintentional, without even an implicit agent in the structure. Four CoS verb classes have been found to be eligible, although the semantics

---

[6] Thanks to Agustín Vicente for this observation.

of the subject/object will be a further constraint (Sect. 3). The next subsection will discuss the semantics of the extra argument subject.

2.2   Role of the Extra Argument

This section will argue for the unique role of the extra argument subject, providing data which show these arguments pattern neither like a typical CoS unaccusative (expected as there is no external argument) nor a non-agentive causer. Below, both the transitive (11-a) and unaccusative (11-b) alternants are different from the extra argument (11-c).

(11)   a.   Jane spilled water [from the bucket].   *transitive*
            ⇒ 'Jane caused the water to spill.'
       b.   Water spilled.   *unaccusative*
            no entailment of cause by anybody/-thing
       c.   The bucket spilled water.   *extra argument*
            ⇏ 'the bucket caused the water to spill.'

Examples (11-b) and (11-c) do not provide any sort of inference regarding the cause of the spilling; it would be odd to infer that the *bucket* is the cause, being neither animate or instrumental. This clear difference between (11-a) and (11-c) suggests that analysing these as "anti-causatives" (cf., e.g., Schäfer 2009) would be inadequate, as the alternation is not unaccusative–transitive.

There are also non-agent causers among the unaccusative constructions. Take, for example, "ambient condition causers" (Rappaport Hovav and Levin 2012), where the CoS is internal, like in the alternations in (12).

(12)   a.   The flowers withered.   *unaccusative*
       b.   *The farmer withered the flowers.   *agentive*
       c.   The heat withered the flowers.   *ambient condition*

At first blush, (12-c) may look similar to the APPEARANCE verbs with an extra argument subject (13). There is, however, a difference visible in the boldfaced prepositional phrase, infelicitous in the ambient condition sentence (14).

(13)   a.   The lizard$_i$ grew a tail **on itself$_i$**.   *extra argument*
       b.   Harry$_i$ developed breasts/a wart/a tail **on himself$_i$**.

(14)   The heat$_i$ withered the flowers #**on itself$_i$**.   *ambient condition*

The extra argument constructions of (13) are felicitous with a phrase localising the CoS on the subject, illustrated by way of a reflexive pronoun. It is perfectly fine to interpret the tail-growing to be on the lizard and the appendage-developing to be on Harry. In contrast—and possibly due to the fact that ambient condition predicates are impossible to combine with AGENTS, as seen in (12)—the CoS in (14) cannot be localised so easily on the *heat* (even with a *heat lamp*, the utterance is still not as good as in (13)). This difference is a defining one for the extra argument construction. In other words, what defines the extra argument is its interpretation as a LOCATION.

While an instrumental causer reading is available for some examples (15), this is not the relevant reading (cp. the irrelevant agentive reading of the BODILY HARM predicates on p. 4). And while it is possible to trigger an instrumental causer reading, the relevant reading for the extra argument construction is that the CoS happened on the subject; see examples in (16).

(15)   The boat broke a rudder **on another boat**.

(16)   a.   The window cracked a pane **#on another window**.
        b.   The runner tore a muscle **#in somebody else's calf**.

Compare the continuations of (16) with (17), where a sublocation of the extra argument subject is identified.[7] These are the relevant readings.

(17)   a.   The boat$_i$ tore a sail [**on its$_i$ mast**].
        b.   The runner$_i$ tore a muscle [**in her$_i$ calf**].

These sentences show that the extra argument localises the CoS on the argument itself. Hole (2006: 415) argues that these subregions "must be semantically active, even if they are not pronounced", as the subjects are "functional" entities. I take this to mean that the extra argument subjects contain relevant sub-regions (boldfaced PP), which are expected from our knowledge of the world and that these sub-regions are necessarily activated when the extra argument is conceptualised as a functioning entity; these sub-regions then comprise other entities, including the THEME. That is, it would be odd to interpret the muscle-tearing event (b) as describing a torn muscle in the biceps, as runners primarily activate the leg muscles. Similarly, it is infelicitous to utter (a) if the torn sail was one simply lying about on the ship, as a functioning ship has sails on its mast—and it is these sails which are most salient. These semantically-active PPs are further motivation that the extra argument subject is a LOCATION.

Building on the observations from above, I submit preliminary criteria for the extra argument construction: (i) an unintentional CoS can occur and (ii) an unintentional CoS which happened on/at the subject. If the CoS is both unintentional and occurred on/at the subject, the construction contains an extra argument which is a LOCATION. Although these two points are necessary for the extra argument structure to be felicitous, this is not the entire picture. The next section will supplement these two criteria.

3   The Part-Whole Relation

This section makes a connection with possession for the relation of the LOCATION subject and the THEME object. It is important to be more specific about the relation between these two arguments, because FIGURE-GROUND (Talmy 1972) is not specific enough. Here, I show that not only must the extra argument be a LOCATION, but that the direct object must be a part of the subject—and that

---

[7] The co-indexation subscripts indicate co-reference; whether or not the possessive pronoun is a bound variable is not central to this paper.

this relation is lexically defined. First, I present a brief introduction to possession theory and how this is reflected in the extra argument construction, followed by a more detailed look at the part-whole relation.

3.1 Locations and Possession

Possession is often discussed to be a lexicalised figure-ground relation. For example, cross-linguistic work by Slobin (1985) discusses how French and German children use locational terms to describe possession (see also, e.g., Armon and Lotem 2004; Kayne 1993). In the cognitive linguistics literature, Langacker (2006: 173) states that "spatial proximity is a usual concomitant of possession". Following these works, I propose that the interpretation of the extra argument as a LOCATION is also reflective of a part-whole relation, with the THEME as the part on/at the LOCATION whole. The examples in (18) display the extra argument interpretation, with subscripts to make the part-whole relation explicit; note that these subscripts are not indicative of a binding relation.

(18) a. The boat$_i$ broke {a/its}$_i$ rudder.
      b. The runner$_i$ tore {a/her}$_i$ muscle.
      c. The lizard$_i$ grew {a/its}$_i$ tail.
      d. The bucket$_i$ spilled (its$_i$) water.

These sentences additionally suggest that the possession relation must be "inalienable". Typically, inalienable possession comprises kinship relations (pragmatically odd for this construction), body parts (such as in the BODILY HARM subclass; (b) and (c) above), and entities in a part-whole relation (*sail-boat* in (a), *water-bucket* in (d)). Kinship and body part are then subcases of inalienable possession. The part-whole relation of sail and boat is different than, say, a person owning a pet or some article of clothing, in that the latter is an alienable relation.[8] So, while the verbs involved in the extra argument construction belong to varying classes, all such constructions describe an inalienable possession relation.

As mentioned in the introduction, there are cross-linguistic parallels to the extra argument possession relation, with what is called an "external possession" relation (see Deal 2017 for an overview); the "external"/"internal" distinction concerns the structure and is not to be confused with the terms "alienable"/"inalienable". These structures are only possible for subjects if the verb is unaccusative (cf. Baker 1988). The sentences in (19) demonstrate the difference between external and internal possession in French.

(19) a. *Je lui   ai   pris   la   main.*
          I   3SG.DAT have taken the hand
          'I took his hand.'   *external possession*

---

[8] An apparent counterexample to the inalienable possession criteria would be *Alex lost his button/keys*, as one reviewer pointed out. I agree that this type of utterance fits into the proposed generalisations, and it would even be translated into a dative possessive in, e.g., German. Not being very productive in English, I will regard it as a minor exception to the inalienable possession constraint.

b. *J'ai   pris   sa   main.*
I-have taken his hand
'I took his hand.'   *internal possession*

[FRENCH; Deal [2017]]

In (19-a), the possessor, the dative clitic *lui*, is syntactically dependent on the
verb, but semantically associated with another verbal argument, *main* 'hand'.
That is to say, the contextually relevant hand is associated with the referent of
the dative clitic and not the subject. In contrast, the possessor, *sa*, of (19-b) is
syntactically dependent on the possessum, *main* 'hand', not the verb; semantically, *sa* is also dependent on *main*. The English extra argument construction is
similar to the external possession structures. For instance, Cuervo (2003) claims
that the Spanish version requires the possessor to be in a part-whole relation.
Regarding the external possession example earlier, the broken leg is Juan's, not
somebody else's.

A major difference between the constructions in English and Romance is that
only English allows inanimate arguments.[9] In the English construction, animates
are only possible if the THEME is a body part (and the CoS is unintentional; see
discussion in Sect. 2.1). For some of the eligible predicates, it is even infelicitous
to include an animate subject, as seen in (20-b). The transitive (20-b) slightly
improves when it is clear it was on somebody else's clothing.

(20)   a. The shirt popped a button.   *extra argument*
        b. #Alex popped a button.   *transitive*

In the literature on possession (cf. Barker 1995, 2011; Partee et al. 2011; Loebner
1985; Vikner and Jensen 2002; a.o.), RELATIONAL NOUNS are explicitly associated
with inalienable possession and are defined in contrast to SORTAL NOUNS: the
former have an argument structure that is inherently two-place, as it specifies
the possession relation, while the latter denotes a one-place relation and can only
express a possession relation after type-shifting (although see, e.g., Le Bruyn et
al. 2013 on all nouns being sortal). In this way, the inalienable possession relation
is lexical for relational nouns and an alienable one can be pragmatically defined
for sortals. A diagnostic to differentiate the two can be seen in (21) and (22),
where only the former can appear with a post-nominal genitive *of*.

(21)   RELATIONAL NOUN   (22)   SORTAL NOUN
        a. John's sister   a. John's lake
        b. The sister of John   b. ?The lake of John

Just like body parts and other inalienable relations, kinship-denoting nouns
can appear with a post-nominal *of* phrase (21). For sortal nouns like *lake*,

---

[9] The animacy of the extra argument in these other languages allows for an AFFECTEE
thematic role, which also accounts for the observation that the act on the possessum
physically affects the possessor (see, e.g., Barnes 1985). However, affecteehood is a
concept only available for animates, which is in sufficient for the construction under
investigation here.

without such a lexically defined relation, it is more difficult to successfully combine the two. The examples in (23)–(24) demonstrate the diagnostic as applied to the extra argument construction; (a) paraphrases the original sentence with the Saxon genitive; (b) with the extra argument and a potential possessor; (c) with the THEME, extra argument, and the post nominal *of* phrase.

(23)   The cello popped a string.   DESTRUCTION
     a.   The cello's string was plucked.
     b.   ?The cello of John ...
     c.   A/the string of the cello was plucked.

(24)   The cake oozed caramel.   SUBSTANCE EMISSION
     a.   The cake's caramel is delicious.
     b.   ?The cake of John ...
     c.   (?)The caramel of the cake is delicious.

The genitive *of* in both (b) sentences is bad, as both nouns are sortal and there is no inherent inalienable possession relation. For the first three subclasses, BODILY HARM, DESTRUCTION, and APPEARANCE, an *of* phrase following the THEME of the original sentence is okay. Where this pattern falters is the SUBSTANCE EMISSION class (24); as liquids are not typically relational nouns—and these predicates take a liquid or other substance as their THEME argument—this is not unexpected. If we take a look at some other examples without an inherently relational noun (25), a pattern emerges: the THEME is a liquid-like substance and the extra argument a container.[10]

(25)   a.   The cake oozed caramel.
     b.   The tunnel seeped water.

Although it might be marginally odd to say *the caramel of the cake* or *the water of the tunnel*, there is a relation akin to possession in these sentences. So, for those predicates which do not combine with a relational noun as the THEME, a necessary condition is that the extra argument representing the 'whole' is a CONTAINER, and the THEME standing for the 'part' is a liquid or other substance. Here, I will assume that the CONTAINER relation is parallel to the part-whole one, and henceforth use "part-whole" to refer to both. Additionally, I am claiming here that the part-whole relation is lexically required by the relational noun. In the next subsection, I will discuss how this restriction is realised.

3.2   Relational Parts and Locational Wholes

In the rest of this subsection, I will discuss the part-whole relation in more detail, including what happens to the interpretation when it is not present. First, let us discuss the nature of the part-whole relation and conditions on its semantics. The examples below illustrate the difference between the extra argument construction

---

[10] Note that, due to lexical semantics, only the APPEARANCE and SUBSTANCE EMISSION classes permit a liquid-container configuration.

with an explicit, concrete, part-whole relation (26) and similar configurations without such a relation (27); the first possible interpretation regards the part-whole relation and the second, the causer of the CoS, if any.

(26)   The car burst a/its tire.   *extra argument*
⇒ 'the tire is a **part** of the car.'
⤳ 'CoS is **unintentional**.'

The object, *a/its tire*, in (26) is conceivably a part of the subject, *the car*, supporting the above claim about a part-whole relation. The felicity of the possessive pronoun, *its*, is further indication. Similarly, the felicity of the indefinite article, plus world knowledge, suggests that UNIQUENESS of the part is irrelevant.

(27)   a.   The pin burst a/(#its) bubble.   *instrumental causer*
⇏ 'the bubble is a part of the pin.'
⤳ 'causer is the pin.'
        b.   The car hurt (#a)/its driver.
⇏ 'the driver is a concrete part of the car.'
⤳ 'causer is the car.'

In contrast, in (27-a), *bubble* is a sortal noun and not a part of the pin, which can be seen by the infelicity of the possessive pronoun; thus, the interpretation is that the subject, the pin caused the damage to the object, the bubble. Interestingly, the *driver* of (27-b) is a relational noun, but the relation is an abstract one with two concrete components. Like with the instrumental *pin*, the most salient interpretation is that the car caused the damage, instead of it being an uncontrolled CoS. The sentences in (27) show that without an explicit, concrete, relation between the components, the surface subject is a causer, not an "extra argument", i.e., not a LOCATION.

A further constraint on the components of the part-whole relation builds on Hole's (2006: 414–415) observations about a holistic condition. This condition requires the subject to be a "functional or organic whole". This would mean that the extra argument itself cannot be a relational noun. The sentences below, Hole's (2006) example (52), illustrate his holistic constraint.

(28)   a.   The ship tore one of its sails.
        b.   #The mast tore one of its sails.

As a native speaker, I agree that (28-b) is deviant; the intended interpretation is that the mast is an instrumental causer, but that is difficult when the sail belongs to the mast. However, if *one of its sails* is changed to the indefinite article, like the other examples discussed in this paper, the sentence improves (29); other native speakers agree on this judgment.

(29)   The mast tore a sail.

Additionally, native judgements suggest that the holistic constraint is not consistent across verbal classes (although I think the construction *is* better when the whole is an organic one). Examples of the four classes are in (30-a)–(30-d).

(30) a. #The skater's mouth chipped a tooth.   BODILY HARM
      b. #The brownie's insides oozed caramel.   SUBSTANCE EMISSION
      c. The ship's mast tore a sail.   DESTRUCTION
      d. Harry's foot developed warts.   APPEARANCE

Examples (30-a) and (30-d) are consistent with the holistic constraint: it is pragmatically odd to say these, presumably because an instrumental causer reading is triggered, and such subjects are implausible instruments. I propose that the felicity depends on whether the subwhole (e.g., mouth, roof) is inside the whole or not. That is, it is impossible to conceptually separate the inside of an entity from its whole, whereas a mast (30-c) is not on the inside of the whole; the sentence in (30-c) is therefore felicitous.[11] So, I propose here that the English construction allows any type of noun as the extra argument, but requires a relational noun as the THEME, and tends to not have an "inside" THEME.

Even though my generalisation differs from Hole's, in that the extra argument can be a relational noun, I agree that there is an inference about functionality in the extra argument construction.[12] The inference is that the THEME is a functioning part of the extra argument; I add to this that the THEME is interpreted to be in a "functional location" on the extra argument. This can be seen in a comparison to the unaccusative variant (with the LOCATION in a postverbal PP) discussed above. Namely, there are differing inferences about where the part is located on the whole—although the CoS is localised on the subject in both cases and even though both alternants have a relational noun as the THEME. This is seen in (31)/(32); the extra argument (a) sentences have the inference that the THEME is located in the functional location.

(31) a. The car burst a tire.   *extra argument*
          ↝ 'in functional location, i.e., one of four.'
      b. A tire burst on the car.   *unaccusative*
          ↛ 'in functional location, i.e., could be the spare tire.'

(32) a. The lizard grew a tail.   *extra argument*
          ↝ 'in functional location, i.e., backside.'
      b. A tail grew on the lizard.   *unaccusative*
          ↛ 'in functional location, i.e., could be from its head.'

---

[11] One sentence not fitting in with the generalisation is (i), even though a brain is inside a person. It is possible that the animacy of the whole, *Harry* in (i), assumes an affectee-type role, which allows the inside sub-whole.

(i)   Harry's brain developed a tumour.

[12] Other argument alternations, such as the *swarm*- or *spray/load*-alternations exhibit a similar effect, most often dubbed "holistic" effect (see Levin 1993 for references). The inference here is different, in that there is no space being filled—to any degree.

The tire that burst in (31) is one of the four used to drive, and the tail that grew in (32) is on the reptile's rear-end, the typical place a tail grows—unless the context otherwise specifies that there was a mutation, which would allow for an interpretation of a tail growing elsewhere on the lizard's body. In contrast to the extra argument construction in (a), the unaccusative variants in (b) lack this inference. Unlike the (a) sentences, it is possible to have the interpretation that the burst tire is the spare (31) or that the tail grew from the lizard's head as the result of an experiment (32). This difference indicates that not only must the subject and object be in a part-whole relation, but that this relation must be functional. The exact status of this inference, however, will have to be set aside for now.

It is possible that this inference is triggered by the configuration of the arguments; that is, it is possible that the functionality inference is triggered by the asymmetrical c-command relation of the extra argument and the THEME. If we look at the locative-inversion parallel of these sentences, the inference of a functional location is not very strong, if there at all; see (34).

(33)   (34)   a.   On the ship tore a sail.
                    ⇏ 'in functional location, i.e., could be a sail not in use.'
               b.   On the lizard grew a tail.
                    ⇏ 'in functional location, i.e., could be from its head.'

Similar to the unaccusative variant, though with a less stark contrast, the locative-inversion variant does not necessarily encode that the THEME is in a functional location on the extra argument. Even though the location is in a PP, and therefore slightly different in structure than the extra argument one, it does not seem that a sentence-initial location triggers the functionality inference. If we look at another variant, with an explicitly stated part-whole relation, however, this functionality inference is present. The sentences in (35) exemplify.[13]

(35)   a.   The car has a burst tire.
        b.   The lizard has a (newly) grown tail.
              ⤳ 'in functional location.'

In both these sentences, the THEME is interpreted to belong to the subject at a "functional" location. For example, the burst tire of (35-a) was one of the four spinning. In the example with an animate subject (35-b), it is possible though not plausible to interpret the subject as being in possession of a newly grown tail. In this case, the CoS would have had occurred elsewhere than the subject. This is a possibility, although not the most salient interpretation. The main difference between these sentences and their extra argument counterpart is that the CoS is not an active part of the sentence's meaning; the HAVE sentences in (35) describe state of possession, as a result of a CoS at some point. In contrast, the extra argument structure describes a CoS having occurred, in addition to possession.

---

[13] Note that there are no SUBSTANCE EMISSION examples, because these verbs do not allow this alternation: ??*The tunnel has seeping water.*

As a summary of the previous subsections, I will describe the necessary ingredients of the extra argument structure. First, there must be an unintentional or uncontrolled CoS. Second, the subject is not a causer or agent, but the LOCATION of the CoS. Third, there is a THEME, which is a relational noun and in an inalienable part-whole relation to the LOCATION.

4 Proposal

This section proposes an account of the extra argument structure, based on insights from the genitive of negation in Russian (Partee and Borschev 2004) and existential unaccusatives in English (Irwin 2012, 2018). Before the proposal, however, a brief note is in order on benefactives and applicatives.

Namely, the extra argument construction is not a benefactive construction. Typically, benefactive/malefactive arguments are animate ones, "since normally only animate participants are capable of making use of the benefit bestowed upon them" (Kittilä and Zúñiga 2010: 2; see also Blake 1994, Lehmann et al. 2008). In a similar vein, the extra argument structure cannot be analysed as an applicative construction, either high or low (Pylkkänen 2002). First, English is not considered to be a high applicative language, nor does this type necessarily involve possession. Second, low applicatives concern a possession relation, but it is only defined as "intended" possession, and does not necessitate possession of the THEME. In contrast, the extra argument construction encodes possession of the THEME by the subject. On the basis of necessary possession of the THEME and possible inanimacy of the extra argument, the extra argument construction is not a low applicative structure.

For the first part of my proposal, I compare the semantics of the extra argument structure to "genitive of negation" in Russian. My main claim will be that in the extra argument construction, the location subject is inferred to exist and the possession relation is asserted, while the CoS is backgrounded.

An alternation in Russian, called the "genitive of negation" (GenNeg), has a nominative (36-a) and genitive variant (36-b). As the name suggests, the genitive construction is only possible under negation and is characteristically associated with non-agentivity.

(36) a. *Petja   na koncerte ne   byl.   Koncerta ne   bylo.*
          Petja-NOM at concert   NEG was-M.SG Concert   NEG was-N.SG
          Petja was not at the concert. There was no concert.
      b. *Peti   na koncerte ne   bylo.   #Koncerta ne   bylo.*
          Petja-GEN at concert   NEG was-N.SG Concert   NEG was-N.SG
          Petja was not at the concert. #There was no concert.

                                              (Partee and Borschev 2004: 218)

As can be seen in the felicity of the continuations, the (b) utterance with the genitive comes with an inference that the location, *the concert* exists. There is no existence inference for the subject Petja. Partee and Borschev use examples like these to argue that the genitive of negation structure has a different "Perspectival

Center" than the nominative counterparts. More precisely, they claim that when a LOCATION is the Perspectival Center, "it turns the predication around: saying of the LOC that it has THING in it" (p. 217). They propose a principle (ex. 18) which states that the Perspectival Center is presupposed to exist. In the GenNeg sentences, the perspectival center is the LOCATION ("LOC"); the LOCATION's existence in a GenNeg sentence is presupposed, while there is no presupposition of the THEME ("THING"). This is what we saw above in (36-b). In contrast, the LOCATION's existence in nominative negated sentences is not presupposed; there is a presupposition of only the THEME's existence. This is what we saw in (36-a). And, this is what we see for extra argument structures, like in (37).

(37)   a.   The shirt popped a button. #The button was on another shirt.
       b.   The lizard grew a tail. #The tail was on another lizard.

For the non-instrumental-causer reading of the extra argument construction, it is infelicitous to deny that the THEME is not on/at the LOCATION. In terms of the Perspectival Center principle, both the shirt and the lizard are presupposed to exist, and the button and tail are in/at this presupposed entity.[14] In these utterances, the CoS is not being denied, but it is not in the foreground informationally. In contrast, the unaccusative variants, e.g., *A tail grew*, the COS event is the main information being expressed; the tail's existence is presupposed, but there is nothing presupposed about a location.

In Irwin's (2012, 2018) dissertation, these observations are taken a step further. The structure she calls existential unaccusative (38) is analysed as having a small clause structure with a necessary locative argument.

(38)   A unicorn wandered over.   (Irwin 2018: 6)

Although Irwin's proposal concerns an existential construction, there are insights relevant to extra arguments. In particular, the discussion on discourse referents can shed light on the unavailability of a definite determiner for the THEME. In (39), indefinite and definite determiners are presented for both arguments.

(39)   A/the ship tore a/its/*the sail.

Even though the LOCATION can appear with both an indefinite and definite determiner, the former still carries the interpretation that the entity is part of a contextually-given subset. Considering the presupposition of the location's existence from the Perspectival Center Principle above, this makes sense. In contrast, CoS unaccusatives, like we have seen as comparison throughout the present paper, have a different Perspectival Center and therefore do not presuppose any location. However, as this principle does not say anything about the THEME, it is not necessarily expected that a "given" discourse referent would be prohibited. Like Partee and Borschev (2004), Irwin also argues that the LOCA-

---

[14] Although I will not go through the Family of Sentences test here, the presupposition of existence does project. Consider, e.g., *Did the shirt pop a button?*, where the shirt is inferred to exist; the CoS, however is not.

TION is already established in the context at the time of utterance (see especially Sect. 3.2.2.), whereas the THEME is a new discourse referent. Following this reasoning, the definite article is unable to combine with the THEME, as we see in (39). At least in terms of information structure, the extra argument construction patterns like GenNeg and the existential unaccusatives. There is more work to be done, particularly with respect to determining how the Perspectival Center is reflected in the (non-existential) syntax of the extra argument construction, but the Perspectival Center Principle already is able to account for the possession relation inherent to the semantics.

5   Conclusion

This paper has described the extra argument construction in English, detailing which predicates are eligible; namely, specific CoS predicates with unintentional or uncontrolled interpretations. It also argued that the extra argument subject is a locational whole with a relational part as the THEME, representative of a part-whole relation; this specific configuration and the ensuing relation is crucial for the construction. For example, without the part-whole relation, the utterance takes on an instrumental causer interpretation. The specific ingredients of the extra argument construction are argued to be (i) unintentional CoS (ii) a LOCA-TION subject argument (iii) a THEME object argument which is a relational noun, i.e., which is part of the LOCATION. Although it does not account for all aspects of the semantics, the Perspectival Center Principle of Partee and Borschev (2004) sheds light on the information structure of extra arguments.

Further work needs to be done with respect to the 'functionality' inference. Namely, the inference that the relational noun THEME is in a functional LOCA-TION is something that needs to be investigated in further detail. Cross-linguistic data could be helpful in this domain, as could an empirical study with native speaker judgments on tests similar to the Family-of-Sentences one. Additionally, it would be interesting to investigate whether there are syntactic parallels to the existential unaccusative construction proposed in Irwin (2012, 2018). Even though the intuitions line up, the tests she uses to diagnose the difference between a simple and a complex structure, namely *there*-insertion and *re*-prefixation, do not pattern consistently with the extra argument.

References

Ahn, B., Sailor, C.: The emerging middle class. In: The Proceedings of CLS, vol. 46 (2012)

Alexiadou, A., Anagnostopoulou, E., Schäfer, F.: The properties of anticausatives crosslinguistically. In: Frascarelli, M. (ed.) Phases of Interpretation. Mouton, Berlin (2006)

Armon-Lotem, S., Crain, S., Varlokosta, S.: Interface conditions in child language: a crosslinguistic look at some aspects of possession. Lang. Acq. **12**, 171–217 (2004)

Baker, M. : Incorporation: a theory of grammatical function changing. University of Chicago Press (1988)

Barker, C.: Possessive Descriptions. CSLI Publications, Stanford (1995)

Barker, C. : Possessives and relational nouns. In: Semantics: An International Handbook of Natural Language Meaning, vol. 2, pp. 1109–1130. De Gruyter, Berlin (2011)

Barnes, B.: A functional explanation of French non-lexical datives. Stud. Lang. **9**, 159–195 (1985)

Beavers, J.: Scalar complexity and the structure of events. In: Dölling, J., Heyde-Zybatow, T., Schäfer, M. (eds.) Event Structures in Linguistic Form and Interpretation, pp. 245–265. Mouton de Gruyter, Berlin (2008)

Beavers, J.: The structure of lexical meaning: why semantics really matters. Language **86**, 821–864 (2010)

Beavers, J.: Aspectual classes and scales of change. Linguistics **51**, 681–706 (2013)

Bhatt, R.: Obligation and possession. Papers from the UPenn/MIT roundtable on argument structure and aspect, MITWPL 32, 21–40 (1998)

Bhatt, R., Pancheva, R.: Implicit arguments. In: Everaert, M., van Riemsdijk, H. (eds.) The Blackwell Companion to Syntax, vol. 2, pp. 554–584. Blackwell Publishing, Malden (2006)

Collins, C.: A smuggling approach to the passive in English. Syntax **8**, 81–120 (2005)

Coppock, E., Beaver, D.: Definiteness and determinacy. L&P **38**, 377–435 (2015)

Cuervo, M.C. : Datives at Large. Doctoral dissertation. MIT (2003)

Deal, A.: External possession and possessor raising. In: Everaert, M., van Riemsdijk, H. (eds.) The Wiley Companion to Syntax, 2nd edn. Wiley, Boca Raton (2017)

Dowty, D.: Word Meaning and Montague Grammar. Springer, Dordrecht (1979). https://doi.org/10.1007/978-94-009-9473-7

Embick, D.: Unaccusative syntax and verbal alternations. In: Alexiadou, A., Anagnostopoulou, E., Everaert, M. (eds.) The Unaccusativity Puzzle: Explorations of the Syntax-Lexicon Interface, pp. 137–158. Oxford University Press, Oxford (2004)

Guéron, J.: On the syntax and semantics of PP extraposition. LI **11**(4), 637–678 (1980)

Harley, H.: Possession and the double object construction. In: Pica, P., Rooryck, J. (eds.) Linguistic Variation Yearbook, vol. 2, pp. 31–70. Benjamins, Amsterdam (2003)

Hay, J., Kennedy, C., Levin, B.: Scalar structure underlies telicity in degree achievements. In: Proceedings of SALT IX, pp. 127–144 (1999)

Hole, D.: Extra argumentality - affectees landmarks and voice. Linguistics **44**, 383–424 (2006)

Hole, D.: Dativ, Bindung und Diathese. Studia grammatica, vol. 78. De Gruyter (2014)

Irwin, P.: Unaccusativity at the interfaces. Doctoral Dissertation. NYU (2012)

Irwin, P.: Existential unaccusativity and new discourse referents. Glossa **3**(1), 24 (2018)

Kayne, R.: Towards a modular theory of auxiliary selection. Stud. Linguist. **47**, 3–31 (1993)

Kennedy, C., Levin, B.: Measure of change: the adjectival core of verbs of variable telicity. In: McNally, L., Kennedy, C. (eds.) Adjectives and Adverbs in Semantics and Discourse, pp. 156–182. OUP, Oxford (2008)

Kennedy, C., McNally, L.: Scale structure, degree modification, and the semantics of gradable predicates. Language **81**, 345–381 (2005)

Langacker, R.: Foundations of Cognitive Grammar, Band I. Stanford (1987)

Langacker, R.: Subjectification, grammaticalization, and conceptual archetypes. In: Athanasiadou, A., Canakis, C., Cornillie, B. (eds.) Subjectification: Various Paths to Subjectivity, pp. 17–40. Mouton de Gruyter, Berlin (2006)

Le Bruyn, B., de Swart, H., Zwarts, J.: Have', 'with' and 'without. Semant. Linguist. Theory **23**, 535–548 (2013)

Levin, B. : English verb classes and alternations. A preliminary investigation (1993)

Löbner, S.: Definites. J. Semant. **4**, 279–326 (1985)

Löbner, S. : Polarity in natural language: Predication, quantification and negation in particular and characterizing sentences. L&P **23**, 213–308 (2000)

McCawley, J.: Remarks on what can cause what. In: Shibatani, M. (ed.) The Grammar of Causative Constructions, pp. 117–129. Academic Press, New York (1976)

Partee, B., Borschev, V.: The semantics of Russian genitive of negation: The nature and role of perspectival structure. In: Young, R.B. (ed.) Proceedings from SALT XIV, pp. 212–234. CLC Publications (2004)

Partee, B., Borschev, V., Paducheva, E.V., Testelets, Y., Yanovich, I.: Russian genitive of negation alternations: the role of verb semantics. Scando-Slavica **52**(1), 135–159 (2011)

Pesetsky, D.: Paths and Categories. Doctoral Dissertation. MIT (1982)

Pylkkänen, L. : Introducing Arguments. Doctoral Dissertation. MIT (2002)

Rappaport Hovav, M.: Lexicalized meaning and the internal temporal structure of events. In: Rothstein, S. (ed.) Theoretical and Crosslinguistic Approaches to the Semantics of Aspect, pp. 13–41. John Benjamins, Amsterdam (2008)

Rappaport Hovav, M., Levin, B.: Change of state verbs: Implications for theories of argument projection. In: Erteschik-Shir, N., Rapoport, T. (eds.) The Syntax of Aspect, pp. 274–286. Oxford University Press, Oxford (2005)

Ritter, E., Rosen, S.T.: Possessors as external arguments: Evidence from Blackfoot. In: Papers of the Forty-Second Algonquian Conference: Actes du Congrés des Algonquinistes. SUNY Press (2014)

Rohdenburg, G.: Sekundäre Subjektivierungen im Englischen und Deutschen: Vergleichende Untersuchungen zur Verb- und Adjektivsyntax. PAKS-Arbeitsbericht 8 (1974)

Schäfer, F.: The causative alternation. Lang. Ling. Compass **3**, 641–681 (2009)

Schäfer, F., Vivanco, M.: Reflexively marked anticausatives are not semantically reflexive. The causative alternation. In: Aboh, E.O., Schaeffer, J., Sleeman, P. (eds.) Romance Language and Linguistic Theory 2013, pp. 203–220, Benjamins (2015)

Slobin, I. Cross-linguistic evidence for the language-making capacity. In: Slobin, D. (ed.) The Cross-Linguistic Study of Language Acquisition, vol. 2: Theoretical issues, pp. 1157–1256. Lawrence Erlbaum Associates, Hillsdale (1985)

von Stechow, A.: Lexical decomposition in syntax. In: Egli, U., Pause, P.E., Schwarze, C., von Stechow, A., Wienold, G. (eds.) Lexical Knowledge in the Organization of Language, pp. 81–117. Benjamins, Amsterdam (1995)

von Stechow, A.: The different readings of wieder: a structural account. J. Semant. **13**, 87–138 (1996)

Vikner, C., Jensen, P.: A semantic analysis of the English genitive: interaction of lexical and formal semantics. Stud. Linguist. **56**, 191–226 (2002)

Szabolcsi, A.: The noun phrase: the syntactic structure of Hungarian. In: Kiefer, F., Kiss, K. (eds.) Syntax and Semantics, vol. 27, pp. 179–274. Academic Press, New York (1994)

Williams, E.: PRO and the subject of NP. NLLT **3**, 297–315 (1985)

An Axiomatization of the d-logic of Planar Polygons

David Gabelaia[1,2(✉)], Kristina Gogoladze[1], Mamuka Jibladze[1,2],
Evgeny Kuznetsov[2], and Levan Uridia[1,2]

[1] Ivane Javakhishvili Tbilisi State University, Tbilisi, Georgia
[2] TSU Razmadze Mathematical Institute, Tbilisi, Georgia
gabelaia@gmail.com

Abstract. We introduce the modal logic of planar polygonal subsets of the plane, with the modality interpreted as the Cantor-Bendixson derivative operator. We prove the finite model property of this logic and provide a finite axiomatization for it.

1 Introduction

There is a separate direction in modal logic, dealing with the specific phenomena related to logical reasoning about various objects of geometric nature with the aid of modal operators. In the literature there are quite a few alternative approaches to the way one interprets modalities in the context of space, shape, dimension, contiguity, etc. When applying logical calculi to reason about planar or spatial regions one often chooses some particular properties one wants to express, and restricts the kind of regions considered to those for which it makes sense to ask whether they have these properties. To point to an example, in a certain context one may find useful to investigate mereological relationships between *regular* subsets of a space—roughly, those which feature "filled up" areas without cracks, hairs or punctures. There also are many other, entirely different approaches. Let us limit ourselves to naming some sources—[1,2,9,10]; Let us also mention a recent paper [4], whose approach is most similar in spirit to ours.

In a recent paper [8] we introduced one more version of such an approach: instead of restricting topologically invariant properties of regions, we have severely restricted their shapes, namely we considered *polygonal regions*—subsets of the plane obtainable as boolean combinations of (either open or closed) halfplanes—equivalently, these are subsets that may be determined by (either strict or non-strict) linear inequalities. In that paper we interpreted modalities as topological closure/interior operators acting on such polygonal regions.

In this paper, our target is another interpretation of modal operators frequently studied in context of topological semantics. Namely, here we interpret ◊

D. Gabelaia, M. Jibladze, E. Kuznetsov and L. Uridia—Supported by Shota Rustaveli National Science Foundation grant #DI-2016-25.

A. Silva et al. (Eds.): TbiLLC 2018, LNCS 11456, pp. 147–165, 2019.
https://doi.org/10.1007/978-3-662-59565-7_8

as the *Cantor-Bendixson derivative* operator. For a polygonal region this roughly means to take its closure but also throw out any isolated points that the region might have.

We thus obtain a certain modal logic $\mathbf{PL}_2^d$: all formulæ that hold true under this interpretation about arbitrary polygonal subsets in the Euclidean plane. Algebraically, we study the variety of modal algebras generated by the Boolean algebra P_2 of polygonal subsets of the plane equipped with the derivative operator $\Diamond$.

We are going to prove that the logic $\mathbf{PL}_2^d$ has the finite model property, and provide five axioms that axiomatize it.

Our main approach is to employ a link to Kripke semantics with the aid of certain maps, called (partial, polygonal) *d-morphisms*, from the plane to various finite Kripke frames. Essentially these are exactly the maps that preserve validity of modal formulas. Algebraically, a *d*-morphism f is a map with the property that the induced Boolean algebra homomorphism f^{-1} from the powerset of the frame to the powerset of the plane (a) lands in the subalgebra consisting of polygonal subsets and (b) is a modal homomorphism with respect to the standard ("R^{-1}") interpretation of the modality $\Diamond$ on the Kripke side, and its above derivative interpretation on the polygonal side.

Using such morphisms helps us in applying a mixture of geometric and relational intuitions to find various finite modal algebras among subquotient algebras of P_2, or, on the contrary, prove that some other finite modal algebras cannot occur as such subquotients. Specifically, we introduce a sequence of finite Kripke frames, called *ir-crown frames*, well suited to be "test objects" for P_2—namely, any point in any polygonal configuration on the plane admits a (partial) polygonal *d*-morphism, in the above sense, onto some ir-crown frame. This allows us to prove that $\mathbf{PL}_2^d$ is the logic of all ir-crown frames, which gives as a result the finite model property for it.

On the other hand, we gather a sufficient but finite supply of certain other "forbidden" finite Kripke frames—those not admitting any partial polygonal *d*-morphisms onto them from the plane. Then using techniques similar to that of Jankov-De Jongh formulas we manage to express the fact of "forbiddenness" of these frames in the modal language, thus providing an axiom system for $\mathbf{PL}_2^d$.

2   Preliminaries

2.1   Syntax and Semantics

Our aim is to set up the reasoning paradigm where we use the basic modal language as our formalism and interpret its formulas as polyhedra in the Euclidean space of fixed dimension, while interpreting the modal diamond as the topological derivative operator and the boolean connectives as their set-theoretic counterparts. In this section we provide the necessary definitions of the relevant notions.

Syntax. We consider the basic modal language $\mathcal{ML}$. The alphabet of $\mathcal{ML}$ consists of a countable set PROP of letters for propositional variables and the symbols $\bot$, $\vee$, $\neg$, $\Diamond$. Formulas of $\mathcal{ML}$ are given by the recursive definition:

$$\varphi ::= \bot \mid p \mid \neg\varphi \mid \varphi \vee \psi \mid \Diamond\varphi.$$

The other connectives, such as $\wedge$, $\rightarrow$, $\leftrightarrow$, $\top$ and $\Box$, will be used as standard shorthand notations. A *normal modal logic* Λ in the modal language $\mathcal{ML}$ is a set of formulas of $\mathcal{ML}$ that contains all propositional tautologies, the formula $\Box(p \rightarrow q) \rightarrow (\Box p \rightarrow \Box q)$, and is closed under the inference rules of *modus ponens* (i.e. if $\varphi \in \Lambda$ and $\varphi \rightarrow \psi \in \Lambda$, then $\psi \in \Lambda$), *uniform substitution* (i.e if φ is in Λ, then so are all of its substitution instances) and *necessitation* (if $\varphi \in \Lambda$, then $\Box\varphi \in \Lambda$). All the logics we consider will be normal modal logics.

Kripke Semantics. We recall the basic notions from the Kripke semantics of modal logic.

A *Kripke frame* $\mathfrak{F}$ consists of a nonempty set W together with a binary relation $R \subseteq W \times W$. Such a pair is denoted by $\mathfrak{F} = (W, R)$ with the set W called the *underlying set* of the frame, and the relation R called the *accessibility relation* on W. To indicate that $(x, y) \in R$ holds we often write xRy (and say that x sees y by R); in such a case the element y is called a *successor* of the element x, and x a *predecessor* of y.

A relation is said to be *transitive* if for any three points $x, y, z \in W$, whenever xRy and yRz, then xRz holds as well. All frames we will consider are transitive.

We say that a point $r \in W$ is a *root* of a transitive Kripke frame $\mathfrak{F} = (W, R)$ if any point y of W distinct from r is a successor of r. A Kripke frame is said to be *rooted* if it has a root. A point in a Kripke frame is said to be *irreflexive* if $(x, x) \notin R$.

If R is a relation on W, and $A \subseteq W$, then the set $\{y \in W \mid \exists\, x \in A(xRy)\}$ of all the successors of elements of A is denoted by $R(A)$; the set $\{y \in W \mid \exists\, x \in A(yRx)\}$ of all the predecessors of elements of A is denoted by $R^{-1}(A)$.

A subset $U \subseteq W$ is called *upwards closed* (or simply an *up-set*) if it contains all successors of all of its elements, i.e. if $R(U) \subseteq U$ holds. Dually, a *down-set* or a *downwards closed* set D is defined as a set containing all predecessors of its elements, i.e. satisfying $R^{-1}(D) \subseteq D$. It is easy to see that the complement of an up-set is a down-set and vice versa.

A subset $A \subseteq W$ is called a *cluster* iff for any distinct $w, v \in A$ both wRv and vRw hold. In words, any two distinct points of a cluster are successors of each other.

Note that with any accessibility relation R on a Kripke frame $\mathfrak{F}$ we can associate a partially ordered set $S_{\mathfrak{F}}$ with elements the equivalence classes with respect to the equivalence relation $\sim_R$ defined by

$$x \sim_R y \iff xR^*y \text{ and } yR^*x,$$

where R^* is the transitive-reflexive closure of R. On these equivalence classes we define the partial order $\leq_R$ via

$$[x] \leq_R [y] \iff xR^*y.$$

A partial order S has *finite height* h if the maximum of the cardinalities of chains in S is equal to h. Then the *height of a Kripke frame* $\mathfrak{F}$ is the height of the partial order $S_{\mathfrak{F}}$ corresponding to $\mathfrak{F}$ as above.

Kripke frames provide semantics for modal logic in the following way. A *valuation* of propositional letters on a Kripke frame $\mathfrak{F} = (W, R)$ is a map $\nu : \text{PROP} \to \mathcal{P}(W)$ assigning a subset of W to each propositional letter. Such valuation is then extended to the valuation of all well-formed formulas of the language $\mathcal{ML}$,

$$
\begin{array}{llll}
x \not\models \bot & & \forall x \in W; \\
x \models p & \text{iff} & x \in \nu(p); \\
x \models \neg\varphi & \text{iff} & x \not\models \varphi; \\
x \models \varphi \vee \psi & \text{iff} & x \models \varphi \text{ or } x \models \psi; \\
x \models \Diamond\varphi & \text{iff} & x \in R^{-1}(\nu(\varphi)).
\end{array}
$$

The pair $\mathcal{M} = (\mathfrak{F}, \nu)$ is called a *Kripke model*, where $\mathfrak{F} = (W, R)$ is a Kripke frame and ν is a valuation as above. We write $\mathcal{M}, x \models \phi$ if a formula φ holds at the point x of a model $\mathcal{M}$.

For a subset $A \subseteq$ we write $\mathcal{M}, A \models \phi$ if $\mathcal{M}, x \models \phi$ holds for all $x \in A$. Further, $\mathcal{M} \models \phi$ (ϕ is valid in $\mathcal{M}$) means that $\mathcal{M}, x \models \phi$ for all $x \in W$. We write $\mathfrak{F} \models \phi$ (ϕ is valid on $\mathfrak{F}$) whenever $(\mathfrak{F}, \nu) \models \phi$ for an arbitrary valuation ν on $\mathfrak{F}$. If $\mathbf{K}$ is a class of Kripke frames we write $\mathbf{K} \models \phi$ when $\mathfrak{F} \models \phi$ for each $\mathfrak{F} \in \mathbf{K}$. By $\mathsf{Log}(\mathbf{K})$ we denote the set of all modal formulas valid in all members $\mathfrak{F} \in \mathbf{K}$ of $\mathbf{K}$. It is a basic fact of Kripke semantics for modal logic that $\mathsf{Log}(\mathbf{K})$ is always a normal modal logic.

Certain operations on frames preserving the modal validity will be useful in our considerations. Notably those of taking a generated subframe of a frame, taking a p-morphic image of a frame and the combination of these two called *an up-reduction* of one frame to another. We proceed to define these.

Definition 1. *Let* $\mathfrak{F}_1 = (W_1, R_1)$ *and* $\mathfrak{F}_2 = (W_2, R_2)$ *be Kripke frames. We say that* $\mathfrak{F}_1$ *is a* subframe *of* $\mathfrak{F}_2$ *if* $W_1 \subseteq W_2$ *and* $R_1 = R_2 \cap (W_1 \times W_1)$. *If in addition* W_1 *is an up-set in* W_2, *then we say that* W_1 *is a* generated subframe *of* W_2.

Proposition 1. *Let* $\mathfrak{F}_1$ *be a generated subframe of* $\mathfrak{F}_2$. *Then* $\mathfrak{F}_2 \models \phi$ *implies* $\mathfrak{F}_1 \models \phi$ *for any modal formula* ϕ.

The proof can be seen in e.g. [5].

Let $\mathfrak{F}_1 = (W_1, R_1)$ and $\mathfrak{F}_2 = (W_2, R_2)$ be Kripke frames. A map $f : \mathfrak{F}_1 \to \mathfrak{F}_2$ is said to be *monotone* if whenever $(x, y) \in R_1$, then $(f(x), f(y)) \in R_2$. A map $f : \mathfrak{F}_1 \to \mathfrak{F}_2$ between Kripke frames is said to be a *p-morphism* if it is monotone, and whenever $(f(x), z) \in R_2$, there exists $y \in W_1$ such that $(x, y) \in R_1$ and $f(y) = z$.

$$
\begin{array}{ccc}
y & \dashrightarrow & z \\
{\scriptstyle R_1}\uparrow & & \uparrow{\scriptstyle R_2} \\
x & \longmapsto & f(x)
\end{array}
$$

The onto p-morphisms also preserve validity of formulas, i.e. the following holds (see [5]):

Proposition 2. *Let $\mathfrak{F}_1$ and $\mathfrak{F}_2$ be Kripke frames and $f : \mathfrak{F}_1 \to \mathfrak{F}_2$ be an onto p-morphism. Then $\mathfrak{F}_1 \models \phi$ implies $\mathfrak{F}_2 \models \phi$ for any modal formula ϕ.*

Hence, $\mathsf{Log}(\mathfrak{F}_1) \subseteq \mathsf{Log}(\mathfrak{F}_2)$ whenever $\mathfrak{F}_2$ is a p-morphic image of $\mathfrak{F}_1$.

Combining the above two constructions, we say that $\mathfrak{F}_2$ is an *up-reduction*[1] of $\mathfrak{F}_1$, if there exists a generated subframe (i.e. an up-set) $\mathfrak{G}_1$ of $\mathfrak{F}_1$, such that $\mathfrak{F}_2$ is a p-morphic image of $\mathfrak{G}_1$. We immediately infer:

Proposition 3. *Let $\mathfrak{F}_1$ and $\mathfrak{F}_2$ be Kripke frames and let $\mathfrak{F}_2$ be an up-reduction of $\mathfrak{F}_1$. Then $\mathfrak{F}_1 \models \phi$ implies $\mathfrak{F}_2 \models \phi$ for any modal formula ϕ.*

Topological Semantics. There are two standard topological semantics for modal logic, depending on whether the modal diamond is interpreted as the closure operator or as the derivative operator. In this paper we focus on the latter. For the corresponding definitions in the closure semantics see [8].

A *topological space* is a pair $\mathfrak{X} = (X, \tau)$, where $X \neq \emptyset$ is a set and $\tau \subseteq \mathcal{P}(X)$ is a family of subsets of X, such that $\emptyset \in \tau$, $X \in \tau$, and τ is closed under finite intersections and arbitrary unions. X is called the underlying set of $\mathfrak{X}$, and τ a topology on X. When there is no danger of ambiguity, we often write just X instead of (X, τ). The elements of τ are called *open* subsets of $\mathfrak{X}$, or simply opens. Set-theoretic complements of opens are called *closed* subsets. Clearly finite unions and arbitrary intersections of closed sets are closed. For a set $A \subseteq X$ the *closure of A* is the intersection of all closed sets containing A,

$$\mathbb{C}(A) = \bigcap \{F \mid A \subseteq F \subseteq X \text{ and } F \text{ is closed}\}.$$

A point x of a topological space $\mathfrak{X}$ is called a *limit point* of a subset $A \subseteq X$ if $x \in \mathbb{C}(A - \{x\})$. The set of all limit points of A is called the *derived set* of A and is denoted by $d(A)$. It is easy to see that $x \in d(A)$ if and only if for any open neighborhood U of x the intersection $U \cap A$ contains at least one point distinct from x, i.e. $U \cap A - \{x\} \neq \emptyset$ (see e.g. [7]). The derived set operator has the following properties:

(i) $\mathbb{C}(A) = A \cup d(A)$;
(ii) If $A \subseteq B$, then $d(A) \subseteq d(B)$;
(iii) $d(A \cup B) = d(A) \cup d(B)$;
(iv) $\bigcup_{i \in I} d(A_i) \subseteq d(\bigcup_{i \in I} A_i)$.

Elements of the set $A - d(A)$ are called *isolated points* of A. A set $A \subseteq X$ is called *dense in itself* (dii for short) if $A \subseteq d(A)$. Thus a subset is dii iff it has no isolated points.

---

[1] We introduce this terminology extending the terminology of [6] where *reduction* means taking a p-morphic image and *subreduction* means taking a p-morphic image of a subframe of the frame. Thus, *up-reductions* are special cases of subreduction, where the subframe under question is an up-set.

At the outset, we slightly generalize the derivative semantics by the following definition.

Definition 2 (General derivative topological space). *A general derivative topological space is a tuple* $\mathfrak{X} = (X, \tau, d, \mathcal{D})$, *where* (X, τ) *is a topological space, d is the derived set operator on* (X, τ), *and* $\mathcal{D} \subseteq \mathcal{P}(X)$, *such that* $X \in \mathcal{D}$ *and $\mathcal{D}$ is closed under the set-theoretic operations (e.g. taking unions and taking complements), as well as under the derivative operation. Subsets from $\mathcal{D}$ will be called the* admissible sets *for* $\mathfrak{X}$.

For further use we mention that a collection $\mathcal{D}$ of subsets in a topological space which is closed under the boolean set-theoretic operations and the derivative operation is said to be a *derivative algebra* of subsets. If the derivative algebra is not specified in a general derivative topological space, we assume that it is equal to the collection of all subsets.

Suppose $\mathfrak{X}$ is a general derivative topological space. A valuation on $\mathfrak{X}$ is a map $\nu : \mathrm{PROP} \to \mathcal{D}(X)$. Each such valuation extends uniquely from propositional variables to all well-formed formulas of the language $\mathcal{ML}$ in the following way:

$$
\begin{aligned}
x &\not\models \bot & & \forall x \in X; \\
x &\models p & \text{iff} & \quad x \in \nu(p); \\
x &\models \neg\varphi & \text{iff} & \quad x \not\models \varphi; \\
x &\models \varphi \vee \psi & \text{iff} & \quad x \models \varphi \text{ or } x \models \psi; \\
x &\models \Diamond\varphi & \text{iff} & \quad x \in d(\nu(\varphi)).
\end{aligned}
$$

The notions of truth in a subset, validity in a model and validity on a space are defined like in the case of Kripke semantics. We just point out that according to the above definition, for a modal formula of the form $\varphi = \Diamond\psi$ we have that $\nu(\varphi) = \nu(\Diamond\psi) = d\nu(\psi)$, i.e. the truth set of $\Diamond\psi$ is the derivative of the truth-set of ψ.

If $\mathbf{K}$ is a class of general derivative spaces, by $\mathrm{Log}(\mathbf{K})$ we denote the set of all modal formulas valid in all members $\mathfrak{X} \in \mathbf{K}$. In case $\mathbf{K}$ consists of a single member $\mathfrak{X}$ we write $\mathrm{Log}(\mathfrak{X})$ to denote the modal logic of $\mathfrak{X}$. It is well-known that the modal logic of T_1 topological spaces (in which each singleton subspace is closed) is the modal logic $K4$, the logic of all transitive Kripke frames. We will only be dealing with T_1 spaces in this paper.

To connect the two semantics described above we need a definition of maps which preserve the validity of modal formulas when the domain of a map is a topological space and the target a Kripke frame (see [3]).

Definition 3. *A map* $f : \mathfrak{X} \to \mathfrak{F}$ *where* $\mathfrak{X} = (X, \tau)$ *is a topological space and $\mathfrak{F} = (W, R)$ is a transitive Kripke frame, is called a d-morphism if the following properties are satisfied:*

(i) *For each open* $U \in \tau$ *it holds that* $R(f(U)) \subseteq f(U)$, *i.e. images of opens are up-sets;*

(ii) For each $V \subseteq W$ such that $R(V) \subseteq V$, it holds that $f^{-1}(V) \in \tau$, i.e. the pre-images of up-sets are open;

(iii) For each irreflexive point $w \in W$ it holds that $f^{-1}(w)$ is a discrete space w.r.t. subspace topology;

(iv) For each reflexive point $w \in W$ it holds that $f^{-1}(w) \subseteq d(f^{-1}(w))$, i.e. $f^{-1}(w)$ is dense in itself (dii).

To synchronize the terminology, notice that the up-sets of a transitive Kripke frame always form a topology. Thus we may occasionally refer to up-sets as open sets and down-sets as closed sets. In this terminology, the first two conditions of a d-morphism amount to the map being *open* and *continuous* (such maps are often called *interior maps*).

It is well-known that d-morphisms preserve modal validity [3]. It is also known that taking open subspaces preserves modal validity in derivative semantics. These facts motivate the following definition, which is similar to that of up-reduction for frames, and serves as a bridge between the Kripke semantics and the derivative semantics.

Let $f : \mathfrak{X} \rightarrow \mathfrak{F}$ denote a *partial* map from the topological space $\mathfrak{X}$ to the Kripke frame $\mathfrak{F}$. In case the domain of f is an open subset of $\mathfrak{X}$, and f satisfies the conditions of being a d-morphism, we say that f is a *partial d-morphism*.

Proposition 4. *Let $\mathfrak{X}$ be a topological space, $\mathfrak{F}$ be a Kripke frame and $f : \mathfrak{X} \rightarrow \mathfrak{F}$ be a partial onto d-morphism. Then for an arbitrary modal formula ϕ we have $\mathfrak{F} \models \phi$ whenever $\mathfrak{X} \models \phi$.*

2.2   The Polygonal Plane

We are interested in specific general models defined over Euclidean spaces $\mathbb{R}^n$. Let P_n be the boolean algebra of the n-dimensional *polyhedra* in $\mathbb{R}^n$. *Basic*, or *elementary* polyhedra can be described as sets that are intersections of finitely many halfspaces in $\mathbb{R}^n$, and are known to be those polyhedra which are convex subsets of $\mathbb{R}^n$. General polyhedra are then the finite unions of the latter.

To be more precise, a *polyhedron* is any subset of $\mathbb{R}^n$ of the form $P = \{\bar{x} \mid \bigvee \bigwedge (\ell_i(\bar{x}) \bowtie a_i)\}$ where ℓ_i are linear forms on $\mathbb{R}^n$ and a_i are real numbers, $\bowtie$ denotes any of the inequality symbols $\geq, >, \leq, <$, while $\bigvee$ and $\bigwedge$ denote finite disjunction and finite conjunction. The sets of the form $P = \{\bar{x} \mid \bigwedge (\ell_i(\bar{x}) \bowtie a_i)\}$ we call *basic* or *elementary* polyhedra.

Then it is clear from definitions that the set P_n of all the n-dimensional polyhedra forms a boolean subalgebra of the powerset $\mathscr{P}(\mathbb{R}^n)$ (note that the negation of an inequality is again an inequality). Moreover, the following holds:

Proposition 5. *The boolean algebra P_n is a modal subalgebra of the powerset derivative algebra $\mathscr{P}(\mathbb{R}^n)$ equipped with the derivative operator for the Euclidean topology on $\mathbb{R}^n$.*

Proof. To show that P_n is closed under the derivative operator, first note that the derivative operator distributes over finite unions. Since every polyhedron is a finite union of basic polyhedra, it suffices to point out that given a nonempty basic polyhedron P defined by a finite conjunction of inequalities, its derivative is either the closed basic polyhedron obtained by turning all strict inequalities into non-strict ones, or is the empty set in case P happens to consist of a single point.

Definition 4. Let $\mathfrak{P}_n = (\mathbb{R}^n, P_n, d)$ be the general derivative space defined by means of the derivative algebra of polyhedra in the Euclidean space $\mathbb{R}^n$. We call such a space the n-dimensional Euclidean polyhedral derivative space. The modal logic $\mathbf{PL}_n^d$ of the n-dimensional Euclidean polyhedral derivative space is defined to be the set of all modal formulas which are valid on $\mathfrak{P}_\mathbf{n}$.

$$\mathbf{PL}_n^d ::= \mathsf{Log}^d(\mathfrak{P}_n)$$

In this paper we concentrate on the 2-dimensional polyhedral modal logic $\mathbf{PL}_2^d$. We call this logic the polygonal modal logic for simplicity and the corresponding general space $\mathfrak{P}_2 = (\mathbb{R}^2, P_2, d)$ is called the *derivative polygonal plane*.

The admissible sets in $\mathfrak{P}_2$ are finite unions of *generalized planar polygons*, where under a generalized planar polygon we understand a (possibly unbounded) region in the plane which is an intersection of finitely many (closed or open) half-planes. It is clear that any point, line, ray or segment also falls under this definition, as do triangles, pentagons and n-gons in general.

We consider a class $\mathbf{IC}$ of finite frames which are called *crown frames with irreflexive root*, ir-crown frames for short (not to be confused with the crown graphs); this class will play crucial role of finite models for $\mathbf{PL}_2^d$ as we will see later.

Definition 5. *A crown frame with irreflexive root, or ir-crown frame is a frame* $\mathfrak{S}_n = (S_n, Q_n)$ *such that* $S_n = \{r, s_1, \cdots, s_{2n}\}$ *and* Q_n *is defined as follows:*

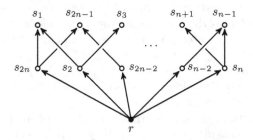

Fig. 1. An ir-crown frame. Here and in further pictures we depict irreflexive points by black circles and the reflexive ones by circles with white interior.

$(r,r) \notin Q_n;$
$(r,s_i) \in Q_n$ $\qquad$ *for all $s_i \in S_n$;*
$(s_i,s_i) \in Q_n$ $\qquad$ *for all $s_i \in S_n$;*
$(s_i,s_j) \in Q_n$ $\qquad$ *when $i < 2n$ is even and $j = i-1, i+1$;*
$(s_{2n},s_1) \in Q_n;$
$(s_{2n},s_{2n-1}) \in Q_n.$

We are going to show that the modal logic of ir-crown frames and the modal logic of derivative polygonal plane coincide. To show that the formulas valid on the derivative polygonal plane are valid on ir-crown frames as well, first we prove the following theorem:

Theorem 1. *For an arbitrary ir-crown frame $\mathfrak{F} = (W, R)$ there is an onto d-morphism $f : \mathfrak{P}_2 \to \mathfrak{F}$ from the derivative polygonal plane to $\mathfrak{F}$ such that for any point $w \in W$ the inverse image $f^{-1}(w)$ belongs to P_2, i.e. is an element of the derivative algebra P_2 of planar polygons.*

Proof. Let $\mathfrak{F} = (W,R)$ be an ir-crown frame. According to the construction of ir-crown frames, $\mathfrak{F}$ is a finite frame with the equal numbers of points on the second (middle) and the third (maximum) layer; let us denote by n the number of maximal points of $\mathfrak{F}$. The construction for obtaining an onto mapping from the polygonal plane to a crown frame introduced in Proposition 3.2 of [8] works here too. Namely, consider any point x and arbitrary distinct rays $l_1, \ldots, l_n$ emanating from x and enumerated in the counter-clockwise direction (Fig. 2).

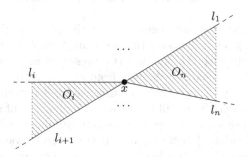

Fig. 2. Polygonal partition of the Euclidean plane corresponding to a d-morphism onto an ir-crown frame.

For $i = 1, \ldots, n-1$ let O_i be the open regions between l_i and l_{i+1}, and let O_n be the open region between l_n and l_1.
Define the map $f : \mathfrak{P}_2 \to \mathfrak{F}$ by putting $f(x) = r$, $f(l_i) = s_{2i}$ and $f(O_i) = s_{2i-1}$ where r is the irreflexive root of the ir-crown frame while s_{2i} and s_{2i-1} are as in Fig. 1. We only need to check that this f satisfies conditions from Definition 3. Conditions (i) and (ii) are easily checked. By the definition of ir-crown frames, the only irreflexive point is the root. Hence we see that condition (iii) holds as well since the preimage of r is the single point x which obviously is a discrete subspace of $\mathbb{R}^2$. Now since by construction of f for any $i \leq n$ the pre-image $f^{-1}(s_i)$ has no isolated points, the condition (iv) holds as well.

Thus by Proposition 4, if $\mathfrak{F}$ is an ir-crown frame with root r and ϕ is a modal formula, it holds that $\mathfrak{F}, r \models \phi$ whenever $\mathfrak{P}_2, x \models \phi$ for some point $x \in \mathbb{R}^2$. Hence $\mathbf{PL}_2^d \subseteq \mathsf{Log}(\mathbf{IC})$.

Let us now prove the converse inclusion.

Theorem 2. *If a formula φ is satisfiable on the derivative polygonal plane $\mathfrak{P}_2$ then it is satisfiable on some ir-crown frame $\mathfrak{F}$.*

Proof. Let ν be a valuation and x be a point such that $\mathfrak{P}_2, \nu, x \models \phi$, where ϕ depends on propositional variables $p_1, \ldots, p_k$. Our strategy is to find a small enough open neighborhood U around x such that the partial d-morphism could be built from U onto one of the ir-crown frames, in such a way that the pre-images of points from the ir-crown frame have constant valuation for propositional variables occurring in ϕ. We follow the construction introduced in Theorem 3.1 of [8].

Suppose ϕ depends on propositional variables $p_1, \ldots, p_k$. It is clear that the truth of ϕ will not be affected if we assume that all the other propositional variables are mapped to the empty set. Let $A_i = \nu(p_i)$ for $i \in \{1, \ldots, k\}$. Then each A_i is a finite union of simple polygons (two-dimensional cases of basic, or elementary polyhedra, as described above towards the beginning of Sect. 2.2). Let S be the collection of all the simple polygons occurring in the A_is. Let E be the collection of all lines or line segments occurring as an edge of one of the simple polygons in S. It is obvious that E is finite. Furthermore, we observe the following:

Key observation: For any segment I on the plane, if the valuation of a propositional letter p_i changes along this segment, then I must intersect with a member of E, namely with the one that is represented as a border of A_i which I must cross in order to change valuation from one point to the other.

Now, for each line in E, calculate the distance from x to that line and to its endpoints (if it has such). This will produce a finite number of non-negative real numbers. Let α be the least *positive* number thus obtained and let $B = B(x, \frac{\alpha}{2})$ be the open ball of radius $\frac{\alpha}{2}$ centered at x. It is straightforward that only those lines from E that pass through x (or have it as an endpoint) will intersect with B. Let us label the intersection points of lines from E with the boundary of B

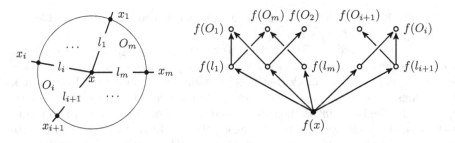

Fig. 3. The open n-gon and its partition corresponding to an ir-crown frame.

in the clockwise direction as $x_1, x_2, \ldots, x_m$, with $m \leq k$. Let l_i denote the open segments (x, x_i) and let O_i denote the open triangles $x_i x x_{i+1}$ inside B bounded by l_i and l_{i+1} for $i \in \{1, \ldots, m-1\}$. Let O_m be the remaining open triangle $x_1 x x_m$ confined between l_m and l_1. Then the open n-gon $x_1 \ldots x_m$ inside B breaks down into the sets $\{x\}$, l_i and O_i (Fig. 3).

The desired d-morphism is built in the same way as in Theorem 1. Then the valuation μ on F defined by putting $\mu(p) = f(\nu(p))$ for each p is such that

$$f(x) \in \mu(p) \text{ iff } x \in \nu(p);$$

and hence $\mathfrak{F}, \mu, f(x) \models \phi$.

Thus we have proved that the two logics $\mathbf{PL}_2^d$ and $\mathsf{Log}(\mathbf{IC})$ coincide.

Corollary 1. *The logic* $\mathbf{PL}_2^d$ *is determined by the class* $\mathbf{IC}$. *Hence this logic has the finite model property* (fmp).

3   Axiomatization

In this section we give a complete axiomatization of the logic $\mathsf{Log}(\mathbf{IC}) = \mathbf{PL}_2^d$.

Let Λ_2^d be the modal logic axiomatized by the following formulas:

$$\begin{aligned}
\theta_1 &= \Diamond\top \\
\theta_2 &= \Diamond\Diamond p \leftrightarrow \Diamond p \\
\theta_3 &= (\Diamond p \wedge \Diamond\neg p) \rightarrow \Diamond\big((p \wedge \Diamond\neg p) \vee (\neg p \wedge \Diamond p)\big) \\
\theta_4 &= \Box\big(p \rightarrow \Box(\neg p \rightarrow \Box\neg p)\big) \\
\theta_5 &= \big[r \wedge \gamma \wedge \Box(r \rightarrow \gamma)\big] \rightarrow \Diamond(\neg r \wedge \Diamond\Box p \wedge \Diamond\Box\neg p)
\end{aligned}$$

where γ is the formula $\Diamond\Box(p \wedge q) \wedge \Diamond\Box(p \wedge \neg q) \wedge \Diamond\Box(\neg p \wedge q)$.

In this section we aim to show that $\Lambda_2^d = \mathsf{Log}(\mathbf{IC}) = \mathbf{PL}_2^d$. To this end, we (a) associate a semantic condition to each of the axioms; (b) demonstrate that each of the five axioms is valid on any ir-crown frame; and (c) show that any rooted frame validating all of these axioms is an up-reduction of some ir-crown frame.

First we associate semantical conditions to each of these formulas. Some of them are well-known - e.g. validating θ_1 is equivalent to the frame being *serial*, i.e. each point having a successor, while validating θ_2 is equivalent to the frame being both *transitive* (successor of a successor is a successor) and *dense* (any successor is a successor of a successor).

In our description we will employ the notions of a subframe, generated subframe and *convex subframe*. We say that $\mathfrak{F} = (W, R)$ is a convex subframe of $\mathfrak{G} = (V, S)$, if $\mathfrak{F}$ is a subframe of $\mathfrak{G}$ and additionally, for arbitrary $w, u \in W$ and $v \in V$, if $wSvSu$ holds, then $v \in W$. In words, a convex subframe must contain whatever is "in between" its two points, akin to the geometric meaning of the term "convex". Clearly, generated subframes are always convex (Fig. 4).

The following picture lists some of the frames that will be useful in the next theorem. Here, as well as in subsequent pictures, hollow circles denote reflexive points, filled circles denote irreflexive points, while the symbol $\times$ is used to denote a point that is either reflexive or irreflexive, so in effect e.g. $\mathfrak{B}_3$ below is a generic name denoting any of the four distinct frames obtained from the corresponding picture by substituting reflexive or irreflexive points in place of ×:

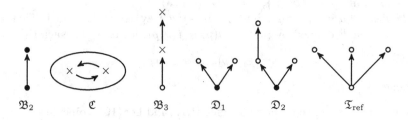

Fig. 4. Frames

Theorem 3. *Let $\mathfrak{F} = (W, R)$ be an arbitrary Kripke frame. Then:*

(i) $\mathfrak{F} \models \theta_1$ *iff $\mathfrak{F}$ is serial, i.e. $\forall w \in W \ \exists v \in W(wRv)$. Moreover, $\mathfrak{F}$ contains no generated subframe consisting of a single irreflexive point.*

(ii) $\mathfrak{F} \models \theta_2$ *iff $\mathfrak{F}$ is transitive and dense, i.e. $\forall w, u, v \in W(wRv \wedge vRu \rightarrow wRv)$ and $\forall w, v \in W(wRv \rightarrow \exists u \in W(wRu \wedge uRv))$. Moreover, $\mathfrak{F} \models \Diamond p \rightarrow \Diamond\Diamond p$ iff $\mathfrak{F}$ contains no convex subframe isomorphic to $\mathfrak{B}_2$.*

(iii) *If $\mathfrak{F} \models \theta_3$ then $\mathfrak{F}$ cannot be up-reduced to any of the frames $\mathfrak{D}_1, \mathfrak{D}_2$.*

(iv) *If $\mathfrak{F} \models \theta_4$, then $\mathfrak{F}$ is of height ≤ 3 and does not contain subframes isomorphic to either $\mathfrak{C}$ or $\mathfrak{B}_3$. If in addition $\mathfrak{F}$ is transitive, the converse holds as well.*

(v) *If $\mathfrak{F} \models \theta_5$ then $\mathfrak{F}$ has no generated subframes isomorphic to $\mathfrak{T}_{\mathrm{ref}}$.*

Proof. (i) This is well-known and easy to check. We note in addition that for finite transitive frames seriality means that all maximal points are reflexive.

(ii) That $\Diamond\Diamond p \rightarrow \Diamond p$ corresponds to transitivity and $\Diamond p \rightarrow \Diamond\Diamond p$ corresponds to density is well-known and easy to check. We only show the second part of the claim.

Suppose $\mathfrak{F}$ has a convex subframe consisting of two irreflexive points xRy as in $\mathfrak{B}_2$. Consider a valuation ν such that $\nu(p) = \{y\}$. It is obvious that $\mathfrak{F}, x \models \Diamond p$ and $\mathfrak{F}, x \not\models \Diamond\Diamond p$, hence $\mathfrak{F}, x \not\models \Diamond p \rightarrow \Diamond\Diamond p$.

To show the converse, suppose $\mathfrak{F} \not\models \Diamond p \rightarrow \Diamond\Diamond p$. Then there exists a point x of $\mathfrak{F}$ such that $\mathfrak{F}, x \models \Diamond p$ and $\mathfrak{F}, x \models \neg\Diamond\Diamond p$. Thus there exists a further point y with xRy and for any point z of $\mathfrak{F}$, either $(x, z) \notin R$ or $(z, y) \notin R$. It clearly follows that both x and y are irreflexive. Hence there is a copy of $\mathfrak{B}_2$ as a convex subframe of $\mathfrak{F}$.

(iii) It is immediate that θ_3 is refuted on the frames $\mathfrak{D}_1$ and $\mathfrak{D}_2$. Just take the valuation shown on the picture below (Fig. 5).

Since $\mathfrak{F}$ validates θ_3 by assumption, and validity is preserved by up-reductions and taking generated subframes, the claim follows.

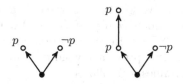

Fig. 5. Refutation of θ_3

(iv) Suppose $\mathfrak{F} \models \theta_4$. Assume $\mathfrak{F}$ contains $\mathfrak{C}$ as a subframe. The frame $\mathfrak{F}$ contains at least two points. Take any two points u and v from $\mathfrak{C}$. Take on it a valuation ν such that $\nu(p) = \{u\}$. Since u and v are interrelated, clearly $v \not\models \Box\neg p$, which means that $v \not\models \neg p \to \Box\neg p$. Hence $u \not\models \Box(\neg p \to \Box\neg p)$ and equally $u \not\models p \to \Box(\neg p \to \Box\neg p)$. The latter implies that $v \not\models \Box(p \to \Box(\neg p \to \Box\neg p))$ which contradicts the assumption. Now assume that $\mathfrak{F}$ contains $\mathfrak{B}_3$ as a subframe. The same reasoning as in the previous case goes through. Again let us show that $\Box(p \to \Box(\neg p \to \Box\neg p))$ is refuted on $\mathfrak{F}$. Clearly $\mathfrak{F}$ contains at least three distinct points u, v and w with uRv and vRw, where R is the accessibility relation of $\mathfrak{F}$. Take a valuation ν such that $\nu(p) = \{u, w\}$. Then $v \not\models \neg p \to \Box\neg p$. Hence $u \not\models \Box(\neg p \to \Box\neg p)$ and $u \not\models p \to \Box(\neg p \to \Box\neg p)$. Since uRu we conclude that $u \not\models \Box(p \to \Box(\neg p \to \Box\neg p))$ which is again in contradiction with the assumption. Now in case $\mathfrak{F}$ contains a frame of strict height more than 3 as a subframe, which means that there are at least four distinct points u, v, w and z with uRv, vRw and wRz, then we choose a valuation V as follows: $V(p) = \{v, z\}$, and the formula is refuted at u. We therefore obtain that if $\mathfrak{F} \models \theta_4$ then $\mathfrak{F}$ is of height ≤ 3 and does not contain subframes isomorphic to $\mathfrak{C}$ and $\mathfrak{B}_3$. To show the converse, assume $\mathfrak{F}$ is transitive and suppose $\mathfrak{F} \not\models \theta_4$, i.e. there is a point x with $x \models \neg\Box(p \to \Box(\neg p \to \Box\neg p))$. This is the same as $x \models \Diamond(p \wedge \Diamond(\neg p \wedge \Diamond p))$, i.e. there exist points y, z, u with $xRyRzRu$ and $y \models p$, $z \models \neg p$, $u \models p$. To wrap up, if $y = u$ or uRx, then $\mathfrak{F}$ contains a copy of $\mathfrak{C}$ as a subframe; if $y \neq u$ and $\neg(uRx)$, then either $\mathfrak{F}$ contains subframes isomorphic to $\mathfrak{B}_3$ in case $x = y$, or is of height greater than 3 in case $x \neq y$.

(v) It is easy to see that the frame $\mathfrak{T}_{\text{ref}}$ refutes the formula θ_5. Indeed, consider the valuation on $\mathfrak{T}_{\text{ref}}$ as follows (Fig. 6):

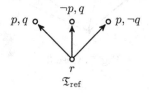

Fig. 6. Refutation of θ_5

Consider the root of $\mathfrak{T}_{\mathrm{ref}}$. We see that r is true only at the root, which models γ as well. Hence the root models $(r \wedge \gamma \wedge \Box(r \rightarrow \gamma))$. It is also clear that the root refutes $\Diamond(\neg r \wedge \Diamond\Box p \wedge \Diamond\Box\neg p)$. Hence θ_5 is refuted at the root.

Since $\mathfrak{F}$ validates θ_5 by assumption, and validity is preserved by up-reductions and taking generated subframes, (v) follows.

Next we show that ir-crown frames validate all of the axioms $\theta_1 - \theta_5$.

Theorem 4. *Let $\mathfrak{G}_n = (S_n, Q_n)$ be an arbitrary ir-crown frame. Then for each $i \le 5$ we have $\mathfrak{G}_n \models \theta_i$.*

Proof. We rely partially on the semantic characterizations afforded by Theorem 3. That ir-crown frames are serial (have reflexive maximum), transitive and dense is trivial to check. Thus $\mathfrak{G}_n \models \theta_1 \wedge \theta_2$. It is also clear that ir-crown frames contain no non-trivial clusters, have height ≤ 3 and contain no subframe isomorphic to $\mathfrak{B}_3$. Thus $\mathfrak{G}_n \models \theta_4$. We show in detail below that $\mathfrak{G}_n \models \theta_3$ and $\mathfrak{G}_n \models \theta_5$.

To show that $\mathfrak{G}_n \models \theta_3$, take an arbitrary ir-crown frame $\mathfrak{S}_n = (S_n, Q_n)$ where $S_n = \{r, s_1, \cdots, s_{2n}\}$ and an arbitrary valuation ν. Take an arbitrary point $u \in S_n$.

Let us distinguish three cases:

Case 1: $u = s_{2k-1}$ for some $1 \le k \le n$. This by definition means that u belongs to the maximal layer. Then $u \not\models \Diamond p \wedge \Diamond\neg p$, which implies that $u \models \theta_3$.

Case 2: $u = s_{2k}$ where $0 \le k \le n$. This by definition means that u belongs to the middle layer. Without loss of generality we can assume that $u \models p$. Assume that $u \models \Diamond p \wedge \Diamond\neg p$. Then either $s_{2k-1} \models \neg p$ or $s_{2k+1} \models \neg p$. Hence $u \models p \wedge \Diamond\neg p$. We thus conclude that $u \models \theta_3$ since s_{2k} is reflexive.

Case 3: $u = r$. Assume that $u \models \Diamond p \wedge \Diamond\neg p$. Then there exist w and v with $w \models p$ and $v \models \neg p$.

Let us consider the case when both w and v belong to the middle layer; other cases follow in a similar way. Assume $w = s_{2k}$ and $v = s_{2k+2l}$ (Fig. 7).

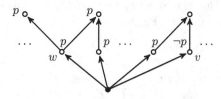

Fig. 7. Case 3.

In case $s_{2k+1} \not\models p$ we are done since $rQ_n w$ and $wQ_n s_{2k+1}$ which yields $r \models \Diamond(p \wedge \Diamond\neg p)$. In case $s_{2k+1} \models p$ we proceed by looking at an immediate (distinct from s_{2k}) predeccessor of s_{2k+1} which is s_{2k+2}. If $s_{2k+2} \not\models p$ we are done since $rQ_n s_{2k+2}$ and $s_{2k+2} Q_n s_{2k+1}$ which means that $r \models \Diamond(\neg p \wedge \Diamond p)$. If

$s_{2k+2} \models p$ we proceed by the same reasoning and either we arrive at an s_m satisfying $s_m \not\models p$ for some m with $2k < m \leq 2k + 2l - 1$, or $s_{2k+2l-1}$ also models p. In the last case since $rQ_n s_{2k+2l-1}$ and $s_{2k+2l-1}Q_n v$ we have that $r \models \Diamond(p \wedge \Diamond \neg p)$. We omit the details for the other cases.

To show that $\mathfrak{S}_n \models \theta_5$, suppose the antecedent of the formula is true at a point w in an ir-crown frame $\mathfrak{S}_n = (S_n, Q_n)$ for some valuation ν. Note that making γ true forces w to have at least three distinct successors. It follows that w is the irreflexive root of $\mathfrak{S}_n$, and is the only point making r true. Moreover, $w \models \gamma$ also implies that both p and $\neg p$ are true at some maximal points of $\mathfrak{S}_n$. Then there exists an $i < n$ such that the maximal points s_{2i-1} and s_{2i+1} differ on the value of p (otherwise all maximal points would agree on the value of p). Since $wQ_n s_{2i}$, $s_{2i}Q_n s_{2i-1}$ and $s_{2i}Q_n s_{2i+1}$, it follows that w makes the consequent of the formula true as well.

Theorem 5. *The logic Λ_2^d has the finite model property.*

Proof. Note that since frames from $B_{\geq 3}$ are not admitted by Λ_2^d, the logic is of finite depth. By Segerberg's Theorem (see e.g. Theorem 8.85 of [6]) any logic of finite depth is characterized by its finite frames. It follows that Λ_2^d has the finite model property.

Since each ir-crown frame validates all the axioms of Λ_2^d, we have the inclusion $\mathbf{IC} \subseteq \mathbf{Fr}(\Lambda_2^d)$. The other inclusion follows from the following theorem.

Let $\mathfrak{F}$ be a rooted transitive frame of height 3 with the root irreflexive, and let $\mathfrak{F}^*$ be the frame obtained from $\mathfrak{F}$ by making the root reflexive. Then the following holds:

Lemma 1. *Let $\mathfrak{S}_n$ be an ir-crown frame, let $\mathfrak{F}$ be a rooted frame with irreflexive root and with height equal to 3. Then $\mathfrak{S}_n$ up-reduces to $\mathfrak{F}$ if and only if $\mathfrak{S}_n^*$ up-reduces to $\mathfrak{F}^*$.*

Proof. Assume that $\mathfrak{S}_n$ up-reduces to $\mathfrak{F}$. Then $\mathfrak{F}$ is obtainable as a p-morphic image of some generated subframe of $\mathfrak{S}_n$. Let us fix U and f to be the mentioned generated subframe and p-morphism. If U is a strict subframe, i.e. $U \subset \mathfrak{S}_n$, then U is a generated subframe of $\mathfrak{S}_n^*$ as well and we take the same function f which is a p-morphism from U to $\mathfrak{F}^*$. In case $U = \mathfrak{S}_n$ it is clear that the irreflexive root is mapped to the irreflexive root since f cannot map reflexive points to irreflexive points due to the monotonicity condition. It follows that the restriction of f to $U \setminus \{r\}$ is a p-morphism, so that f is a p-morphism from $\mathfrak{S}_n^*$ to $\mathfrak{F}^*$.

Conversely, assume that $\mathfrak{S}_n^*$ up-reduces to $\mathfrak{F}^*$. Again let us fix U and f to be the generated subframe and the p-morphism doing the up-reduction. The case $U \subset \mathfrak{S}_n^*$ is exactly the same as for the previous direction. Now assume that $U = \mathfrak{S}_n^*$ and let us show that the preimage of the root $w \in \mathfrak{F}^*$ is exactly the root r of $\mathfrak{S}_n^*$. Clearly $r \in f^{-1}(w)$. Assume that some other point $x \in f^{-1}(w)$. Since the height of $\mathfrak{F}^*$ is three, there are at least two distinct points $v, v' \in \mathfrak{F}^*$ with $v \neq w \neq v'$ and $wRvRv'$. Since f is a p-morphism, there exist u and u' in U with $xQ_n uQ_n u'$ and $f(u) = v$, $f(u') = v'$. But this is a contradiction, since ir-crown frames do not contain a chain of four distinct points such as $rQ_n xQ_n uQ_n u'$.

Theorem 6. *If a finite rooted frame validates all five axioms of Λ_2^d, then it is an up-reduction of some ir-crown frame.*

Proof. Assume that $\mathfrak{F} = (W, R)$ is a finite rooted frame validating the logic Λ_2^d. By Theorem 3 (iv) we know that $\mathfrak{F}$ does not contain clusters and is of height ≤ 3. Let us distinguish three cases.

Case 1 (height = 1). This means that there are no arrows between distinct points in $\mathfrak{F}$. Since $\mathfrak{F}$ is a rooted serial frame, it can only consist of a single reflexive point. Clearly we can obtain one reflexive point by taking a generated subframe of the ir-crown frame $\mathfrak{S}_1$. Hence $\mathfrak{F}$ is an up-reduction of an ir-crown frame (Fig. 8).

Case 2 (height = 2). By Theorem 3 (v), $\mathfrak{F}$ cannot have width > 2. Additionally, by seriality, we know that the maximum of $\mathfrak{F}$ must be reflexive. Let us picture all frames with no clusters, with reflexive maximal points, having width < 3 and height 2. These are:

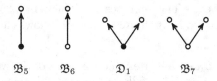

$\qquad \mathfrak{B}_5 \qquad \mathfrak{B}_6 \qquad \mathfrak{D}_1 \qquad \mathfrak{B}_7$

Fig. 8. Frames with height 2, width < 3 and reflexive maximum.

By Theorem 3 (iii), $\mathfrak{F}$ cannot be $\mathfrak{D}_1$. The other three frames are up-reductions of ir-crown frames. $\mathfrak{B}_5$ is a p-morphic image of the ir-crown frame $\mathfrak{S}_1$ below; $\mathfrak{B}_6$ is a generated subframe of $\mathfrak{S}_1$ and $\mathfrak{B}_7$ is a generated subframe of $\mathfrak{S}_2$, as pictured below (Fig. 9).

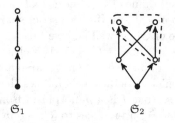

$\qquad\qquad \mathfrak{S}_1 \qquad\qquad\qquad \mathfrak{S}_2$

Fig. 9. Frames $\mathfrak{S}_1$ and $\mathfrak{S}_2$.

Case 3 (height = 3). Let us define a relation $\overline{R}$ by $(x, y) \in \overline{R}$ iff $xRy \wedge x \neq y$. Denote the root of $\mathfrak{F}$ by r_1. We define a partition of W as follows:

   (i) $G_0 = \{r_1\}$

   (ii) $G_1 = \{x \in W \mid r_1\overline{R}x \wedge \neg\exists y(r_1\overline{R}y \wedge y\overline{R}x)\}$

   (iii) $G_2 = \{x \in W \mid \exists y \in G_1, y\overline{R}x\}$

Since height $= 3$, we know that $G_2 \neq \emptyset$. We investigate the structure of $\mathfrak{F}$ in a series of claims below.

   Claim 1. Every point of G_2 is reflexive. Clearly every point of G_2 is a maximal point of $\mathfrak{F}$. The claim then follows from seriality of $\mathfrak{F}$.

   Claim 2. The root r_1 is irreflexive. Since the height of $\mathfrak{F}$ is exactly 3, there exist distinct points $u, v \in W$ with r_1RuRv. Recall that by Theorem 3 (iv) the frame $\mathfrak{B}_3$ cannot be a subframe of $\mathfrak{F}$. It follows that r_1 is irreflexive.

   Claim 3. Every point of G_1 is reflexive. Indeed, as r_1 is irreflexive and $\mathfrak{B}_2$ is not a convex subframe of $\mathfrak{F}$, we deduce that all points in G_1 are reflexive.

   Claim 4. Each point in G_1 sees at most 2 points from G_2. Indeed, otherwise the frame $\mathfrak{T}_{\mathrm{ref}}$ would be a generated subframe of $\mathfrak{F}$, which contradicts Theorem 3 (v).

   Claim 5. G_1 does not contain maximal points. Assume it does. Since the height of $\mathfrak{F}$ is 3 it can be divided into three nonempty parts—the root r_1, $A = \{w \in G_1 \mid w \text{ is maximal}\}$ and $B = W \setminus (\{r_1\} \cup A)$. We construct a p-morphism $f : \mathfrak{F} \to \mathfrak{D}_1$ by sending the root to the root, A to the left successor of the root and B to the right successor of the root.

   Given the properties of $\mathfrak{F}$ revealed by these claims, we further claim that $\mathfrak{F}^*$ cannot be up-reduced to any of the following five frames taken from the paper [8] (Fig. 10).

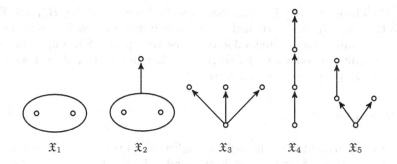

Fig. 10. Posets $\mathfrak{X}_1 - \mathfrak{X}_5$.

   Indeed, that $\mathfrak{F}^*$ is reflexive, transitive, without non-trivial clusters and of height 3, means precisely that it cannot be up-reduced to any of $\mathfrak{X}_1, \mathfrak{X}_2$ and $\mathfrak{X}_4$.

   Suppose that $\mathfrak{F}^*$ is up-reduced to $\mathfrak{X}_3$ using a generated subframe U and a p-morphism f. If $U \subsetneq W$, then U is also a generated subframe of $\mathfrak{F}$ and the same up-reduction works for reducing $\mathfrak{F}$ to $\mathfrak{X}_3$, which is isomorphic to $\mathfrak{T}_{\mathrm{ref}}$, and this is forbidden by Theorem 3 (v). Suppose then $U = W$. It is clear, that $f(r_1)$ is the root of $\mathfrak{X}_3$. If any other point from W maps to the root of $\mathfrak{X}_3$, we can reason as for the case $U \subsetneq W$. Otherwise, the same f works as a p-morphism from $\mathfrak{F}$ onto $\mathfrak{T}_{irr}$ below, which clearly maps p-morphically onto $\mathfrak{D}_1$ thus contradicting Theorem 3 (iv) (Fig. 11).

$$\mathfrak{T}_{Irr}$$

Fig. 11. The frame $\mathfrak{T}_{irr}$.

Therefore, $\mathfrak{F}^*$ cannot be up-reduced to $\mathfrak{X}_3$.

Suppose, finally, that $\mathfrak{F}^*$ can be up-reduced to $\mathfrak{X}_5$. A reasoning similar to that in Lemma 1 suffices to deduce that $\mathfrak{F}$ can be up-reduced to $\mathfrak{D}_2$, in contradiction to Theorem 3 (iv).

It follows that none of the frames $\mathfrak{X}_1 - \mathfrak{X}_5$ is an up-reduction of $\mathfrak{F}^*$. By Lemma 4.2 in [8] we can now claim that $\mathfrak{F}^*$ is an up-reduction of some crown frame (i.e. a frame similar to an ir-crown frame, but with reflexive root). Using Lemma 1 above, we can conclude that $\mathfrak{F}$ is an up-reduction of an ir-crown frame, as required.

Theorem 7. *The d-logic of the polygonal plane is axiomatized by the axioms* $\theta_1 - \theta_5$, *i.e.* $\mathbf{PL}_2^d = \Lambda_2^d$.

Proof. By Theorem 4 any ir-crown validates Λ_2^d, hence $\Lambda_2^d \subseteq \mathsf{Log}(\mathbf{IC})$. By Theorem 5 the logic Λ_2^d is determined by its finite frames and by Theorem 6 each such rooted frame is an up-reduction of an ir-crown frame. Since up-reductions preserve validity, it follows that $\mathsf{Log}(\mathbf{IC}) \subseteq \Lambda_2^d$. Since by Corollary 1 we have $\mathbf{PL}_2^d = \mathsf{Log}(\mathbf{IC})$, the proof is completed.

4   Conclusion

We have axiomatized the modal logic of the Euclidean plane when propositional letters denote planar polygons, while the modal diamond is interpreted as the standard derivative operator on the plane. The obvious question is how these results can be generalized to higher dimensions. The research is under way to determine the C-logic of the polyhedra in the Euclidean space of dimension 3. We are convinced that the approach taken in this paper to determine the d-logic of planar polygons given the knowledge of their C-logic can be lifted to higher dimensions as well. That for each $n < \omega$ the C-logic and the d-logic of n-dimensional polyhedra are Kripke complete can be proved using the Segerberg's theorem on transitive logics of finite height, since a formula like θ_4 can be written in each dimension utilizing the fact that the border $\mathbf{C}A - A$ of a polyhedron is a polyhedron of strictly lower dimension. The link between Kripke semantics for the C-logic and the d-logic seems to be that admissible rooted frames of maximal possible height for these logics are very similar, with the only difference of having an irreflexive root in the case of d-logic and the reflexive one in the case of C-logic. Some further general observations can be made for such frames, like all of them necessarily being without trivial clusters, but the precise details have to be postponed until a more in-depth investigation.

Some final words about possible applications of the formalism and semantic interpretation studied in this paper. Modal language is often praised for its fine balance between simplicity and expressivity. Thus it is desirable to find ways of interpreting it on mathematical structures modeling phenomena of particular interest. Many spatial phenomena and their interrelations can be modelled with arbitrary precision using polyhedra in the Euclidean space. Our approach in this paper interprets the modal language on such structures and studies the emerging basic reasoning mechanisms. We believe this prepares the ground for fruitful applications in the area of spatial knowledge representation and reasoning.

References

1. Aiello, M., van Benthem, J., Bezhanishvili, G.: Reasoning about space: the modal way. J. Log. Comput. **13**, 889–920 (2003)
2. van Benthem, J., Bezhanishvili, G.: Modal logics of space. In: Aiello, M., Pratt-Hartmann, I., van Benthem, J. (eds.) Handbook of Spatial Logics. Springer, Dordrecht (2007). https://doi.org/10.1007/978-1-4020-5587-4_5
3. Bezhanishvili, G., Esakia, L., Gabelaia, D.: Some results on modal axiomatization and definability for topological spaces. Stud. Logica **81**, 325–355 (2005)
4. Bezhanishvili, N., Marra, V., McNeill, D., Pedrini, A.: Tarski's theorem on intuitionistic logic, for polyhedra. Ann. Pure Appl. Logic **169**, 373–391 (2017)
5. Blackburn, P., de Rijke, M., Venema, Y.: Modal Logic. Cambridge University Press, Cambridge (2001)
6. Chagrov, A., Zakharyaschev, M.: Modal Logic. Oxford University Press, Oxford (1997)
7. Engelking, R.: General Topology. Polish Scientific Publishers, Warszawa (1977)
8. Gabelaia, D., Gogoladze, K., Jibladze, M., Kuznetsov, E., Marx, M.: Modal logic of planar polygons. http://arxiv.org/abs/1807.02868
9. Kontchakov, R., Pratt-Hartmann, I., Zakharyaschev, M.: Interpreting topological logics over euclidean spaces. In: Proceedings of Twelfth International Conference on the Principles of Knowledge Representation and Reasoning, pp. 534–544. AAAI Press (2010)
10. Kontchakov, R., Pratt-Hartmann, I., Zakharyaschev, M.: Spatial reasoning with RCC8 and connectedness constraints in euclidean spaces. Artif. Intell. **217**, 43–75 (2014)

An Ehrenfeucht-Fraïssé Game
for Inquisitive First-Order Logic

Gianluca Grilletti[1]([⊠]) and Ivano Ciardelli[2]

[1] Institute for Logic, Language and Computation, Amsterdam, The Netherlands
grilletti.gianluca@gmail.com
[2] Munich Center for Mathematical Philosophy, LMU, Munich, Germany
ivano.ciardelli@lmu.de

Abstract. Inquisitive first-order logic, InqBQ, is an extension of classical first-order logic with questions. From a mathematical point of view, formulas in this logic express properties of sets of relational structures. In this paper we describe an Ehrenfeucht-Fraïssé game for InqBQ and show that it characterizes the distinguishing power of the logic. We exploit this result to show a number of undefinability results: in particular, several variants of the question *how many individuals have property P* are not expressible in InqBQ, even in restriction to finite models.

1 Introduction

According to the traditional view, the semantics of a logical system specifies truth-conditions for the sentences in the language. This focus on truth restricts the scope of logic to a special kind of sentences, namely, statements, whose semantics can be adequately characterized in terms of truth-conditions. In recent years, a more general view of semantics has been developed, which goes under the name of *inquisitive* semantics (see [4] for a language-oriented introduction, and [2] for a logic-oriented one). In this approach, the meaning of a sentence is laid out not by specifying when the sentence is true relative to a state of affairs, but rather by specifying when it is supported by a given state of information. This view allows us to interpret in a uniform way both statements and questions: for instance, the statement *it rains* will be supported by an information state s if the information available in s implies that it rains, while the question *whether it rains* will be supported by s if the information available in s determines whether or not it rains.

In its first-order version, referred to as InqBQ, inquisitive logic can be seen as a conservative extension of classical first-order logic with formulas expressing questions. Thus, in addition to standard first-order formulas like Pa and $\forall x Px$, we also have formulas like $?Pa$ (*whether a has property P*), $\exists x Px$ (*what is an instance of property P*), and $\forall x ?Px$ (*which individuals have property P*). A

Financial support by the European Research Council under the EU's Horizon 2020 research and innovation programme (grant No. 680220) is gratefully acknowledged.

© Springer-Verlag GmbH Germany, part of Springer Nature 2019
A. Silva et al. (Eds.): TbiLLC 2018, LNCS 11456, pp. 166–186, 2019.
https://doi.org/10.1007/978-3-662-59565-7_9

model for this logic is based on a set W of possible worlds, each representing a possible state of affairs, corresponding to a standard first-order structure. An information state is modeled as a subset $s \subseteq W$. The idea, which goes back to the work of Hintikka [11], is that a set of worlds s stands for a body of information that is compatible with the actual world being one of the worlds $w \in s$, and incompatible with it being one of the worlds $w \notin s$. The semantics of the language takes the form of a support relation holding between information states in a model and sentences of the language.

From a mathematical point of view, a sentence of InqBQ expresses a property of a set s of first-order structures. The crucial difference between statements and questions is that statements express *local* properties of information states—which boil down to requirements on the individual worlds $w \in s$—while questions express *global* requirements, having to do with the way the worlds in s are related to each other. Thus, for instance, the formula $?Pa$ requires that the truth-value of Pa be the same in all worlds in s; the formula $\exists x Px$ requires that there be an individual that has property P uniformly in all worlds in s; and the formula $\forall x?Px$ requires that the extension of property P be the same across s. Global properties can also take the form of *dependencies*: thus, e.g., $?Pa \rightarrow ?Qa$ requires that the truth-value of Qa be functionally determined by the truth-value of Pa in s, while $\forall x?Px \rightarrow \forall x?Qx$ requires that the extension of property Q be functionally determined by the extension of property P in s. Thus, inquisitive first-order logic provides a language that can be used to talk about both local and global features of an information state.

In contrast to propositional inquisitive logic, which has been thoroughly investigated (see, among others, [1,3,5,9,16–18]), first-order inquisitive logic has received comparatively little attention [2,10]. In particular, a detailed investigation of the expressive power of the logic has so far been missing. This paper is a first step in this direction.

In the classical context, a powerful tool to study the expressiveness of first-order logic is provided by Ehrenfeucht-Fraïssé games (also known as EF games or back-and-forth games), introduced in 1967 by Ehrenfeucht [6], developing model-theoretic results presented by Fraïssé [8]. These games provide a particularly perspicuous way of understanding what differences between models can be detected by means of first-order formulas of a certain quantifier rank. Reasoning about winning strategies in this game, one can prove that two first-order structures are elementarily equivalent, or one can find a formula telling them apart. As an application, EF games provide relatively easy proofs that certain properties of first-order structures are not first-order expressible.

The basic idea of EF games has proven to be very flexible and adaptable to a wide range of logical settings, including fragments of first-order logic with finitely many variables [13]; extensions of first-order logic with generalized quantifiers [14]; monadic second order logic [7]; modal logic [20]; and intuitionistic logic [15,21]. In each case, the game provides an insightful characterization of the distinctions that can and cannot be made by means of formulas in the logic.

Our aim in this paper is to introduce an EF-style game for InqBQ and to show that this game provides a characterization of the expressive power of InqBQ. As an application, we show that certain natural questions are not expressible in InqBQ: in particular, the question *how many individuals have property P* (supported in s if the extension of P has the same cardinality in all the worlds in s) and the question *whether there are only finitely many individuals satisfying P* (supported in s if the extension of P is finite in all the worlds in s, or infinite in all the worlds in s).

The paper is structured as follows: in Sect. 2 we provide some technical background on the logic InqBQ. In Sect. 3 we describe the game and prove our main result, linking winning strategies for bounded versions of the game to the distinguishing power of fragments of InqBQ. In Sect. 4 we use this result to show that certain questions are not expressible in InqBQ. In Sect. 5 we summarize our findings and mention some directions for future work.

2   Inquisitive First-Order Logic

In this section we provide a basic introduction to inquisitive first-order logic. For a more comprehensive introduction, the reader is referred to [2].

Let Σ be a predicate logic signature. For simplicity, we first restrict to the case in which Σ is a *relational* signature, i.e., contains no function symbols. The extension to an arbitrary signature, which involves some subtleties familiar from the classical case [12], is discussed in Sect. 3.4. The set $\mathcal{L}$ of formulas of InqBQ over Σ is defined as follows, where $R \in \Sigma$ is an n-ary relation symbol:

$$\varphi ::= R(x_1, \dots, x_n) \mid (x_1 = x_2) \mid \bot \mid \varphi \wedge \varphi \mid \varphi \to \varphi \mid \forall x \varphi \mid \varphi \vee\!\!\!\vee \varphi \mid \exists\!\!\!\exists x \varphi$$

We will take negation to be a defined operator:

$$\neg \varphi := \varphi \to \bot$$

Formulas without occurrences of $\vee\!\!\!\vee$ and $\exists\!\!\!\exists$ are referred to as *classical* formulas, and can be identified with standard first-order logic formulas. If α and β are classical formulas, then we can define $\alpha \vee \beta := \neg(\neg \alpha \wedge \neg \beta)$ and $\exists x \alpha := \neg \forall x \neg \alpha$. The set of classical formulas is denoted $\mathcal{L}_c$.[1]

The connective $\vee\!\!\!\vee$ and the quantifier $\exists\!\!\!\exists$, referred to respectively as *inquisitive disjunction* and *inquisitive existential quantifier*, allow us to form questions. For instance, if α is a classical formula then the formula $?\alpha := \alpha \vee\!\!\!\vee \neg \alpha$ represents the question *whether α*; the formula $\exists\!\!\!\exists x \alpha(x)$ represents the question *what is an individual satisfying $\alpha(x)$*; and the formula $\forall x ?\alpha(x)$ represents the question *which individuals satisfy $\alpha(x)$*.

A model for InqBQ consists of a set W of worlds—representing possible states of affairs—a set D of individuals, and an interpretation function I which determines at each world the extension of all relation symbols, including identity.

---

[1] We could in principle define $\vee$ and $\exists$ for arbitrary formulas; however, these operators are only natural and useful when applied to classical formulas, on which they yield the standard disjunction and existential quantifier of classical logic.

Definition 1 (Models). *A model for the signature Σ is a tuple $M = \langle W, D, I \rangle$ where W and D are sets and I is a function mapping each world $w \in W$ and each n-ary relation symbol $R \in \Sigma \cup \{=\}$ to a corresponding n-ary relation $I_w(R) \subseteq D^n$ —the extension of R at w. The interpretation of identity is subject to the following condition:*

$I_w(=)$ is an equivalence relation $\sim_w$ which is a congruence, i.e., if $R \in \Sigma$ and $d_i \sim_w d_i'$ for $i \leq n$, then $\langle d_1, \ldots, d_n \rangle \in I_w(R) \iff \langle d_1', \ldots, d_n' \rangle \in I_w(R)$.

As discussed in the introduction, in inquisitive logic the semantics of the language specifies when a formula is supported at an information state $s \subseteq W$, rather than when a formula is true at a possible world $w \in W$. As usual, to handle open formulas and quantification, the support relation is defined relative to an assignment, which is a function from variables to the set D of individuals; if g is an assignment and $d \in D$, then $g[x \mapsto d]$ is the assignment which maps x to d and behaves like g on all other variables.

Definition 2 (Support). *Let $M = \langle W, D, I \rangle$ be a model and let $s \subseteq W$.*

$$M, s \models_g R(x_1, \ldots, x_n) \iff \forall w \in s : \langle g(x_1), \ldots, g(x_n) \rangle \in I_w(R)$$
$$M, s \models_g x_1 = x_2 \iff \forall w \in s : g(x_1) \sim_w g(x_2)$$
$$M, s \models_g \bot \iff s = \emptyset$$
$$M, s \models_g \varphi \wedge \psi \iff M, s \models_g \varphi \text{ and } M, s \models_g \psi$$
$$M, s \models_g \varphi \lor\!\!\lor \psi \iff M, s \models_g \varphi \text{ or } M, s \models_g \psi$$
$$M, s \models_g \varphi \rightarrow \psi \iff \forall t \subseteq s : M, t \models_g \varphi \text{ implies } M, t \models_g \psi$$
$$M, s \models_g \forall x\varphi \iff M, s \models_{g[x \mapsto d]} \varphi \text{ for all } d \in D$$
$$M, s \models_g \exists x\varphi \iff M, s \models_{g[x \mapsto d]} \varphi \text{ for some } d \in D$$

As usual, if $\varphi(x_1, \ldots, x_n)$ is a formula whose free variables are among $x_1, \ldots, x_n$, then the value of g on variables other than $x_1, \ldots, x_n$ is irrelevant. If $d_1, \ldots, d_n \in D$, we can therefore write $M, s \models \varphi(d_1, \ldots, d_n)$ to mean that $M, s \models_g \varphi$ holds with respect to an assignment g that maps x_i to d_i. In particular, if φ is a sentence we can drop reference to g altogether. Moreover, we write $M \models \varphi$ as a shorthand for $M, W \models \varphi$ and we say that M supports φ.

The support relation has the following two basic features:

– Persistency: if $M, s \models_g \varphi$ and $t \subseteq s$ then $M, t \models_g \varphi$;
– Empty state property: if $M, \emptyset \models_g \varphi$ for all φ.

In restriction to classical formulas, the above definition of support gives a non-standard semantics for classical first-order logic. To see why, let us associate to each world $w \in M$ a corresponding relational structure $\mathcal{M}_w$, having as its domain the quotient $D/ \sim_w$, and with the interpretation of relation symbols induced by $I_w(R)$. Given $\alpha \in \mathcal{L}_c$ a classical formula, by a simple induction on the structure of α we obtain:

$$M, s \models_g \alpha \iff \forall w \in s : \mathcal{M}_w \models_{\bar{g}} \alpha \text{ holds in first-order logic}$$

where $\bar{g}$ is the assignment mapping x to the $\sim_w$-equivalence class of $g(x)$. Thus, as far as the standard fragment of the language is concerned, the relation of support is essentially a recursive definition of global truth with respect to a set of structures sharing the same domain. Notice that the standard definition of truth can be recovered as a special case of support by taking s to be a singleton. We will also write $M, w \models_g \alpha$ as an abbreviation for $M, \{w\} \models_g \alpha$.

Thus, evaluating a classical formula on an information state s amounts to evaluating it at each world in s and determining whether it is satisfied at each world. The same is not true for formulas that contain the operators $\lor\hspace{-0.5em}\lor$ and $\exists\hspace{-0.5em}\exists$; typically, such formulas allow us to express global requirements on a state, which cannot be reduced to requirements on the single worlds in the state. We will illustrate this point by means of some examples. First take a classical sentence α, and consider the formula $?\alpha := \alpha \lor\hspace{-0.5em}\lor \neg\alpha$. We have:

$$M, s \models ?\alpha \iff (\forall w \in s : M, w \models \alpha) \text{ or } (\forall w \in s : M, w \models \neg\alpha)$$

Thus, in order for s to support $?\alpha$, all the worlds in s must agree on the truth-value of α. In other words, $?\alpha$ is supported at s only if the information available in s determines whether or not α is true.

Next take $\alpha(x)$ to be a classical formula having only the variable x free, and consider the formula $\exists\hspace{-0.5em}\exists x \alpha(x)$. We have:

$$M, s \models \exists\hspace{-0.5em}\exists x \alpha(x) \iff \exists d \in D \text{ s.t. } \forall w \in s : M, w \models \alpha(d)$$

Thus, in order for s to support $\exists\hspace{-0.5em}\exists x \alpha(x)$ there must be an individual d which satisfies $\alpha(x)$ at all worlds in s. In other words, $\exists\hspace{-0.5em}\exists x \alpha(x)$ is supported at s if the information available in s implies for some specific individual that it satisfies $\alpha(x)$—i.e., gives us a specific witness for $\alpha(x)$.

Finally, let again $\alpha(x)$ be a classical formula having only x free, and let us denote by α_w the extension of $\alpha(x)$ at w, i.e., $\alpha_w := \{d \in D \mid M, w \models \alpha(d)\}$. Consider the formula $\forall x ?\alpha(x) := \forall x (\alpha(x) \lor\hspace{-0.5em}\lor \neg\alpha(x))$. We have:

$$M, s \models \forall x ?\alpha(x) \iff \forall w, w' \in s : \alpha_w = \alpha_{w'}$$

Thus, in order for s to support $\forall x ?\alpha(x)$, all the worlds in s must agree on the extension that they assign to $\alpha(x)$. In other words, $\forall x ?\alpha(x)$ is supported at s if the information available in s determines exactly which individuals satisfy $\alpha(x)$.

An aspect of InqBQ which is worth commenting on is the interpretation of identity. In InqBQ, the interpretation of identity may differ at different worlds. This allows to deal with uncertainty about the identity relation: e.g., one may have information about two individuals, a and b (say, one knows Pa and Qb) and yet be uncertain whether a and b are distinct individuals, or the same. This also allows for uncertainty about how many individuals there are. Indeed, although a model is based on a fixed set D of epistemic individuals—objects to which information can be attributed—the domain of actual individuals at a world w is given by the equivalence classes modulo $\sim_w$; the number of actual individuals that exist at w is the number of such equivalence classes, i.e., the cardinality of the quotient $D/\sim_w$. Of course, as a special case we could take $\sim_w$ to be the actual relation of identity on D at each world. A model in which identity is treated in this way is called an id-model.

This section is only intended as a summary of the relevant notions and as a quick illustration of the features of information states that can be expressed by means of formulas of InqBQ. With these basic notions in place and hopefully some grasp of how the logic works, we are now ready to turn to the novel contribution of the paper: an Ehrenfeucht-Fraïssé game for InqBQ.

3   An Ehrenfeucht-Fraïssé Game for InqBQ

The EF game for InqBQ is played by two players, S (Spoiler) and D (Duplicator), using two inquisitive models M_0, M_1 as a board. As in the classical case, the game proceeds in turns: at each turn, S picks an object from one of the two models and D must respond by picking a corresponding object from the other model. At the end of the game, a winner is decided by comparing the atomic formulae supported by the sub-structures built during the game.

However, there are two crucial differences with the classical EF game. First, the objects that are picked during the game are not just individuals $d \in D_i$, but also information states $s \subseteq W_i$. This is because the logical repertoire of InqBQ contains not only the operators $\forall$ and $\exists$, which quantify over individuals, but also the operator $\rightarrow$, which quantifies over information states. Second, the roles of the two models in the game are not symmetric. This is connected to the absence of a classical negation in the language of InqBQ; unlike in classical logic, it could be that a model M_0 supports all the formulas supported by a model M_1, but not vice versa. This directionality is reflected by the game.

3.1   The Game

A position in an EF game for InqBQ is a tuple $\langle M_0, s_0, \bar{a}_0; M_1, s_1, \bar{a}_1 \rangle$ where:

- $M_0 = \langle W_0, D_0, I_0 \rangle$ and $M_1 = \langle W_1, D_1, I_1 \rangle$ are models for InqBQ;
- s_0 and s_1 are information states in the models M_0 and M_1 respectively;
- $\bar{a}_0$ and $\bar{a}_1$ are tuples of equal length of elements from D_0 and D_1 respectively.

If not otherwise specified, a game between the models M_0 and M_1 starts from position $\langle M_0, W_0, \varepsilon; M_1, W_1, \varepsilon; \rangle$, where ε indicates the empty tuple.

Starting a round from a position $\langle M_0, s_0, \bar{a}_0; M_1, s_1, \bar{a}_1 \rangle$, the following are the possible moves:[2]

- $\exists$-move: S picks an element $b_0 \in D_0$; D responds with an element $b_1 \in D_1$; the game continues from the position $\langle M_0, s_0, \bar{a}_0 b_0; M_1, s_1, \bar{a}_1 b_1 \rangle$;
- $\forall$-move: S picks an element $b_1 \in D_1$; D responds with an element $b_0 \in D_0$; the game continues from the position $\langle M_0, s_0, \bar{a}_0 b_0; M_1, s_1, \bar{a}_1 b_1 \rangle$;
- $\rightarrow$-move: S picks a sub-state $t_1 \subseteq s_1$; D responds with a sub-state $t_0 \subseteq s_0$; S picks $i \in \{1, 0\}$. The game continues from $\langle M_i, t_i, \bar{a}_i; M_{1-i}, t_{1-i}, \bar{a}_{1-i} \rangle$.

---

[2] In the following, the notation $\bar{a}b$ indicates the sequence obtained by adding the element b at the end of the sequence $\bar{a}$.

Notice the asymmetry between the roles of the two models: by performing an
$\rightarrow$-move, S can pick an information state from M_1, but not a state in M_0.

With respect to termination condition, we consider different versions of the
game. In the bounded version of the game, a pair of numbers $\langle i, q \rangle \in \mathbb{N}^2$ is fixed
in advance. This number constrains the development of the game: in total, S
can play only i implication moves and only q quantifier moves (i.e., $\overline{\exists}$-move
or a $\forall$-move). When there are no more moves available, the game ends. If
$\langle M_0, s_0, \overline{a}_0; M_1, s_1, \overline{a}_1 \rangle$ is the final position, the game is won by Player D if the
following condition is satisfied, and by player S otherwise:

- Winning condition for D: for all atomic formulas $\alpha(x_1, \ldots, x_n)$ where n is the
  size of the tuples $\overline{a}_0$ and $\overline{a}_1$, we have:

$$M_0, s_0 \models \alpha(\overline{a}_0) \implies M_1, s_1 \models \alpha(\overline{a}_1) \tag{1}$$

In the unbounded version of the game, no restriction is placed at the outset
on the number of moves to be performed. Instead, player S has the option to
declare the game over at the beginning of each round: in this case, the winner
is determined as in the bounded version of the game. If the game never stops,
then D is the winner.[3]

Example 1. Take the signature $\Sigma = \{P\}$, where P is a unary predicate symbol.
Given the models M_0 and M_1 in Fig. 1, in the table we show a run of the bounded
game with $\langle i, q \rangle = \langle 1, 2 \rangle$ between M_0 and M_1. At the end of the run, the position
is $\langle M_1, \{v_1\}, \langle e_2, e_1 \rangle; M_0, \{w_0\}, \langle d_1, d_2 \rangle \rangle$. The winner is Spoiler, since

$$M_1, \underbrace{\{v_1\}}_{t_1} \models P(\underbrace{e_2}_{b_1}) \qquad \not\Longrightarrow \qquad M_0, \underbrace{\{w_0\}}_{s_1} \models P(\underbrace{d_1}_{a_1})$$

As usual, a winning strategy for a player is a strategy which guarantees victory
to them, no matter what the opponent plays. If D has a winning strategy in the
EF game of length $\langle i, q \rangle$ starting from position $\langle M_0, s_0, \overline{a}_0; M_1, s_1, \overline{a}_1 \rangle$ we write:

$$(M_0, s_0, \overline{a}_0) \preceq_{i,q} (M_1, s_1, \overline{a}_1)$$

We write $\approx_{i,q}$ for the relation $\preceq_{i,q} \cap \succeq_{i,q}$. Notice that the game is finite (since
the number of turns is bounded by $i + q$), zero-sum (as can be seen from the
winning condition) and has perfect information. Therefore, if $(M_0, s_0, \overline{a}_0) \preceq_{i,q}$
$(M_1, s_1, \overline{a}_1)$ does not hold, then it follows from the Gale-Stewart Theorem that S
has a winning strategy in the EF game of length $\langle i, q \rangle$ starting from the position
$\langle M_0, s_0, \overline{a}_0; M_1, s_1, \overline{a}_1 \rangle$.

We write $M_0 \preceq_{i,q} M_1$ as a shorthand for $(M_0, W_0, \varepsilon) \preceq_{i,q} (M_1, W_1, \varepsilon)$.

The following two propositions follow easily from the definition of the game.
For Proposition 2, notice that the three conditions amount precisely to the fact

---

[3] Here we consider games in which the rounds of play are indexed by natural numbers.
To define games of transfinite length, one would have to specify how to determine
the game position corresponding to a limit ordinal. We leave this for future work.

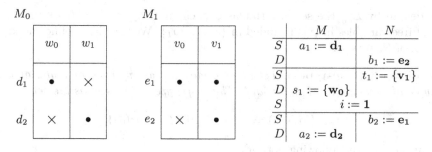

Fig. 1. On the left, two inquisitive models in the signature $\Sigma = \{P\}$. The top row represents the set of worlds of the models (e.g., $\{w_0, w_1\}$ for the model M_0) and the left column represents the domain (e.g., $\{d_1, d_2\}$ for M_0); $I(P)$ is encoded by the entries of the table: a • indicates that P holds, while a × indicates that P does not hold (e.g., $d_1 \in I_{w_0}(P)$ and $d_2 \notin I_{w_0}(P)$). On the right, a run of the Ehrenfeucht-Fraïssé game.

that, for every move available to Spoiler, there is a corresponding move for Duplicator that leads to a sub-game in which Duplicator has a winning strategy. This is precisely what is needed for Duplicator to have a winning strategy in the original game.

Proposition 1. *If* $(M_0, s_0, \overline{a}_0) \preceq_{i,q} (M_1, s_1, \overline{a}_1)$ *then* $(M_0, s_0, \overline{a}_0) \preceq_{i',q'} (M_1, s_1, \overline{a}_1)$ *for all* $i' \leq i$ *and* $q' \leq q$.

Proposition 2. *Suppose* $\langle i, q \rangle \neq \langle 0, 0 \rangle$. $(M_0, s_0, \overline{a}_0) \preceq_{i,q} (M_1, s_1, \overline{a}_1)$ *iff the following three conditions are satisfied:*

- *If* $i > 0$, *then* $\forall t_1 \subseteq s_1 \; \exists t_0 \subseteq s_0 \; : (M_0, t_0, \overline{a}_0) \approx_{i-1,q} (M_1, t_1, \overline{a}_1)$
- *If* $q > 0$, *then* $\forall b_0 \in D_0 \; \exists b_1 \in D_1 : (M_0, s_0, \overline{a}_0 b_0) \preceq_{i,q-1} (M_1, s_1, \overline{a}_1 b_1)$
- *If* $q > 0$, *then* $\forall b_1 \in D_1 \; \exists b_0 \in D_0 : (M_0, s_0, \overline{a}_0 b_0) \preceq_{i,q-1} (M_1, s_1, \overline{a}_1 b_1)$

3.2 IQ Degree and Types

We define the *implication degree* (Ideg) and *quantification degree* (Qdeg) of a formula by the following inductive clauses, where p stands for an atomic formula:

$\text{Ideg}(p)$	$= 0$	$\text{Qdeg}(p)$	$= 0$
$\text{Ideg}(\bot)$	$= 0$	$\text{Qdeg}(\bot)$	$= 0$
$\text{Ideg}(\varphi_1 \wedge \varphi_2)$	$= \max(\text{Ideg}(\varphi_1), \text{Ideg}(\varphi_2))$	$\text{Qdeg}(\varphi_1 \wedge \varphi_1)$	$= \max(\text{Qdeg}(\varphi_1), \text{Qdeg}(\varphi_2))$
$\text{Ideg}(\varphi_1 \vee\!\!\!\vee \varphi_2)$	$= \max(\text{Ideg}(\varphi_1), \text{Ideg}(\varphi_2))$	$\text{Qdeg}(\varphi_1 \vee\!\!\!\vee \varphi_1)$	$= \max(\text{Qdeg}(\varphi_1), \text{Qdeg}(\varphi_2))$
$\text{Ideg}(\varphi_1 \rightarrow \varphi_2)$	$= \max(\text{Ideg}(\varphi_1), \text{Ideg}(\varphi_2)) + 1$	$\text{Qdeg}(\varphi_1 \rightarrow \varphi_1)$	$= \max(\text{Qdeg}(\varphi_1), \text{Qdeg}(\varphi_2))$
$\text{Ideg}(\forall x \varphi)$	$= \text{Ideg}(\varphi)$	$\text{Qdeg}(\forall x \varphi)$	$= \text{Qdeg}(\varphi) + 1$
$\text{Ideg}(\exists x \varphi)$	$= \text{Ideg}(\varphi)$	$\text{Qdeg}(\exists x \varphi)$	$= \text{Qdeg}(\varphi) + 1$

The combined degree of a formula is defined as $\text{IQdeg}(\varphi) = \langle \text{Ideg}(\varphi), \text{Qdeg}(\varphi) \rangle$. We define an order relation $\leq$ on such degrees by setting:

$$\langle a, b \rangle \leq \langle a', b' \rangle \iff a \leq a' \text{ and } b \leq b'$$

We denote by $\mathcal{L}^l_{i,q}$ the set of formulas φ such that $\mathrm{IQdeg}(\varphi) \leq \langle i, q \rangle$ and the set of free variables in φ is included in $\{x_1, \ldots, x_l\}$. We can then define the key notion of $\langle i, q \rangle$-type.

Definition 3 ($\langle i, q \rangle$-**types**). *Let M be a model, s an information state, and $\bar{a}$ a tuple of elements in M of length l. The $\langle i, q \rangle$-type of $\langle M, s, \bar{a} \rangle$ is the set*

$$\mathrm{tp}_{i,q}(M, s, \bar{a}) := \{\varphi \in \mathcal{L}^l_{i,q} \mid M, s \models \varphi(\bar{a})\}$$

We also define the following notation:

$$(M_0, s_0, \bar{a}_0) \sqsubseteq_{i,q} (M_1, s_1, \bar{a}_1) \stackrel{def}{\Longleftrightarrow} \mathrm{tp}_{i,q}(M_0, s_0, \bar{a}_0) \subseteq \mathrm{tp}_{i,q}(M_1, s_1, \bar{a}_1)$$
$$(M_0, s_0, \bar{a}_0) \equiv_{i,q} (M_1, s_1, \bar{a}_1) \stackrel{def}{\Longleftrightarrow} \mathrm{tp}_{i,q}(M_0, s_0, \bar{a}_0) = \mathrm{tp}_{i,q}(M_1, s_1, \bar{a}_1)$$

Example 2. Consider the models M_0, M_1 in Fig. 1. We have $(M_1, \{v_0, v_1\}, \langle e_1 \rangle)$ $\not\sqsubseteq_{0,0}$ $(M_0, \{w_0, w_1\}, \langle d_1 \rangle)$ since $M_1, \{v_0, v_1\} \models P(e_1)$, while $M_0, \{w_0, w_1\} \not\models P(d_1)$.

Notice that, if the signature is finite, there are only a finite number of non-equivalent formulas of combined degree at most $\langle i, q \rangle$, and consequently only a finite number of $\langle i, q \rangle$-types. This can be shown inductively as follows:

- up to logical equivalence, there are only finitely many formulas in $\mathcal{L}^l_{0,0}$—atoms and their boolean combinations. Notice the hypothesis of working with a finite signature is necessary here.
- formulas in $\mathcal{L}^l_{i,q}$ are equivalent to Boolean combinations of formulas in $A \cup B$, for $A = \{\varphi \to \psi \mid \varphi, \psi \in \mathcal{L}^l_{i-1,q}\}$ and $B = \{\exists x.\varphi, \forall x.\varphi \mid \varphi \in \mathcal{L}^{l+1}_{i,q-1}\}$—where we impose by definition $\mathcal{L}^l_{i,q} = \emptyset$ if $i < 0$ or $q < 0$. By induction hypothesis, A and B contain only finitely many non-equivalent formulas.

3.3   The EF Theorem

What follows is the main result of the paper: the relations $\preceq_{i,q}$ and $\sqsubseteq_{i,q}$ coincide.

Theorem 1. *Suppose the signature Σ is finite. Then*

$$(M_0, s_0, \bar{a}_0) \preceq_{i,q} (M_1, s_1, \bar{a}_1) \quad \Longleftrightarrow \quad (M_0, s_0, \bar{a}_0) \sqsubseteq_{i,q} (M_1, s_1, \bar{a}_1)$$

Proof. We will prove this by well-founded induction on $\langle i, q \rangle$. For the basic case, $\langle i, q \rangle = \langle 0, 0 \rangle$, we just have to verify that, if Condition (1) holds for all atomic formulas, then it holds for all formulas $\varphi \in \mathcal{L}^l_{0,0}$. This is straightforward. Next, suppose $\langle i, q \rangle > \langle 0, 0 \rangle$ and suppose the claim holds for all $\langle i', q' \rangle < \langle i, q \rangle$. For the left-to-right direction, proceed by contraposition. Suppose that for some $\varphi \in \mathcal{L}^l_{i,q}$

$$M_0, s_0 \models \varphi(\bar{a}_0) \qquad M_1, s_1 \not\models \varphi(\bar{a}_1)$$

We proceed by induction on the structure of φ; some cases are easy to consider:

- If φ is an atom, it follows $(M_0, s_0, \bar{a}_0) \not\preceq_{0,0} (M_1, s_1, \bar{a}_1)$; so, by Proposition 1 also $(M_0, s_0, \bar{a}_0) \not\preceq_{i,q} (M_1, s_1, \bar{a}_1)$. Thus, in this case the conclusion follows.
- If φ is a conjunction $\psi \wedge \chi$ then we have:

$$\begin{cases} M_0, s_0 \models \psi(\bar{a}_0) \wedge \chi(\bar{a}_0) \implies M_0, s_0 \models \psi(\bar{a}_0) \text{ and } M_0, s_0 \models \chi(\bar{a}_0) \\ M_1, s_1 \not\models \psi(\bar{a}_1) \wedge \chi(\bar{a}_1) \implies M_1, s_1 \not\models \psi(\bar{a}_1) \text{ or } M_1, s_1 \not\models \chi(\bar{a}_1) \end{cases}$$

So, either ψ or χ is a less complex witness of $(M_0, s_0, \bar{a}_0) \not\sqsubseteq_{i,q} (M_1, s_1, \bar{a}_1)$.
- If φ is a disjunction $\psi \vee \chi$, we can reach a conclusion analogous to the one we reached for conjunction.

So the cases left are φ of the form $\psi \to \chi$, $\forall x \psi$ and $\exists x \psi$ (cases $\Rightarrow^1$, $\Rightarrow^2$, $\Rightarrow^3$ respectively). Let us consider the three cases separately.

Case $\Rightarrow^1$: φ is an implication $\psi \to \chi$. In this case we have

$$M_1, s_1 \not\models \psi(\bar{a}_1) \to \chi(\bar{a}_1) \implies (\exists t_1 \subseteq s_1) \, [M_1, t_1 \models \psi(\bar{a}_1) \text{ and } M_1, t_1 \not\models \chi(\bar{a}_1)]$$
$$M_0, s_0 \models \psi(\bar{a}_0) \to \chi(\bar{a}_0) \implies (\not\exists t_0 \subseteq s_0) \, [M_0, t_0 \models \psi(\bar{a}_0) \text{ and } M_0, t_0 \not\models \chi(\bar{a}_0)]$$

Thus there exists a state $t_1 \subseteq s_1$ with a different $\langle i-1, q \rangle$-type than every $t_0 \subseteq s_0$—either because it supports ψ or because it does not support χ. So by induction hypothesis, if S performs a $\to$-move and chooses t_1, for every choice t_0 of D we have $(M_0, t_0, \bar{a}_0) \not\approx_{i-1,q} (M_1, t_1, \bar{a}_1)$. It follows by Proposition 2 that $(M_0, s_0, \bar{a}_0) \not\preceq_{i,q} (M_1, s_1, \bar{a}_1)$ as wanted.

Case $\Rightarrow^2$: φ is a universal $\forall x \psi$. In this case we have

$$M_1, s_1 \not\models \forall x \psi(\bar{a}_1, x) \implies (\exists b_1 \in D_1) \, M_1, s_1 \not\models \psi(\bar{a}_1, b_1)$$
$$M_0, s_0 \models \forall x \psi(\bar{a}_0, x) \implies (\forall b_0 \in D_0) \, M_0, s_0 \models \psi(\bar{a}_0, b_0)$$

Thus if S performs a $\forall$-move and chooses b_1, for every choice b_0 of D, by induction hypothesis we have

$$(M_0, s_0, \bar{a}_0 b_0) \not\preceq_{i,q-1} (M_1, s_1, \bar{a}_1 b_1)$$

It follows by Proposition 2 that $(M_0, s_0, \bar{a}_0) \not\preceq_{i,q} (M_1, s_1, \bar{a}_1)$ as wanted.

Case $\Rightarrow^3$: φ is an inquisitive existential $\exists x \psi$. This case is similar to the previous one: S can perform an $\exists$-move and pick an element b_0 in D_0 with no counterpart in D_1, and by Proposition 2 we get the result.

This completes the proof of the left-to-right direction of the inductive step. Now consider the converse direction. Again, we proceed by contraposition. Suppose that S has a winning strategy in the EF game of length $\langle i, q \rangle$ starting from $(M_0, s_0, \bar{a}_0; M_1, s_1, \bar{a}_1)$. We consider again three cases, depending on the first move of the winning strategy (cases $\Leftarrow^1$, $\Leftarrow^2$, $\Leftarrow^3$ respectively).

Case $\Leftarrow^1$: the first move is a $\to$-move. Suppose S starts by choosing $t_1 \subseteq s_1$. As this is a winning strategy for S, for every choice $t_0 \subseteq s_0$ of D we have

$$(M_0, t_0, \bar{a}_0) \not\preceq_{i-1,q} (M_1, t_1, \bar{a}_1) \quad \text{or} \quad (M_1, t_1, \bar{a}_1) \not\preceq_{i-1,q} (M_0, t_0, \bar{a}_0)$$

By inductive hypothesis, this translates to

$$\exists \psi_{t_0} \in \mathrm{tp}(t_0) \setminus \mathrm{tp}(t_1) \quad \text{or} \quad \exists \theta_{t_0} \in \mathrm{tp}(t_1) \setminus \mathrm{tp}(t_0)$$

where $\mathrm{tp}(t_0) := \mathrm{tp}_{i-1,q}(M_0, t_0, \overline{a}_0)$ and $\mathrm{tp}(t_1) := \mathrm{tp}_{i-1,q}(M_1, t_1, \overline{a}_1)$.

Given this, there exist two families $\{\psi_{t_0} \mid t_0 \subseteq s_0\}$ and $\{\theta_{t_0} \mid t_0 \subseteq s_0\}$ s.t.:

$$\begin{cases} \psi_{t_0} \in \mathrm{tp}(t_0) \setminus \mathrm{tp}(t) \text{ if } \mathrm{tp}(t_0) \setminus \mathrm{tp}(t) \neq \emptyset \\ \psi_{t_0} := \bot \qquad \text{otherwise} \end{cases}$$

$$\begin{cases} \theta_{t_0} \in \mathrm{tp}(t) \setminus \mathrm{tp}(t_0) \text{ if } \mathrm{tp}(t) \setminus \mathrm{tp}(t_0) \neq \emptyset \\ \theta_{t_0} := \top \qquad \text{otherwise} \end{cases}$$

Moreover, we can suppose the two families to be finite, as there are only a finite number of formulas of degree $\langle i-1, q \rangle$ up to logical equivalence (see Sect. 3.2). Define now

$$\varphi := \bigwedge_{t_0 \subseteq s_0} \theta_{t_0} \rightarrow \bigvee_{t_0 \subseteq s_0} \psi_{t_0}$$

We have: (i) $\mathrm{IQdeg}(\varphi) \leq \langle i, q \rangle$, (ii) $\varphi \notin \mathrm{tp}_{i,q}(M_0, s_0, \overline{a}_0)$ (since by construction φ is falsified at $t_1 \subseteq s_1$) and (iii) $\varphi \in \mathrm{tp}_{i,q}(M_1, s_1, \overline{a}_1)$ (since by construction φ holds at every state $t_0 \subseteq s_0$). Thus we have $(M_0, t_0, \overline{a}_0) \not\equiv_{i-1,q} (M_1, t_1, \overline{a}_1)$, as we wanted.

Case $\Leftarrow^2$: the first move is a $\forall$-move. Suppose S starts by choosing $b_1 \in D_1$. As this is a winning strategy for S, for every choice $b_0 \in D_0$ of D we have

$$(M_0, s_0, \overline{a}_0 b_0) \not\equiv_{i,q-1} (M_1, s_1, \overline{a}_1 b_1)$$

By induction hypothesis, the above translates to

$$\exists \psi_{b_0} \in \mathrm{tp}(b_0) \setminus \mathrm{tp}(b_1)$$

where $\mathrm{tp}(b_0) := \mathrm{tp}_{i,q-1}(M_0, s_0, \overline{a}_0 b_0)$ and $\mathrm{tp}(b_1) := \mathrm{tp}_{i,q-1}(M_1, s_1, \overline{a}_1 b_1)$.

Now the formula

$$\varphi := \forall x \bigvee_{b_0 \in D_0} \psi_{b_0}$$

has IQ-degree at most $\langle i, q \rangle$, and by construction we have $\varphi \in \mathrm{tp}_{i,q}(M_0, s_0, \overline{a}_0)$ and $\varphi \notin \mathrm{tp}_{i,q}(M_1, s_1, \overline{a}_1)$. Thus, we have $(M_0, t_0, \overline{a}_0) \not\equiv_{i-1,q} (M_1, t_1, \overline{a}_1)$.

Case $\Leftarrow^3$: the first move is a $\exists$-move. Reasoning as in the previous case, we find that there exists a $b_0 \in D_0$—the element chosen by S—s.t. for every $b_1 \in D_1$

$$\exists \theta_{b_1} \in \mathrm{tp}(t_0) \setminus \mathrm{tp}(t_1)$$

In particular, it follows that the formula

$$\varphi := \exists x \bigwedge_{b_1 \in D_1} \psi_{b_1}$$

is a formula of complexity at most $\langle i, q \rangle$ such that $\varphi \in \mathrm{tp}_{i,q}(M_0, s_0, \overline{a}_0)$ and $\varphi \notin \mathrm{tp}_{i,q}(M_1, s_1, \overline{a_1})$. Again, it follows that $(M_0, t_0, \overline{a}_0) \not\equiv_{i-1,q} (M_1, t_1, \overline{a}_1)$. $\quad\square$

As a corollary, we also get a game-theoretic characterization of the distinguishing power of formulas in the $\langle i, q\rangle$-fragment of InqBQ.

Corollary 1. *For a finite signature Σ, we have:*

$$(M_0, s_0, \overline{a}_0) \approx_{i,q} (M_1, s_1, \overline{a}_1) \quad \Longleftrightarrow \quad (M_0, s_0, \overline{a}_0) \equiv_{i,q} (M_1, s_1, \overline{a}_1)$$

3.4 Extending the Result to Function Symbols

The results we just obtained assume that the signature Σ is relational. However, it is not hard to extend them to the case in which Σ contains function symbols (including nullary function symbols, i.e., constant symbols). In InqBQ, function symbols are interpreted rigidly: if $f \in \Sigma$ is an n-ary function symbol, then the interpretation function I of a model M must assign to all worlds w in the model the same function $I_w(f) : D^n \to D$.[4]

As in the case of classical logic [12] §3.3, the presence of function requires some care in formulating the EF game. The fundamental reason is that allowing atomic formulas to contain arbitrary occurrences of function symbols allows us to generate with a finite number of choices in the game an infinite sub-structure of the model—which spoils the crucial locality feature of the game. Technically, a simple way to achieve this is to follow [12] §3.3 and work with formulas which are *unnested*.

Definition 4 (Unnested formula). *An* unnested atomic formula *is a formula of one of the following forms:*

$$x = y \qquad c = y \qquad f(\overline{x}) = \overline{y} \qquad R(\overline{x})$$

An unnested formula *is a formula that contains only unnested atoms.*

Examples of *nested formulas*—i.e., non-unnested formulas—are $f(x) = g(y)$, $R(f(x))$ and $f(c) = x$.

We can now make the following amendments to the definition above: (i) the winning conditions for the game are determined by looking at whether Eq. (1) is satisfied for all *unnested* atomic formulas, and (ii) the $\langle i, q\rangle$-types are re-defined as sets of *unnested* formulas of degree at most $\langle i, q\rangle$. Other than that, the statement of the result and the proof are the same as above.

Clearly, using identity we can turn an arbitrary formula into an equivalent unnested one (e.g., replacing $P(f(x))$ with $\forall y(y = f(x) \to Py)$) so the restriction to unnested formula is not a limitation to the generality of the game-theoretic characterization; rather, it can be seen as an indirect way of assigning formulas containing function symbols with the appropriate $\langle i, q\rangle$-degree—making explicit a quantification which is implicit in the presence of a function symbol.

---

[4] In the general case, non-rigid function symbols are also allowed; however, such symbols can be dispensed with as usual in favor of relation symbols constrained by suitable axioms. See §4.3.5 of [2] for the details.

4 An Application: Expressive Limitations of InqBQ

In this section we exploit our game-theoretic characterization of the expressive power of InqBQ to show that certain questions—i.e., certain global properties of information states—are not definable in InqBQ over certain classes of structures.

Let us start by making the relevant notions more precise. If $\mathcal{C}$ is a class of models for InqBQ, by a *property* over $\mathcal{C}$ we mean a sub-class $\mathcal{P} \subseteq \mathcal{C}$. We also write $\mathcal{P}(M)$ instead of $M \in \mathcal{P}$.[5] Given a property $\mathcal{P}$ over $\mathcal{C}$, we say that a formula φ of InqBQ *defines* $\mathcal{P}$ over $\mathcal{C}$ if $\mathcal{P} = \{M \in \mathcal{C} \mid M \models \varphi\}$.

We will focus on the following classes:

- $\mathcal{C}_{\text{tot}}$ the class of all inquisitive models;
- $\mathcal{C}_{\text{f}}$ the class of finite models;
- $\mathcal{C}_{\text{id}}$ the class of id-models;[6]
- $\mathcal{C}_{\text{id,f}} = \mathcal{C}_{\text{id}} \cap \mathcal{C}_{\text{f}}$.

The figure on the right shows these classes ordered by containment. Notice that if a property $\mathcal{P}$ is definable by a formula φ relative to a class $\mathcal{C}$, and $\mathcal{C}' \subseteq \mathcal{C}$, then over $\mathcal{C}'$ the formula φ defines the restriction of $\mathcal{P}$ to $\mathcal{C}'$. Thus, a property which is definable over a class of models is also definable over a sub-class. Contrapositively, if a property $\mathcal{P}$ is not definable over $\mathcal{C}$ and $\mathcal{C}' \supseteq \mathcal{C}$, then no property $\mathcal{P}'$ whose restriction to $\mathcal{C}$ coincides with $\mathcal{P}$ can be definable over $\mathcal{C}'$.

Before delving into our undefinability results, let us give an example of an interesting property that *is* definable in InqBQ. Consider the question "which individuals have property P"—where P is a unary predicate symbol. This question is supported in a state s if the information available in s determines which individuals have property P—i.e., if all worlds $w \in s$ agree on the extension P_w. Thus, the question corresponds to the following property of inquisitive models:

$$\mathcal{P}_0(M) \overset{def}{\iff} \forall w, w' \in W : P_w = P_{w'}$$

It is easy to check that, with respect to any class of inquisitive models, this property is defined by the formula $\forall x ? P x$.

Now let us turn to undefinable properties. The first result that we show is that the question "how many individuals have property P"—for P a unary predicate symbol—is not definable in InqBQ, even in restriction to the class $\mathcal{C}_{id,f}$ of finite id-models. To make this precise, let us denote by $\#P_w$ the cardinality of P_w. In the setting of id-models, the question "how many individuals have property P" is supported at a state s if the information available in s determines the cardinality of the set P_w, i.e., if $\forall w, w' \in s$ we have $\#P_w = \#P_{w'}$. The following result says that no sentence of InqBQ has these support conditions, even in restriction to finite id-models.

Theorem 2. *The following property is not definable over $\mathcal{C}_{id,f}$:*

$$\mathcal{P}_1(M) \stackrel{def}{\Longleftrightarrow} \forall w, w' \in W : \#P_w = \#P_{w'}$$

To make the argument clearer, we will consider only models in the signature $\{P\}$ for P a unary predicate symbol. The argument can be easily generalized to arbitrary signatures containing P.[7]

To show this result, a special class of models will be especially useful.

Definition 5 (switch models). *For $h, k \in \omega \cup \{\omega\}$, define the following model*[8]:

$$M_{h,k} = \langle \{w_0, w_1\}, [1, h+k], I \rangle$$

where $P_{w_0} = [1, h]$ and $P_{w_1} = [h+1, h+k]$.

We will start by showing an interesting property of these models. For two ordinals h, k and a natural number q, write $h =_q k$ in case $h = k$ or both h and k are $\geq q$.

Lemma 1. *Consider two switch models $M_{h,k}$ and $M_{h',k'}$ and $i, q \in \mathbb{N}$. If $h =_q h'$ and $k =_q k'$, then $M_{h,k} \equiv_{i,q} M_{h',k'}$.*

Proof. In this proof we will assume that elements a_i belong to $M_{h,k}$, and the elements b_j belong to $M_{h',k'}$.

By Theorem 1 it suffices to give winning strategies for D in the two games with indexes $\langle i, q \rangle$ played between models $M_{h,k}$ and $M_{h',k'}$. Since the situation is symmetrical in the two cases, one direction suffices. The trick consists in preserving the following invariants through the game.

1. The info states in the current position are the same.
2. Given $\bar{a}$ and $\bar{b}$ the l-sequences of elements picked, then $a_i \leq h \iff b_i \leq h'$ and $a_i = a_j \iff b_i = b_j$ for every $i, j \leq l$.

Maintaining these invariants until the end of the game ensures the victory for D. The invariants can be maintained by the following strategy:

---

[7] One way to generalize the result to an arbitrary signature Σ: extend the models in the proofs that follow by interpreting the symbols in $\Sigma \setminus \{P^{(1)}\}$ as the empty relation. The proofs then follow through.

[8] Where + denotes ordinal sum and $[\alpha, \beta]$ denotes the set of ordinals $\{\gamma \mid \alpha \leq \gamma \leq \beta\}$.

- If S plays an implication move and picks the state s in one of the models, then D picks s in the other model.
- If S plays a quantifier move and picks $a_l = a_i$ an element already chosen, then D picks $b_l = b_i$.
- If S plays a quantifier move and picks a_l a fresh element, then D picks b_l a fresh element such that $a_l \leq h \iff b_l \leq h'$. This is possible since $h =_q h'$ and $k =_q k'$ by hypothesis. □

We are now ready to prove Theorem 2.

Proof (Theorem 2). Suppose for a contradiction that there exists a φ which defines $\mathcal{P}_1$. Let $\langle i, q \rangle$ be its combined degree. Then by Lemma 1 the models $M_{q,q}$ and $M_{q,q+1}$ are $\langle i, q \rangle$-equivalent. In particular, $M_{q,q} \models \varphi \iff M_{q,q+1} \models \varphi$. But this contradicts the assumption that φ defines the property $\mathcal{P}_1$, since $\mathcal{P}_1$ holds for $M_{q,q}$ but not for $M_{q,q+1}$. □

With proofs completely analogous to the one just given, we can show that the question "whether the number of elements that satisfy P is even" is not definable in InqBQ relative to the class of finite id-models, and, more generally, that for any natural number k the question "what is the number of elements that satisfy P, modulo k" is not definable. More formally, we have the following result.

Theorem 3. *The following property is not definable over* $\mathcal{C}_{id,f}$:

$$\mathcal{P}_2(M) \overset{def}{\iff} \forall w, w' \in W : \#P_w \text{ and } \#P_{w'} \text{ are congruent modulo } k$$

Moving away from the restriction to finite domains, we can consider the question *whether the extension of P is finite*. This question is resolved in a state s in case from the information available in s it follows that the extension of P is finite (i.e., $\#P_w$ is finite for all $w \in s$) or it follows that the domain is infinite (i.e., $\#P_w$ is infinite for all $w \in s$). In other words, the question is resolved if all worlds in s agree on whether $\#P_w$ is finite. We can use switch models to show that this question is not definable in InqBQ, even in restriction to the class $\mathcal{C}_{id}$.

Theorem 4. *The following property is not definable over* $\mathcal{C}_{id}$:

$$\mathcal{P}_3(M) \overset{def}{\iff} \forall w, w' \in W : (\#P_w \text{ is finite} \iff \#P_{w'} \text{ is finite})$$

Proof. Suppose for a contradiction that some formula φ defines $\mathcal{P}_3$. Let $\langle i, q \rangle$ be its combined degree. By Lemma 1 the models $M_{q,q}$ and $M_{q,\omega}$ are $\langle i, q \rangle$-equivalent, and so in particular $M_{q,q} \models \varphi \iff M_{q,\omega} \models \varphi$. But this contradicts the assumption that φ defines $\mathcal{P}_3$, since $\mathcal{P}_3$ holds of $M_{q,q}$ but not of $M_{q,\omega}$. □

Notice that, as we remarked above, these undefinability results extend to super-classes of the class for which they were stated: for instance, Theorem 2 implies that InqBQ cannot define any property of states s in arbitrary inquisitive model whose restriction to the class coincides with the property "the cardinality of P_w is constant in s".

Next, let us look at models where the interpretation of identity is not fixed, but varies across worlds. As we discussed in Sect. 2, the actual individuals existing at a world w can be equated with the equivalence classes modulo $\sim_w$, i.e., with the elements of the quotient $D/\sim_w$. Let us denote this quotient by D_w and refer to it as the *essential domain* at w. The number of actual individuals existing at w is the cardinality of this set, $\#D_w$. Now consider the question "how many individuals there are". This question is resolved at a state s if in s there is no uncertainty about the cardinality of the essential domain, i.e., if all the worlds $w \in s$ agree on the number $\#D_w$. The following theorem says that this question is not definable in InqBQ, even in restriction to the class $\mathcal{C}_f$ of finite models.

Theorem 5. *The following property is not definable over $\mathcal{C}_f$:*

$$\mathcal{P}_4(M) \models \varphi \overset{def}{\Longleftrightarrow} \forall w, w' \in W : \#D_w = \#D_{w'}$$

As before we will only consider models in the empty signature—thus we have to deal only with atoms of the form $(x = y)$—but the theorem can be easily generalized to arbitrary signatures.

To prove this result we will introduce another useful class of models.

Definition 6 (Grid models). *For $h, k \in \omega \cup \{\omega\}$, define the model*

$$G_{h,k} = \langle \{w_0, w_1\}, [1, h] \times [1, k], I \rangle$$

where, denoting by $\sim_w$ the equivalence relation $I_w(=)$, we have:

$$\langle a, b \rangle \sim_{w_0} \langle a', b' \rangle \iff a = a'$$
$$\langle a, b \rangle \sim_{w_1} \langle a', b' \rangle \iff b = b'$$

Given two sequences of elements $\overline{a} = \langle a_1, \dots, a_l \rangle$ and $\overline{b} = \langle b_1, \dots, b_l \rangle$, define $\mathbf{zip}(\overline{a}, \overline{b}) = \langle \langle a_1, b_1 \rangle, \dots, \langle a_l, b_l \rangle \rangle$.

Notice that, in a grid model $G_{h,k}$, the essential domain D_{w_0} has cardinality h, while the essential domain D_{w_1} has cardinality k. Thus, the cardinality of the essential domain is constant in $G_{h,k}$ only if $k = h$.

Intuitively, we can play an Ehrenfeucht-Fraïssé game between two grid models on the two components of the grid independently, by considering the projections of the elements chosen. This intuition is formalized in the following Definition and Propositions.

Definition 7. *Given $\overline{a}$ a sequence of natural numbers, we define*

- *its* identity trace: $E_{\overline{a}} = \{\langle i, j \rangle \mid a_i = a_j\}$
- *its* identity l-trace: $E_{\overline{a}}|_l = E_{\overline{a}} \cap ([1, l] \times [1, l])$

The following proposition follows easily from Definition 3.

Proposition 3. *Consider a grid model $G_{h,k}$, a state s and a sequence of elements $\mathbf{zip}(\bar{a},\bar{b})$. Then $\mathrm{tp}_{0,0}(G_{h,k}, s, \mathbf{zip}(\bar{a},\bar{b}))$ is univocally determined by the identity traces $E_{\bar{a}} = \{\langle i,j \rangle \,|\, a_i = a_j\}$ and $E_{\bar{b}} = \{\langle i,j \rangle \,|\, b_i = b_j\}$.*

Proposition 4. *Fix $l, h, k \in \mathbb{N}$ such that $l < \min(h,k)$. Consider a grid model $G_{h,k}$ and two sequences of elements $\mathbf{zip}(\bar{a},\bar{b})$ and $\mathbf{zip}(\bar{c},\bar{d})$ of length l and $l+1$ respectively. Moreover suppose that*

$$E_{\bar{a}} = E_{\bar{c}}|_l \qquad\qquad E_{\bar{b}} = E_{\bar{d}}|_l$$

Then there exists an element $\langle a_{l+1}, b_{l+1} \rangle$ such that

$$E_{\bar{a}a_{l+1}} = E_{\bar{c}} \qquad\qquad E_{\bar{b}b_{l+1}} = E_{\bar{d}}$$

Proof. First we choose a_{l+1} conditionally on $\bar{c}$: if $c_{l+1} = c_i$ for $i < l$, then pick $a_{l+1} = a_i$; otherwise pick an element in $[1, h] \setminus \{a_1, \dots, a_l\}$—notice this is always possible as $l < h$. b_{l+1} is chosen in a similar way conditionally on $\bar{d}$. The identities $E_{\bar{a}a_{l+1}} = E_{\bar{c}}$ and $E_{\bar{b}b_{l+1}} = E_{\bar{d}}$ hold by construction. □

The following lemma is an analogue of Lemma 1 for switch models.

Lemma 2. *Consider $G_{h,k}$ and $G_{h',k'}$ two grid models and $i, q \in \mathbb{N}$. If $h =_q h'$ and $k =_q k'$, then $G_{h,k} \equiv_{i,q} G_{h',k'}$.*

Proof. We will define a winning strategy for D applicable to both the games $\mathrm{EF}_{i,q}(G_{h,k}, G_{h',k'})$ and $\mathrm{EF}_{i,q}(G_{h',k'}, G_{h,k})$. The trick of the strategy consists in preserving the following invariants throughout the game.

1. The info states in the current position are the same.
2. Identity traces coincide in each coordinate: i.e., given $\mathbf{zip}(\bar{a},\bar{b})$ and $\mathbf{zip}(\bar{a}',\bar{b}')$ the sequences in the current position, we have $E_{\bar{a}} = E_{\bar{a}'}$ and $E_{\bar{b}} = E_{\bar{b}'}$.

By Proposition 3, preserving the invariants ensures the victory of player D. The winning strategy for D is the following:

- If S plays a $\twoheadrightarrow$-move and chooses s: D chooses s in the other model. This ensures the invariant is preserved.
- If S plays a $\forall$ or $\exists$ move and chooses $\langle a_k, b_k \rangle$ from one model: Suppose the current sequences of elements are $\mathbf{zip}(\bar{a},\bar{b})$ and $\mathbf{zip}(\bar{a}',\bar{b}')$. Condition 2 ensures that $E_{\bar{a}} = E_{\bar{a}'}$ and $E_{\bar{b}} = E_{\bar{b}'}$. By Proposition 4, D can choose an element $\langle a'_k, b'_k \rangle$ such that $E_{\bar{a}a_k} = E_{\bar{a}'a'_k}$ and $E_{\bar{b}b_k} = E_{\bar{b}'b'_k}$. This ensures the invariant is preserved. □

We are now ready to prove Theorem 5.

Proof. (Theorem 5) Suppose for a contradiction that some formula φ of InqBQ defines $\mathcal{P}_4$. Let $\langle i, q \rangle$ be the combined degree of φ. By Lemma 2 the models $G_{q,q}$ and $G_{q,q+1}$ are $\langle i, q \rangle$-equivalent, so $G_{q,q} \models \varphi \iff G_{q,q+1} \models \varphi$. But this contradicts the assumption that φ defines the property $\mathcal{P}_4$, since $\mathcal{P}_4$ holds of $G_{q,q}$ but not of $G_{q,q+1}$. □

In a similar way we can prove that, relative to finite models, InqBQ cannot express the question "how many individuals there are, modulo k" (and, as a special case, "whether the number of individuals is even"). This is formalized by the following theorem.

Theorem 6. *The following property is not definable over $\mathcal{C}_f$:*

$$\mathcal{P}_5(M) \models \varphi \stackrel{def}{\iff} \forall w, w' \in W : \#D_w \text{ and } \#D_{w'} \text{ are congruent modulo } k$$

Proof. As noted in the proof of Theorem 5, $G_{q,q} \equiv_{i,q} G_{q,q+1}$ for every $q \in \mathbb{N}$, but the models disagree on the property $\mathcal{P}_5$. Reasoning as in the proof of Theorem 5 we conclude that $\mathcal{P}_5$ is not definable.

We can also use grid models to show that, this time relative to the class $\mathcal{C}_{tot}$ of all models, InqBQ cannot express the question "whether the domain is finite", which is supported in a state s in case all worlds in s agree on whether the domain is finite. The proof is analogous to the one of Theorem 4, using grid models instead of switch models.

Theorem 7. *The following property is not definable over $\mathcal{C}_{tot}$:*

$$\mathcal{P}_6(M) \stackrel{def}{\iff} \forall w, w' \in W : (\#D_w \text{ is finite } \iff \#D_{w'} \text{ is finite})$$

Proof. By Lemma 2, we have that $G_{q,q} \equiv_{i,q} G_{q,\omega}$ for every $i, q \in \mathbb{N}$, but only one of these models has property $\mathcal{P}_6$. Again, this implies that $\mathcal{P}_6$ is not definable.

Finally, notice that, in the general case where the interpretation of identity varies across worlds, the number of actual individuals that satisfy P at a world is not given by $\#P_w$, but rather by the number of equivalence classes $\#(P_w/\sim_w)$. So, in the general setting of $\mathcal{C}_{tot}$ (or $\mathcal{C}_f$), the question "how many individuals have property P" corresponds to the following property:

$$\mathcal{P}_7(M) \stackrel{def}{\iff} \forall w, w' \in W : \#(P_w/\sim_w) = \#(P_{w'}/\sim_{w'})$$

However, notice that the restriction of this property to the class $\mathcal{C}_{id,f}$ of finite id-models is just the property $\mathcal{P}_1$. Since $\mathcal{P}_1$ is not definable over $\mathcal{C}_{id,f}$, it follows that $\mathcal{P}_7$ is not definable over $\mathcal{C}_{tot}$ (or $\mathcal{C}_f$). Analogous conclusions can be drawn for the generalized versions of properties $\mathcal{P}_2$ and $\mathcal{P}_3$ which take into account the role of variable identity: in restriction to id-models, the generalized version boils down to the simple version considered above; since the simple version is not definable, the generalized version is not definable either.

The following table summarizes our results in this section. A symbol '×' means that the relevant property is not definable over the relevant class; a symbol '⊤' means that the relevant property is trivial over the relevant class, i.e., satisfied by all models (thus obviously definable by a tautology). Finally, '✓' means that the property is non-trivial and definable over the relevant class.

	$\mathcal{C}_{\text{id},f}$	$\mathcal{C}_f$	$\mathcal{C}_{\text{id}}$	$\mathcal{C}_{\text{tot}}$
$\mathcal{P}_0$	✓	✓	✓	✓
$\mathcal{P}_1$	✗	✗	✗	✗
$\mathcal{P}_2$	✗	✗	✗	✗
$\mathcal{P}_3$	⊤	✗	⊤	✗
$\mathcal{P}_4$	⊤	⊤	✗	✗
$\mathcal{P}_5$	⊤	⊤	⊤	⊤
$\mathcal{P}_6$	⊤	⊤	⊤	✗
$\mathcal{P}_7$	✗	✗	✗	✗

5   Conclusions and Further Work

EF games often provide an insightful perspective on a logic, and a useful characterization of its expressive power. In this paper we have described an EF game for
inquisitive first-order logic, InqBQ. This game presents two novelties with respect
to its classical counterpart. First, the roles of the two models on which the game
is played are not symmetric: certain moves have to be performed mandatorily
in one of the models. This feature reflects the fact that InqBQ lacks a classical negation, and that the theory of a model—unlike in the classical case—can
be properly included in that of another. Secondly, the objects that are picked
in the course of the game are not just individuals $d \in D$, but also information
states, i.e., subsets $s \subseteq W$ of the universe of possible worlds. This feature reflects
the fact that InqBQ contains not only the quantifiers $\forall, \overline{\exists}$ over individuals, but
also the implication $\rightarrow$, which allows for a restricted kind of quantification over
information states. We proved that certain fragments of the language of InqBQ
are preserved from a model M_0 to a model M_1 if and only if a winning strategy
exists for Duplicator in a corresponding finite game played on these models. We
used this result to establish a number of undefinability results. In particular,
we have seen that the question *how many individuals satisfy P* (or any variant
of this question concerning cardinality modulo $k > 1$) is not expressible, even
in restriction to finite models in which '=' is interpreted rigidly as the actual
identity on D. Moreover, the question *whether the extension of P is finite* is
not expressible, even given this restriction on the interpretation of identity. In
the general context where the interpretation of identity is variable across worlds,
analogous results hold concerning the total number of individuals in the domain.

The work presented in this paper can be taken further in several directions.
First, in the context of classical logic, several variants of the EF game have
been studied. For example, [19] presents a *dynamic* EF game, corresponding to
a more fine-grained classification of classical structures. In the inquisitive case,
an analogous refinement could lead to interesting insights into the structure of
inquisitive models.

Second, EF games can be used to study which properties invariant under
automorphism are expressible in InqBQ. This includes, in particular, properties
dependent only on the cardinality of the domain: $\mathcal{P}_4$, $\mathcal{P}_5$ and $\mathcal{P}_6$ are *not* expressible in InqBQ, while, e.g., "there are n individuals" for $n \in \mathbb{N}$ is expressible.

In future work, we aim to use the game-theoretic perspective developed in this paper to study in more generality which automorphism-invariant properties—in particular, which cardinality properties—are expressible in InqBQ.

Finally, EF games can be used also to compare different extensions of a fixed logic, as shown in [14]. In this regard, the results presented in Sect. 4 already yield some interesting corollaries. For example, adding to InqBQ a generalized quantifier corresponding to any of the properties $\mathcal{P}_1, \ldots, \mathcal{P}_7$ discussed in Sect. 4 yields a logic which is strictly more expressive than InqBQ. More generally, the techniques introduced in this paper are likely to provide a useful tool for a systematic study of generalized quantifiers in inquisitive logic.

References

1. Ciardelli, I.: Inquisitive semantics and intermediate logics. M.Sc. thesis, University of Amsterdam (2009)
2. Ciardelli, I.: Questions in logic. Ph.D. thesis, Institute for Logic, Language and Computation, University of Amsterdam (2016)
3. Ciardelli, I.: Questions as information types. Synthese **195**, 321–365 (2018)
4. Ciardelli, I., Groenendijk, J., Roelofsen, F.: Inquisitive Semantics. Oxford University Press, Oxford (2018)
5. Ciardelli, I., Roelofsen, F.: Inquisitive logic. J. Philos. Log. **40**(1), 55–94 (2011)
6. Ehrenfeucht, A.: An application of games to the completeness problem for formalized theories. Fund. Math. **49**(2), 129–141 (1961). http://eudml.org/doc/213582
7. Fagin, R.: Monadic generalized spectra. Zeitschrift für mathematische Logik und Grundlagen der Mathematik **21**, 89–96 (1975)
8. Fraïssé, R.: Sur quelques classifications des systèmes de relations. Publications Scientifiques de l'Université D'Alger **1**(1), 35–182 (1954)
9. Frittella, S., Greco, G., Palmigiano, A., Yang, F.: A multi-type calculus for inquisitive logic. In: Väänänen, J., Hirvonen, Å., de Queiroz, R. (eds.) WoLLIC 2016. LNCS, vol. 9803, pp. 215–233. Springer, Heidelberg (2016). https://doi.org/10.1007/978-3-662-52921-8_14
10. Grilletti, G.: Disjunction and existence properties in inquisitive first-order logic. Stud. Logica (2018). https://doi.org/10.1007/s11225-018-9835-3. ISSN 1572-8730
11. Hintikka, J.: Knowledge and Belief: An Introduction to the Logic of the Two Notions. Cornell University Press (1962)
12. Hodges, W.: A Shorter Model Theory. Cambridge University Press, New York (1997)
13. Immerman, N.: Upper and lower bounds for first order expressibility. J. Comput. Syst. Sci. **25**(1), 76–98 (1982)
14. Kolaitis, P., Väänänen, J.: Generalized quantifiers and pebble games on finite structures. Ann. Pure Appl. Log. **74**(1), 23–75 (1995)
15. Połacik, T.: Back and forth between first-order kripke models. Log. J. IGPL **16**(4), 335–355 (2008)
16. Punčochář, V.: Weak negation in inquisitive semantics. J. Log. Lang. Inf. **24**(3), 323–355 (2015)
17. Punčochář, V.: A generalization of inquisitive semantics. J. Philos. Log. **45**(4), 399–428 (2016)
18. Roelofsen, F.: Algebraic foundations for the semantic treatment of inquisitive content. Synthese **190**(1), 79–102 (2013)

19. Väänänen, J.: Models and Games, 1st edn. Cambridge University Press, New York (2011)
20. van Benthem, J.: Modal correspondence theory dissertation, pp. 1–148. Universiteit van Amsterdam, Instituut voor Logica en Grondslagenonderzoek van Exacte Wetenschappen (1976)
21. Visser, A.: Submodels of Kripke models. Arch. Math. Log. **40**(4), 277–295 (2001)

Computational Model of the Modern Georgian Language and Search Patterns for an Online Dictionary of Idioms

Irina Lobzhanidze[⊠] [iD]

Ilia State University, Kakutsa Cholokashvili Ave 3/5, 0162 Tbilisi, Georgia
irina_lobzhanidze@iliauni.edu.ge

Abstract. This paper describes the results of projects implemented under the financial support of the Shota Rustaveli National Science Foundation (Nos. DP2016_23, LE/17/1-30/13, AR/320/4-105/11, Y-04-10), particularly the use of finite state technology, specifically *lexc* and *xfst*, for the morphological analysis of the Modern Georgian language and the application of a morphological transducer to solve problems of lemmatization and alphabetization noticed in Georgian dictionaries. For instance information on lemmas and the morphological structures of words determined by means of the transducer was used to solve the above-mentioned lexicographic problems in the Online Dictionary of Idioms available at http://idioms.iliauni.edu.ge/.

The Online Dictionary of Idioms is monolingual (Georgian) and bidirectional bilingual (Modern Georgian – Modern Greek and vice versa), and is equipped with a friendly environment to find idioms, their meaning and context of use. Its compilation followed the main principles as stated by Gibbon & Van Eynde (2000) regarding the linguistic specification, database management system, phases of lexicographic database construction and access to lexical information.

Keywords: Modern Georgian language · Lexicography · Finite state transducers (fst)

1 Introduction

Any kind of electronic dictionary can be considered a database; generally, its purpose is to provide adequate explanation or translation of separate words or multi-word expressions (MWEs), to store information and to allow the user to find appropriate language units. Following Atkins and Rundell [2], Gibbon and Van Eynde [10] and others, there are four major prerequisites to the design of any lexicographic database, i.e. dictionary:

- Linguistic specification (of the macrostructure and the microstructure);
- Database management system (DBMS) specification;
- Specification of the phases of lexicographic database construction: input, verification and modification;
- Presentation of and access to lexical information: access, re-formatting, dissemination.

© Springer-Verlag GmbH Germany, part of Springer Nature 2019
A. Silva et al. (Eds.): TbiLLC 2018, LNCS 11456, pp. 187–208, 2019.
https://doi.org/10.1007/978-3-662-59565-7_10

In the case of Modern Georgian, the main problems are associated first with linguistic specification, which corresponds to the types of lexical information involved in the linguistic analysis of Modern Georgian, and in addition with access by end-users to lexical information stored in the database (DB). The Modern Georgian language belongs to the morphologically rich languages (MRLs). The term morphologically rich language refers to languages in which substantial grammatical information, i.e. information concerning syntactic units and relations, is expressed at the word level and a large number of word forms can be obtained from a single root. Descriptions of Georgian morphological structure emphasize the large number of inflectional categories, the large number of elements that verb or noun paradigms can contain, the interdependence in the occurrence of various elements and the large number of regular, semi-regular and irregular patterns. This means that the morphologically rich nature of Georgian expresses different levels of information at the word level and affects the compilation of dictionaries, i.e. lexicographic databases, of the Georgian language. As a result of rich morphology a word can have many forms, making it hard to find them in the dictionary if you don't know the language's morphological structure. Thus, the main problems are associated with:

1. The representation of verbal forms in dictionary entries due to the absence of an infinitive. By comparison there is no infinitive in the Modern Greek language as well, but while in Modern Greek dictionaries verbal forms are represented in the first person singular of the present tense, active voice, in Modern Georgian dictionaries the headwords of verbal entries can be represented in the third person singular of the present or future tenses or in the form of a verbal noun etc. Different kinds of approaches to the headword complicate access to the verbal entries as a whole;

2. The polypersonalism of the Georgian verb allows the subject and object markers to appear in the form of different affixes reflecting subject, direct and indirect objects, and genitive, locative and causative meanings. Specifically, a verb can agree with between one and three arguments, which leads to the inclusion of verbs with different morphological structures in the majority of Georgian printed or electronic dictionaries, without any attention being paid to similarities between their meanings;

3. In electronic and online dictionaries it is completely impossible to arrange verbal headwords in alphabetical order. The problem is caused by the position of the verbal root in the verbal template. The verbal root occupies the fourth slot in the nine-slot template, which makes it impossible to arrange verbs in alphabetical order without paying attention to preverbs, prefixal person markers and version markers and causes difficulties in determining the lemma form.

While the second problem of printed Georgian lexicography generally depends on the size of the printed dictionaries and the author's approach, the electronic dictionaries are not limited to the lemma of inflected words and the size restrictions are of little importance, which means that there is the possibility of representing the different inflected forms of a Georgian verb from the viewpoint of their polypersonalism. But the nature of the first and the third problems is different. Firstly, native and, especially, non-native speakers of Modern Georgian are interested in access to the lexical definition of concrete words. Secondly, it is difficult to find words or MWEs in Modern Georgian dictionaries without knowing the grammatical structure of a concrete verbal form as found in the raw text; for example, if you do not know that (a) *mkonia* 'I had' is an inflected form of *akvs* 'he/she has' or *kona* 'possession' or (b) *viçer* 'I write smth. for myself' is an inflected form of *çers* 'he/she writes' or *çera* 'writing', you simply do not know how to find their meaning in the dictionary. Thus, the inflected forms shown in the dictionaries should meet the requirements of end-users and the morphological-analyzer-based solution is the best choice, which affects not only the end-user, but the compilation of an online dictionary as a whole.

This paper describes a system that overcomes the difficulties stated above by automating the process of inflection in the Online Dictionary of Idioms [24][1] and the morphological analyzer of Modern Georgian[2] used for the above-mentioned purpose.

The paper is divided into the following parts: 1. Introduction; 2. Problems of Georgian Lexicography and a Brief Description of Dictionaries; 3. Morphological Analyzer and Generator of the Modern Georgian Language; 4. Findings; 5. Conclusions.

2 Problems of Georgian Lexicography and a Brief Description of Dictionaries

2.1 Problems of Georgian Lexicography

The compilation of linguistic specification is always associated with the macrostructure and the microstructure of a dictionary. Deciding on the types of entry the dictionary will include and organizing the headword list are macrostructure decisions, but planning the entries in the dictionary and deciding on their structure and components are microstructure decisions [2].

A single lexeme or a MWE can have different senses, but only one lemma of a concrete lexeme or MWE can be used as a head word for all of its forms and, appropriately, stored in the dictionary DB in an alphabetical order. In the case of European languages, the problem of alphabetization relates to print dictionaries, but not to electronic ones, and, generally, arises in the case of MWE headwords. In the case of the Georgian language, the headword list as well as the structure of entries in the dictionary are always under the impact of complex morphological structure and affect the compilation of the dictionary as a whole.

Thus, the issues under consideration are subdivided into four parts: nominal and verbal inflections of Modern Georgian, which cause lemmatization and alphabetization problems, and, correspondingly, the access by end-users to the headwords of dictionary entries.

---

[1] Projects No. Y-04-10, Georgian-Modern Greek Dictionary of Idioms, and No. LE/17/1-30/13, Online Dictionary of Idioms, financed by the Shota Rustaveli National Science Foundation (SRNSF).

[2] Projects No. AR/320/4-105/11, Corpus Annotation and Analysis Software for the Modern Georgian Language, and No. DP2016_23, Digital Humanities: Digital Epigraphy, Computational Linguistics, Digital Prosopography, financed by the Shota Rustaveli National Science Foundation (SRNSF).

2.2   Nominal Inflection

In Modern Georgian, the structures of noun, adjective, numeral and pronoun have something in common, but the quantity of slots and formation models are different. Generally, the formation of nominal inflection is carried out by suffixation; only the degrees of an adjective are formed by means of circumfixation.

Types of stems are a base for different types of inflections, therefore, nouns are subdivided into common and proper nouns with different types of inflections. Pronouns are subdivided into personal, demonstrative, possessive, indefinite, interrogative, relative, reflexive, reciprocal and negative ones; numerals into cardinal, ordinal and fractal ones; and adjectives into gradable and non-gradable ones, i.e. those which can produce degrees of comparison and those which cannot.

The main morphological categories which affect the formation of nominal inflection are as follows: case (nominative, ergative, dative, genitive, instrumental, adverbial and vocative), number (singular and plural), postpositions and clitics. Additional categories are degree for adjectives and person for pronouns. Animacy corresponds to nouns and pronouns, but can be considered only at the level of the sentence as a part of subject-verb Agreement, where an animate noun or pronoun agrees with the verb in both the singular and plural, while an inanimate one agrees in the singular, but in the plural requires the singular form of a verb [18]. The scheme of formation for nouns, numerals and pronouns is as follows:

Type → Number → Declension → Postpositions → Clitics

While for adjectives it is:

Type → Degree → Number → Declension → Postpositions → Clitics

The quantity of slots depends on the part of speech (PoS) under consideration. Thus, the maximum possible quantity of slots in noun, pronoun and numeral templates is equal to 8 and consists of the following units: (1) root, (2) number, (3) case, (4) emphatic vowel, (5) postposition, (6) emphatic vowel, (7) particle and (8) auxiliary verb; e.g. (1).

(1)   *mercxl-Ø-is-a-tvis-a-ca-a*
        swallow-SG-GEN-EMP-POST-EMP-PTCL-AUX
        'is also for swallow'

The maximum possible quantity of slots in the adjective template is equal to 10 slots and consists of the following units: (1) degree marker, (2) root, (3) degree marker, (4) number, (5) case, (6) emphatic vowel, (7) postposition, (8) emphatic vowel, (9) particle and (10) auxiliary verb; see (2).

(2)   *mo-tetr-o-Ø-s-a-vit-a-ca-a*
        COMP-white-CIRC-SG-GEN-EMP-POST-EMP-PTCL-AUX
        'is also whitish'

Thus, from the lexicographic point of view, the initial points for the above-mentioned PoSs, especially nouns, pronouns and numerals, are the initial letters of a root, and the keyword form of a dictionary entry can be easily represented by their lemmatized form, especially the nominative case in the singular. This means that access to these words in a dictionary can easily be gained by typing letters one by one or by simply putting them in an alphabetical order. From this point of view the adjectives in comparative or superlative degrees cannot be easily accessed; however, keeping in mind that in dictionaries they generally are represented by a lemmatized form, especially by a positive degree in the nominative case, singular, the possibility of finding them by means of a morphological analyzer can be considered an additional option for their search in the online dictionary.

2.3   Verbal Inflection

In Modern Georgian, the main morphological categories which affect the formation of verbal inflection are as follows: the TAM (tense-aspect-mood) series, which specify case-marking and linking between participants like agent and patient and grammatical relations like subject and object by means of preverbs, version vowels and thematic suffixes, diathesis/voice subdivided into active, auto active, inactive, passive and auto passive, personality, which covers unipersonal, bipersonal and tripersonal verbs, and number. Following Melikishvili [21], Hewitt [12], Shanidze [30] and others, version, subdivided into subjective, objective and locative, is traditionally described as one of the inflectional categories of the Georgian verb, but keeping in mind that pre-radical vowels, called version markers, are generally used to show grammatical relations between subject and objects and change the lexical meaning of the verb, we have considered them from the viewpoint of the above-mentioned relation.

A Georgian verb contains many morphemes, which from one point of view are typical for agglutinating structures, but from another point of view are characteristic of inflected ones. The Georgian verbal paradigm can be considered a mixed one. The maximum possible quantity of slots in the verbal template varies from 9 to 12 as described by Boeder [4], Hewitt [12] and others and consists of the following units: (1) preverbs, (2) prefixal person markers, (3) version markers, (4) root, (5) passive markers, (6) thematic markers, (7) causative markers, (8) screeve markers and (9) suffixal person markers.

Most verbs have preverbs lexically associated with them, although there is a group of verbs that do not have preverbs. Preverbs can be classified as a closed class always associated with verbs, which are used for the production of future, aorist and perfect screeves and, generally, reflect perfective aspect (3); however, at the same time, there are some preverbs that do not follow this rule (4).

(3)   *da-xaṭ-av-s*
    PV-draw-TS-FUT.3SGSBJ
    'will draw'

(4)   *mo-di-s*
    PV-come-PRES.3SGSBJ
    'comes'

In addition, any kind of morpheme gives information about the grammatical function of the word. A Georgian verb, generally, uses bound morphemes to show its grammatical function. The main types of use are as follows:

1. affixation, e.g. *čmun-av-d-a* 'was worrying'
   worry-TS-PAS-IMPERF.3SGSBJ
2. root vowel alternation, e.g. *grex-s* 'twists smth.', *mo-grix-a* 'twisted smth.'
   a. bend-PRES.3SGSBJ
   b. PV-bend-AOR.3SGSBJ
3. root alternation, e.g. *e-tqv-i-s* 'will tell smb. smth.'
   prv-say-TS-FUT.3SGSBJ

Also, it should be mentioned that if we consider the possibility of stemming of verbal forms, the second person singular in the present indicative may be considered a unit closely associated with a stem. But the focus of our research is closely connected to the existence of word indexes as well as the possibility of their processing by means of finite state automata. This means that the lexicon data should contain forms which can provide generation for object- and subject-based paradigms and provide their further analysis. Keeping in mind that the majority of Georgian dictionaries follow a mixed type of verbal presentation, which covers not only verbal nouns, but also the third person singular forms, we have chosen the third person singular in the present or future indicative as represented by Chikobava [6] and Melikishvili [21].

From the lexicographic point of view, the problems of lemmatization as well as alphabetization are closely connected to the existence of preverbs, personal markers and version markers, which affect the initial point of a verbal entry and change the lexical meaning of the verb. Personal and version markers are generally used to indicate agent, patient and beneficiary and, respectively, subject, direct and indirect object relations, but they do not change the lexical meaning of a verb.

2.4   Lemmatization Problem

Lemmatization is the process of deriving the base form or lemma of a word from one of its inflected forms and, generally, refers to the use of vocabulary and morphological analysis of words. In spite of the fact that for the majority of words inflection and morphological analysis can be considered clear, there are many cases where the normative rules are not sufficient and the morphological analysis and inflection itself can be a rather difficult task. For languages with rich morphology like Modern Georgian, lemmatization may be considered a quite difficult task, keeping in mind that it requires a lexicon consisting of lemmata with a set of rules for creating inflected forms.

According to the Morpho-syntactic Annotation Framework (MAF) a lemma is a lemmatized form class of inflected forms differing only by inflectional morphology. In European languages, the lemma is usually the /singular/ if there is a variation in /number/, the /masculine/ form if there is a variation in /gender/ and the/infinitive/ for all verbs. In some languages, certain nouns are defective in the singular form, in which case the/plural/is chosen. In Arabic, for a verb, the lemma is usually considered to be the third person singular with the accomplished aspect [14].

In the case of Modern Georgian, the lemma for the nominal paradigm is represented with the nominative singular, but the Georgian verb does not have an infinitive. Thus there is no clear rule with regard to the representation of a lemma for the verbal paradigm. That is why the majority of Modern Georgian dictionaries include the following approaches to the base form of dictionary entries:

1. Verbal noun, the so-called masdar form;
2. Root-based form, the so-called abstract root;
3. The third person singular in the presentor future indicative.

All of these approaches can be described within the combination of slots and rules used for the formation of verbal nouns from transitive and intransitive verbs.

The first approach considers a verbal noun to be a base form for a verbal entry [35] and sometimes considers it to be an infinitive of the verbal paradigm, keeping in mind that the extraction of an abstract root from given forms is a simpler task than the other way around [7, 11] etc. This approach can be seen in the dictionaries of MWEs as well [28], where in headwords verbal constituents of idioms or other compounds are represented in the form of verbal nouns in spite of the fact that a verbal constituent is used in the citation form. Such a type of headword representation has a negative influence on the meaning of idiomatic expressions as a whole and the majority of idioms require the fixed grammatical structures of concrete verbs, but not verbal nouns. Following Hewitt [13], Shanidze [30] and others, a verbal noun, a so-called masdar, shares morphological structures and syntactic features of noun and verb. A verbal noun can be produced by adding the marker-*a* to the root of a verb without markers of version, voice, situation and subject or object in the present or future indicative. But the production rules are different; in particular, some forms retain slots for preverbs (PV), but lose slots for pre-radical vowels (prv) and the *a*-vowel in thematic suffixes (TS) -*am* (5) and -*av* (6), e.g.

(5) *a-b-am-s > b-m-a > da-b-m-a*
  a. *a-b-am-s*
    prv-tie_up-TS-PRES.3SGSBJ
    'ties up smth.'
  b. *b-m-a*
    tying_up-TS-NOM
    'tying up'
  c. *da-b-m-a*
    PV-tying_up-TS-NOM
    'tying/binding up'
(6) *xaṭ-av-s > xaṭ-v-a > mo-xaṭ-v-a*
  a. *xaṭ-av-s*
    draw-TS-PRES.3SGSBJ
    'draws'
  b. *xaṭ-v-a*
    drawing-TS-NOM
    'drawing'
  c. *mo-xaṭ-v-a*
    PV-drawing-TS-NOM
    'painting'

Other verbal nouns are derived by means of the suffixes-*n* (*txov-s* 'asks smth.'→ *txov-n-a* 'asking', *i-p' ov-i-s* 'will find smth.'→ *p'ov-n-a* 'finding' etc.), -*om* (*dg-a-s* 'stands' → *dg-om-a* 'standing', *ḳrt-i-s* 'trembles' → *ḳrt-om-a* 'trembling' etc.) and-*ol* (*kr-i-s* 'wind blows' → *kr-ol-a* 'wind blowing', *qar-s* 'stinks'→ *qr-ol-a* 'stinking' etc.). The medials and the indirect verbs, which regardless of series, require the logical subject in the dative case and the logical object in the nominative, donot provide clear rules for the above-mentioned formation, and there are some verbs which never produce verbal nouns at all. Thus, if a dictionary follows this approach, some verbal entries are not represented at all, while others are represented at least twice with and/or without preverbs, and it is rather awkward to use such a list of verbal nouns for the generation of verbs in Natural Language Processing (NLP) systems.

The second approach differs from the previous one by representing the headwords in the form of an abstract verbal root with appropriate paradigms [33]. For dictionary users it is rather difficult to find the appropriate meaning of words by trying to determine their possible verbal roots and to derive them from the existing structures without basic knowledge of Georgian grammar, especially the rules for the formation of verbal paradigms. Thus, if you do not know the word, you may not be able to determine which part of the word is the root. Also, this principle of dictionary entries cannot be shared for the dictionaries of MWEs because idioms or any other kind of phraseological units are groups of words with a fixed lexical composition and grammatical structure.

The third approach considers a verb in the third singular subject form in the present or future indicative [6, 26] etc., which includes forms indicating grammatical categories like version, causation etc. as well. This kind of approach is also shared by the dictionaries of idioms [23], where in headwords verbal constituents are represented in two ways: (a) as required by the fixed grammatical structure of the MWE, and (b) as a verb in the third singular subject form in the present or future indicative if a fixed grammatical structure of the MWE can be violated.

All of the above-mentioned approaches complicate the compilation of monolingual and bilingual dictionaries by making it generally awkward for users, especially learners of Modern Georgian, to find the meaning of appropriate words or MWEs without knowledge of Georgian grammar.

2.5   Access to Headwords in Alphabetical Order

The headword of a dictionary entry has different functions, but, generally, in the broad sense it is a form which the user tries to access first; in the narrow sense it is a representative of inflected forms of the same word, which could be generated by means of affixation. Thus, the separate question which also complicates the compilation of Georgian dictionaries is the alphabetization problem, i.e. access to headwords of dictionary entries in alphabetical order. Following Atkins and Rundell [2], the problem of alphabetization can be considered in relation to print dictionaries, but not to electronic ones, and it does not exist if all the headwords are single words, but does arise if the headword list contains MWEs. At the same time there are two options of alphabetization, namely, word-by-word, where spaces between words take precedence, or letter-by-letter, where spaces and hyphens are disregarded. Both of these options are easily adopted for the Georgian language, but at the same time the position of the verbal root in the verbal template, which occupies the fourth slot in the nine-slot template, makes it impossible to arrange verbs in alphabetical order without paying attention to preverbs, prefixal person markers and version markers.

In the case of Modern Georgian the alphabetization problem affects not only print, but electronic dictionaries as well, especially in the case of verbal idioms consisting of a verb plus one or more constituents. In order to allow the user to access the headword of a dictionary entry, the headwords have to be arranged, from one point of view, in accordance with the alphabet of Modern Georgian and, from another, taking into account the possible generation of inflected forms by means of affixation. Keeping in mind that there is a big difference in meaning between verbal forms formed by means of different affixes and the appropriate verbal noun, it is awkward to appropriately decide the alphabetical order of verbs and of MWEs with verbal constituents. For example, if we consider the third person singular in the present or future indicative of verbs: *u-vl-is* 'looks after, tends to smb.', *a-u-vl-is* 'will make the rounds of smth.', *cha-u-vl-is* 'will go down smth.', *she-u-vl-is* 'will drop in to see smth.' etc. as a main form, the quantity of dictionary entries with affixes attached to the same stem and reflecting different meanings will be more than enough. Otherwise, if we consider the appropriate verbal noun-*svla* 'going, walking' to be the headword of the dictionary entry, different meanings of verbal constituents like those mentioned above will not be represented at the appropriate level. We can put MWEs with verbal nouns like *gverd-is avla* 'avoiding, ignoring', *gverd-is chavla* 'ignoring', but there do not exist forms like * *gverd-is uvla*or * *gverd-is shevla*. Thus, the dictionary of idioms should contain MWEs with verbal constituents to preserve all possible meanings (7, 8, 9, 10):

(7) *gverd-s   a-u-vl-is*
    side-DAT   PV-prv-go-FUT.3SGSBJ
    'He/she will overlook/bypass smb./smth.'

(8) *gverd-s   u-vl-is*
    side-DAT   prv-go-PRES.3SGSBJ
    'He/she disregards, avoids smb./smth.'

(9) *gverd-s   cha-u-vl-i-s*
    side-DAT   PV-prv-go-FUT.3SGSBJ
    'He/she will ignore smb./smth.'

(10) *gverd-s   she-u-vl-i-s*
    side-DAT   PV-prv-go-FUT.3SGSBJ
    'He/she will drop in to see smb./smth.'

Also, it is well known that, generally, idioms treated as words with spaces pose problems to Natural Language Processing applications [8, 27] etc. In particular, there exist flexibility and lexical proliferation problems based on the different structures of MWEs. Different NLP schools have been studying the peculiarities of MWEs and different approaches like Alegria et al. [1], Oflazer et al. [22], Villavicencio et al. [37], Breidt et al. [5], Jacquemin [15], Karttunen et al. [3, 16] etc. have been proposed with regard to treating multi-word units as described and compared by Savary [29], with different results. In the case of Modern Georgian, the cascaded finite state approach by means of *lexc* and *xfst* tools is considered the best choice with regard to the problems already discussed.

As was mentioned above, the main problem of alphabetization as well as lemmatization in the case of Modern Georgian is associated with the position of theverbal root in the verbal template and, appropriately, the position of the verbal constituent in a MWE. There are two types of verbal idioms which should be described separately, specifically those which do not undergo grammatical transformations as given in example (11.a) for the subject and (11.b) for the object verbal paradigms, and those which show morphosyntactic flexibility with regard to the perfective/imperfective aspects (12.a) and tenses (12.b) and allow inflections etc.

(11) *tavze buzs ar isvams*, but not **tavze buzs isvams*; *tavi ara makvs*, but not
    *tavi makvs etc.
  a.   *tav-ze   buz-s   ar   i-sv-am-s*
       head-on   fly-DAT   NEG prv-seat-TS-PRES.3SGSBJ
       'He/she does not let a fly land on one's head; ~He/she is haughty'
  b.   *tav-i   ara   m-akv-s*
       head-NOM   NEG 1SGOBJ-have-PRES.3SGSBJ
       'I have no head; ~I am not able to do smth.'
(12) *tavs igdebs, tavi chaigdo, ǧurs ugdebs, ǧuri vugde* etc.
  a.   *tav-s   i-gd-eb-s*
       head-DAT   prv-throw-TS- PRES.3SGSBJ
       'He/she is throwing a head; ~ He/she is insolent'
       tav-i   ča-i-gd-o
       head-NOM   PV-prv-throw-TS-AOR.3SGSBJ
       'He/she was throwing a head; ~He/she was insolent'
  b.   *ǧur-s   u-gd-eb-s*
       ear-DAT   prv-throw-TS-PRES.3SGSBJ
       'He/she is throwing an ear; ~He/she listens to smb.'
       ǧur-i   v-u-gd-e
       ear-NOM   1SGSBJ-prv-throw-AOR
       'He/she was throwing an ear; ~He/she listened to smb.'

Also, there are idioms with non-fixed wording, e.g. *dushash-imo-s-d-is* vs*mo-s-d-is dushash-i* (13), non-fixed canonical forms, e.g. *dana-s ɣor-is ḳud-zega-ṭex-s*vs*dana-s ɣor-is ḳud-zegada-ṭex-s* (14), or with the possibility of substitution, e.g. *cecxl-s u-ḳid-eb-s* vs *al-cecxl-s u-ḳid-eb-s* (15):

(13) *dushash-i mo-s-d-is* vs *mo-s-d-is dushash-i*

  a.   *dushash-i*   *mo-s-d-is*

      double_sixes-NOM   PV-3SGOBJ-come-TS-PRES.3SGSBJ

  b.   *mo-s-d-i-s*   *dushash-i*

      PV-3SGOBJ-come-TS-3SGSBJ.PRES double_sixes-NOM

      'He/she gets double sixes; ~ He/she is lucky'

(14) *dana-s γor-is ḳud-ze ga-ṭex-s* vs *dana-s γor-is ḳud-ze gada-ṭex-s*

  a.   *dana-s*   *γor-is*   *ḳud-ze*   *ga-ṭex-s*

      knife-DAT   pig-GEN   tail-on   PV-break-FUT.3SGSBJ

  b.   *dana-s*   *γor-is*   *ḳud-ze*   *gada-ṭex-s*

      knife-DAT   pig-GEN   tail-on   PV-break-FUT.3SGSBJ

      'He/she will break knife on tail; ~He/she will leave the job half-done'

(15) *cecxl-s u-ḳid-eb-s* vs *al-cecxl-s u-ḳid-eb-s*

  a.   *cecxl-s*   *u-ḳid-eb-s*

      fire-DAT   prv-light-TS-PRES.3SGSBJ

  b.   *al-cecxl-s*   *u-ḳid-eb-s*

      flame-fire-DAT   prv-light-TS-PRES.3SGSBJ

      'He/she makes smb. burn with love or sells smth. for a fortune'

In the case of Modern Georgian, a word-by-word approach with space precedence does not work correctly for these more flexible examples of MWEs and affects the compilation of dictionaries of Modern Georgian as well, especially in the case of verbal idioms. So, to overcome the problems of lemmatization and alphabetization in the Online Dictionary of Idioms, the use of the morphological analyzer of Modern Georgian compiled as a finite state transducer was considered. And the main focus of our research was to compile an Online Dictionary of Idioms and to provide user-friendly access to lexical information stored in the DB on the basis of the appropriate linguistic specification and the morphological analyzer ofthe Georgian language.

Thus, the methods used are subdivided into two steps. First, following approaches of finite state techniques ([3, 17] etc.), a morphological analyzer of the Georgian language was created using xfst and lexc tools and tested on the Georgian Language Corpus, and, second, following approaches of modern corpus-based lexicography ([2, 25, 31] etc.), the compilation of a dictionary was based on the corpus of the Modern Georgian language [9] and an additional one created in the TLex Suite: Dictionary Compilation system [34].

3 Morphological Analyzer and Generator of the Modern Georgian Language

3.1 Encoding Georgian Morphology

The Modern Georgian language belongs to the MRLs. This means that the morphologically rich nature of Georgian expresses different levels of information at the word level, and the grammatical functions of words are not always determined by their syntactic position. This leads to a high degree of word order variation at the level of the sentence, and the priority of morphosyntactic descriptions in the case of Modern Georgian.

Naturally, there are many Natural Language Processing (NLP) systems used for treating languages with concatenative-type morphology like Georgian. One of the most famous approaches to the morphological analysis of such kinds of languages is a finite state technology as described by Beesley and Karttunen [3]. Finite state technology is used in morphological processing, semantics and discourse modeling.

The standard tools for morphological analysis of this kind are *xfst* and *lexc*. The finite state transducers can be used for both analysis and generation, but their compilation is subdivided into two parts, namely, the lexicons and the regular expressions compiled into finite-state networks. In spite of the fact that the development of a finite state transducer is a time-consuming task, the approach has been successfully used for the processing of typologically different languages and can be considered to be language-independent. Therefore, the morphological analyzer of Modern Georgian was compiled by means of the above-mentioned tools and covers the full inflectional paradigm of Modern Georgian. The morphotactics is encoded in lexicons and alternation rules are encoded in regular expressions.

The lexicon consists of the Lexical Unit and Continuation Classes. The Lexical Unit is the unmarked form of the word, i.e. the root or headword given in a dictionary. The lexical items were taken from the Georgian Explanatory Dictionary by Chikobava [6], and the digitized index of Georgian verbs was taken from the Conjugation System of the Georgian Verb by Melikishvili [21]. At this point the lexicons comprise more than 78,000 nominals and 85,000 verbs.

As for the Continuation Class, it is just a pointer to another lexicon or it can be finished by the end marker '#'. At the same time, the lexical items consist of lexical and surface levels. Thus, where the format does not include a colon, the mapping is provided to the lexical item as (16) demonstrates, while the format with a colon maps lexical levels to surface ones as (17) shows:

> (16) *cixe* 'prison'
> ```
> cixe     N5 ;
> ```
> (17) *kvitini* 'sobbing'
> ```
> kvitin-i:kvitini    N1 ;
> ```

The tags used for the marking of morphological features are described as part of the `Multichar_Symbol` at the start of the text file, e.g. +Noun, +Nom, +Erg etc. Also, the long-distance dependencies expressed by different prefixes linked to other affixes after the root, as occur in the case of adjectives and verbs, are rather difficult to model because of the fact that a finite state network has no memory to constrain the sequence of transitions from one state to another. In such cases, the use of Flag Diacritics allows us to block the illegal paths. And, to avoid overgeneration and overrecognition, we have used flag diacritics – standard multicharacter symbols reflecting feature-based constraints, which do not appear in output strings. For nominal inflection, the quantity of flags is equal to 73, while for verbal inflection it is 167.

Having a high number of affixes and morphemes per word and separate morphemes denoting multiple grammatical features, Modern Georgian belongs to the agglutinating type of languages with weak fusion. This means that most noun and verb forms can be segmented into agglutinatively concatenated morphemes, which sometimes denote multiple grammatical features, and their further processing depends on the grouping of morphemes into affix slots. Inflections also frequently include modification to the stem, including root alternation (18) and root vowel alternation (19). At the same time, when affixes are attached to a root, several phonological processes can take place like metathesis (20), vowel substitution (21), vowel deletion, i.e. syncope (22), etc.

(18) *e-t̨q̇v-i-s* 'will say to smb.'
    prv-say-TS-FUT.3SGSBJ
    Ipfv+eubneb-a+V+Trans+Act+Fut+Subj3Sg+ObjRec3+Obj3

(19) *mo-drik̨-a* 'twisted smb.'
    PV-bend-AOR.3SGSBJ
    Pfv+drek-s+V+Din+Trans+Act+Aor+Subj3Sg+Obj3

(20) *xn-av-s* 'ploughs smth.' and *xvna* 'ploughing'
    a.   plough-TS-PRES.3SGSBJ or plough-TS-FUT.3SGSBJ
        { Ipfv+xnav-s+V+IDt+Din+Trans+Act+Pres+Obj3+Subj3Sg
        | Ipfv+xnav-s+V+IDt+Din+Trans+Act+Fut+Obj3+Subj3Sg }
    b. ploughing<TS>-NOM
        Ipfv+xnav-s+VN+Sg+Nom

(21) *nivr-is* 'of garlic'
    garlic-GEN
    nior-i+N+Com+Sg+Gen

(22) *da-čer-i* 'you cut smth.' / 'cut it'
    PV-cut-AOR.2SGSBJ or PV-cut-AOR.IMP.2SGSBJ
    { Pfv+čri-s+V+IDt+Din+Trans+Act+Aor+Subj2Sg+Obj3
    | Pfv+čri-s+V+IDt+Din+Trans+Act+AorImp+Subj2Sg+Obj3 }

The stem modifications and final mutations are solved by means of special mark-up tags triggering replace rules, e.g. ^S, ^P etc. For instance, it is not sufficient to add the plural suffix -*eb* to vowel-stem nouns like *deda* 'mother', we have to remove -*a* as well. Thus, a replace rule must be used to implement this change as shown in examples (23) and (24):

(23)   a- > [] || _ %^P

(24)   *ded-eb-i* 'mothers'
      mother-PL-NOM
      deda+N+Com+Pl+Nom: deda^Pebi

The system described above is able to do both analysis and generation. The parts of speech (PoS) are subdivided into open and closed inflected classes accordingly. New items are generally added to open classes. The open classes include noun, adjective, verb, adverb and interjection, while the closed classes are pronoun, numeral, conjunction and particle. The lexicons as represented in the morphological analyzer are organized in accordance with declension types and the sequence of slots within concrete PoSs as described above.

Thus, there are 22 declension types for nouns undergoing different phonological processes, 4 for adjectives and numerals and 49 for pronouns. The verbal module of the morphological analyzer consists of 66 standard paradigms with appropriate lexicons and a block of irregular verbs (so-called suppletive verbs, e.g. *q̇opna* 'be', *kmna* 'do' etc.; verbs with root alternation according to number, e.g. *gdeba–q̇ra* 'throw', *jdoma–sxdoma* 'sit' etc.; verbs with root alternation according to animacy *mi/moṭana–mi/moq̇vana* 'take', *koneba–q̇ola* 'have' etc.) as described by Melikishvili [21]. The minimum quantity of forms generated per root is 54, the maximum is 1,076 without preverbs. At the same time, there are differently encoded object- and subject-based paradigms. The verbal paradigm is compiled from *lexc* – for lexicon data including information on stems and affixes and their dependencies, and *xfst* – for alternation rules. Both formats are connected with each other and generate a single two-level transducer. The analyzer uses Unicode, specifically utf8, which allows us to provide testing and error analysis of different Georgian texts.

3.2   Testing and System Analysis

During the compilation of the system, testing and error analysis were based on the following:

a. Rule integrity: the *lexc* tools offer the lookup and lookdown commands, while *xfst* tools apply up and apply down. At the same time, the finite-state calculus has the option of a regression test used within a version-control system. Thus, we have used (a) regression testing comparing two versions to find lost surface words, and (b) regression testing comparing two versions to find added words. The system was run and fixed periodically;

b. Well-formedness of the surface representation of paradigms: the tags appear in accordance with the order defined preliminarily to provide their possible integration into other systems;

c. Language coverage testing: the language coverage of the lexicon in terms of frequency always depends on "zipfian" distributions [38]. Such kinds of distribution mean that in all the languages of the world, a small quantity of words has a high frequency, an average quantity has an intermediate frequency and a large quantity of words has a very low frequency, which varies from 1 to 2. The resource used to evaluate language coverage is the Georgian Language Corpus available at corpora. iliauni.edu.ge [9]. The results are as follows (see Fig. 1):

As can be seen, the transducer recognizes only 92.17% of the 10,000 most frequently used words automatically generated from the corpus [10]. At the same time the quantity of verbs per 1,000 words is equal to 530, including 521 recognized by the

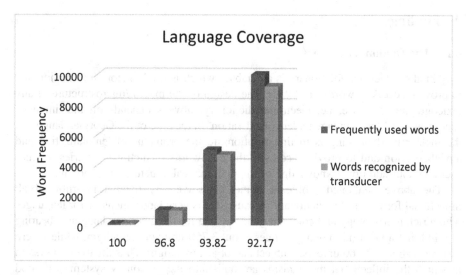

Fig. 1. Language coverage test

transducer and 9 not recognized by the transducer. This means that the recognition rate for verbs is equal to 98.31%.

At the same time, analyzing the forms recognized by the transducer, we have to mention overlapping between different paradigms within the first diathesis, especially between classes No. 19 and No. 26 (25) and overlapping between different diatheses, specifically between classes No.28, No. 29, No. 30 and No. 47 (26).

(25) *ķec-av-s* 'folds smth.'

fold-TS-PRES.3SGSBJ
{Ipfv+ķecav-s+V+IDt+#19+Din Trans+Act+Pres+Subj3Sg+Obj3
| Ipfv+ķecav-s+V+IDt+#26+Din+Trans+Act+Pres+Subj3Sg+Obj3}

(26) *ča-i-gd-o* 'was thrown down'

PV-prv-throw-AOR.3SGSBJ or PV-prv-throw-AOR. SJV.3SGSBJ
{Pfv+agdeb-s+V+IDt+#28+Din+Trans+Act+AorSbj+SubjBen2Sg+Obj3
| Pfv+igdeb-s+V+IDt+#29+Din+Trans+Act+Aor+SubjBen3Sg+Obj3
| Pfv+igdeb-s+V+IDt+#29+Din+Trans+Act+AorSbj+SubjBen2Sg+Obj3
| Pfv+igdeb-s+V+IDt+#30+Din+Trans+Act+Aor+SubjBen2Sg+Obj3
  +IndSpeech3
| Pfv+igdeb-s+V+IDt+#30+Din+Trans+Act+AorSbj+SubjBen2Sg+Obj3
| Pfv+igdeb-a+V+IIDt+#47+Din+Intr+Pass/AutAct+Aor+SubjBen3Sg
| Pfv+igdeb-a+V+IIDt+#47+Din+Intr+Pass/AutAct+AorSbj+SubjBen2Sg}

These classes generally differ in structure and at the same time generate similar forms for aorist and aorist subjunctive screeves.

4 Findings

4.1 The Dictionary System

Any kind of electronic dictionary is a database which is used to store information and to provide access to words or MWEs. The design of the macro-/microstructure of any lexicographic database, i.e. electronic dictionary, always depends on the linguistic specification of the language under consideration. In the case of the Georgian language, the most important thing is to find solutions to the problems of lemmatization and alphabetization and to provide access to the headwords of dictionary entries by end-users without additional knowledge of the grammatical structure of Georgian.

The above-mentioned problems cannot be solved by manual insertion of all inflectional forms into the dictionary, because this is very labor intensive for languages with a rich morphology and completely impossible for the Georgian language, bearing in mind that a Georgian noun generates about 3,750 units per paradigm, while a verb generates about 33,260 units per subject and object paradigm. Thus the use of a system in which the inflectional mechanisms are built into the dictionary system is a good choice. And taking into account the complexity of the task, the morphological analyzer of the Georgian language can be considered a tool that deals with inflection by means of a paradigm-based approach, when a word is associated with a concrete inflectional paradigm, and provides generation of appropriate forms. At the same time the possibility of analyzing a word and determining a lemma form is very important for solving the problems described previously.

Thus, the compilation of any dictionary includes a sequence of stages. In the case of the Online Dictionary of Idioms, we determined the structure of the dictionary and entries, revised the idioms by means of a corpus-based approach (we used the Georgian Language Corpus [9] and one additional corpus compiled in the *TLex* suite [34]), added entries to the system, prepared. xml-s and launched an online version of the dictionary. At the same time we developed advanced search options to allow easy access to entries and provided solutions to the problems mentioned above by means of the morphological analyzer of Modern Georgian.

4.2 The Dictionary Structure and Entries

The Online Dictionary of Idioms available at http://idioms.iliauni.edu.ge/ consists of monolingual and bidirectional bilingual blocks. The bidirectional bilingual dictionary is subdivided into Modern Georgian – Modern Greek and Modern Greek – Modern Georgian parts as described by Lobzhanidze [19]. The monolingual dictionary was compiled on the basis of the Modern Georgian – Modern Greek dictionary of idioms [20]. The quantity of entries, approximately 8,300, allows us to incorporate some changes in the meanings of idioms represented in the works of different authors from the twentieth century up totoday. Citations come not only from the classics of Georgian literature, but from Modern Georgian writers as well. At the beginning each idiom was equipped with essential grammar information, but then it was decided to develop an advanced search option which allows users to find constituents of idioms in any form as they are represented in the text and to avoid the above-mentioned problems of Georgian lexicography.

The dictionary entries for the monolingual and bilingual parts differ in the elements represented. Most entries include information on lemma sign, derivational variants of use, and etymological notes for some entries, definitions, and literary citation with the indication of literary source. The polysemous meanings of idioms are listed as well. Noun phrases are generally used in the singular. Verbal phrases are used in finite forms, i.e. third person singular in the present or future indicative. Also, we have included forms with verbal noun constituents as well, keeping in mind that some idioms require them. Non-flexible idioms are represented in accordance with their fixed forms.

4.2.1 Quick Search and Alphabetical Search Option

A typical search starts with the user entering some letters to choose the direction of the search, specifically, the source language and its correlation with the target language, i.e. Georgian, Georgian – Modern Greek, Modern Greek – Georgian. All idioms from the source language are then available to the user in alphabetical order, who can browse through them and see explanations or translations of MWEs or try to access a specific idiom by entering its headword letter by letter.

In the Online Dictionary of Idioms entries are ordered alphabetically in accordance with the letter-by-letter principle and allow the user to get the complete information on the idioms represented in the dictionary. Also, if a user wants to find any of the constituents, including those which are not the first word of a MWE, this is allowed directly from the web by means of a Quick Search Option, specifically by entering the initial letters of the constituent under consideration; for example, if a user tries to find the second verbal constituent of *gzas daubnevs* 'will confuse smb.', it will be sufficient to print the initial letters like *dau-* or to use a wildcard *, which describes any combination of letters.

But the majority of MWEs in Modern Georgian are morphosyntactically flexible like (27) and (28):

> (27) *gza-s   m-i-čr-is*
> road-DAT   1SGOBJ-prv-cut-PRES.3SGSBJ
> 'He/she blocks my way; ~He/she waits for me'
>
> (28) *gza-s   v-u-čr-i*
> road-DAT   1SGSBJ-prv-cut-PRES
> 'I block his/her way; ~I wait for smb./smth.'

Such kinds of forms can be seen in literary sources and the user whose knowledge of Modern Georgian is not at a high level will not be able to find them at all without the Advanced Search Option, which allows one to find lemma signs for words not listed in the dictionary and to access them in accordance with the letter-by-letter principle.

4.2.2 Advanced Search Option

The possibility of solving the problems of lemmatization and alphabetization is closely connected to the system in which the inflectional mechanism is integrated in the dictionary and which allows the end-user to search for appropriate forms. Thus, the Advanced Search is a solution to the problem of lemmatization associated with the absence of an infinitive form and the problem of alphabetization associated with the impossibility of finding appropriate verbal MWEs by initial letters of headwords. This option performs a search for any kind of word as it is found in the raw text and gives users the possibility of seeing direct translation of its initial form; in our case it is the third person singular in the present or future indicative for verbs and the nominative case singular for nouns.

If we consider for example an idiom *sicocxle aqv-s* 'he/she is happy', we can see that the Georgian language allows for flexibility like *sicocxle m-kon-ia* 'I have been happy' (29), *sicocxle g-kon-ia* 'you have been happy' (30) etc.

(29) *sicocxle   m-kon-ia*
     life-NOM   1SGOBJ-be-PRF.3SGSBJ
     'I have been happy'
(30) *sicocxle   g-kon-ia*
     life-NOM   2SGOBJ-be-PRF.3SGSBJ
     'You have been happy'

Such kinds of idioms can be seen in literary sources as well, and the user whose knowledge of Modern Georgian is not very great will not be able to find them considering that the headword in the dictionary entries for the verbal forms *m-kon-ia* 'I have been' and *g-kon-ia* 'you have been' is *aqv-s* 'has'. Thus, if an end-user does not directly type *sicocxle aqv-s* 'he/she is happy' or the verb *aqvs* 'has', a typical search is of no use. The case is similar for other verbal idioms like those represented in examples (27) and (28). The headword of the dictionary entry is the verb *čr-is* 'cuts', but to determine this we have to use the output of the analyzer, specifically (31) and (32):

(31) *gza-s m-i-čr-i-s* 'He/she blocks my way'
     road-DAT   1SGOBJ-prv-cut-TS-PRES.3SGSBJ
     gza+N+Com+Inanim+Sg+Dat
     Ipfv+čr-is+V+IDt+#23+Din+Trans+Act+Fut+Subj3Sg+Obj3
(32) *gza-s v-u-čr-i* 'I block his/her way'
     road-DAT   1SGSBJ-prv-cut-TS.PRES.3SGOBJ
     gza+N+Com+Inanim+Sg+Dat
     { Ipfv+čr-is+V+IDt+#23+Din+Trans+Act+Pres+Subj1Sg+Obj3+ObjBen3
     | Ipfv+čr-is+V+IDt+#23+Din+Trans+Act+Fut+Subj1Sg+Obj3+ObjBen3 }

Thus, the system available online determines lemmas for the words given above and then based on the lemma *čris* 'cuts' carries out a search in the database and returns all verbal MWEs associated with the above-mentioned verb including *gzas čris* 'He/she blocks the way'. As a result, the system gives the end-user access to the headword of the word under consideration.

5   Conclusions

As shown in this article, the use of two-level morphology and finite-state technology is both theoretically and technologically suitable for the Modern Georgian language if we keep in mind that in spite of long distance dependencies within words the concatenative structure of Modern Georgian can be implemented without difficulty using finite state transducers like *lexc* and *xfst*. The morphological transducer can easily be adapted for different purposes, especially for solving the problems of lemmatization and alphabetization noticed in Georgian dictionaries (in monolingual and bilingual ones).

It became possible to use the morphological analyzer of Modern Georgian for automatic inflection of the lexicon, and its application to the Online Dictionary of Idioms has provided easy access to the headwords of dictionary entries by the end-user in a way which answers the needs of native and non-native speakers of the Georgian language. From one point of view such a kind of search can be considered a naive one, but taking into account the main problems of lemmatization and alphabetization which we face in Georgian dictionaries, at this point in time the morphological-analyzer-based solution can be considered the best choice.

Also, the compilation of the Online Dictionary of Idioms is useful for the further development of computational approaches to the Georgian language. Considering that the compilation of the monolingual and bidirectional bilingual dictionary of idioms is finished, we have the possibility of using the results we have obtained in other dictionaries of the Georgian language and of providing further development of light parsing in the case of noun phrases, verb phrases etc. by means of finite-state techniques.

Acknowledgments. The author is grateful to two anonymous reviewers for their critical remarks on the previous version of this study.

Abbreviations

DB	Database
DBMS	Database management system
ISO	International organization for standardization
MRL	Morphologically rich language
MWE	Multi-word expression
NLP	Natural Language Processing
PoS	Part of speech
TAM	Tense-aspect-mood
TS	Thematic suffix

Glosses

1, 2, 3	1st, 2nd, 3rd person
AOR	aorist
AUX	auxiliary
CIRC	circumfix
COMP	comparative degree

DAT	dative
EMP	emphatic vowel
GEN	genitive
FUT	future
NEG	negation
NOM	nominative
OBJ	object
PAS	passive marker
PL	plural
POST	postposition
PRES	present
PRF	perfect
prv	pre-radical vowel
PTCL	particle
PV	preverb
SG	singular
SBJ	subject
SJV	subjunctive
TS	thematic suffix

Morphological Feature Tags

+A	adjective
+Act	active
+Aor	aorist
+AorImp	aorist imperative
+AorSbj	aorist subjunctive
+AutAct	autoactive
+Aux	auxiliary verb
+Com	common
+Din	dynamic
+Emp	emphatic vowel
+Fut	future
+Gen	genitive case
+IDt	the first diathesis
+IIDt	the second diathesis
+Inanim	inanimate
+Intr	intransitive
+N	noun
+Nom	nominative case
+Pass	passive
+Pl	plural
+Post(for)	postposition for
+Post(like)	postposition like
+Ptcl	particle
+Sg	singular
+Trans	transitive

+V   verb
+VN   verbal noun
Ipfv+   imperfective aspect
Pfv+   perfective aspect

References

1. Alegria, I., Ansa, O., Artola, X., Ezeiza, N., Gojenola, K., Urizar, R.: Representation and treatment of multiword expressions in basque. In: Tanaka, T. (eds.) Second ACL Workshop on Multiword Expressions, pp. 48–55, Spain, Barcelona (2004)
2. Atkins, S.B.T., Rundell, M.: The Oxford Guide to Practical Lexicography. Oxford University Press, Oxford (2008)
3. Beesley, K.R., Karttunen, L.: Finite State Morphology. CSLI Publications, Stanford (2003)
4. Boeder, W.: The South Caucasian Languages. Lingua **115**(1–2), 5–89 (2005)
5. Breidt, E., Segond, F., Valetto, G.: Formal description of multi-word lexemes with the finite-state formalism IDAREX. In: Tsuji, J.(eds.) Proceedings of COLING 1996. 16th International Conference on Computational Linguistics, pp. 1036–1040, Copenhagen, Denmark (1996)
6. Chikobava, A.: Georgian Explanatory Dictionary. Academy of Sciences, Tbilisi (2008–2010; 1950–1964)
7. Chubinashvili, D.: Georgian-Russian-French Dictionary. Imperial Academy of Sciences, Saint-Petersburg (1840)
8. Copestake, A., et al.: Multiword expressions: linguistic precision and reusability. In: Rodriguez, M., Araujo, C. (eds.) Proceedings of the 3rd International Conference on Language Resources and Evaluation (LREC), pp. 1941–1947, Las Palmas, Canary Islands (2002)
9. Georgian Language Corpus. http://corpora.iliauni.edu.ge/. Accessed 20 Mar 2018
10. Gibbon, D., Van Eynde, F.: Lexicon Development for Speech and Language Processing. Kluwer Academic Publishers, London (2000)
11. Gippert, J.: Complex morphology and its impact on lexicology: the kartvelian case. In: Margalitadze, T. (ed.) The XVII EURALEX International Congress, pp. 16–37. IvaneJavakhishvili Tbilisi University Press, Tbilisi (2016)
12. Hewitt, B.: Georgian: A Structural Reference Grammar. John Benjamins, Amsterdam (1995)
13. Hewitt, G.: Georgian: A Learner's Grammar. Routledge, Abingdon (2005)
14. International Organization for Standardization (ISO): Language Resource Management – Morpho-syntactic Annotation Framework (MAF), no. 24611 (2012)
15. Jacquemin, C.: Spotting and Discovering Terms through Natural Language Processing. MIT Press, Cambridge (2001)
16. Karttunen, L., Kaplan, R.M., Zaenen, A.: Two-Level Morphology with Composition. In: Proceedings of COLING-92, Nantes, pp. 141–148 (1992)
17. Koskenniemi, K.: Two-Level Morphology: A General Computational Model for Word-Form Recognition and Production. University of Helsinki, Helsinki (1983)
18. Kvachadze, L.: Syntax of Modern Georgian Language (in Georgian). Rubikon, Tbilisi (1996)

19. Lobzhanidze, I.: Online Dictionary of Idioms. In: Margalitadze, T. (ed.) The XVII EURALEX International Congress, pp. 710–717. Ivane Javakhishvili Tbilisi University Press, Tbilisi (2016)
20. Lobzhanidze, I.: To the compilation of Georgian-modern Greek dictionary of idioms. In: Gavriilidou, Z., Efthymiou, Z. (eds.) 10th International Conference of Greek Linguistics, pp. 899–904. Democritus University of Thrace, Komotini (2012)
21. Melikishvili, D.: Conjugation System of the Georgian Verb (in Georgian). Logos presi, Tbilisi (2001)
22. Oflazer, K., Çetonoglu,Öz., Say, B.: Integrating Morphology with Multi-word Expression Processing in Turkish. In: Tanaka, T. (eds.) Second ACL Workshop on Multiword Expressions, Spain, Barcelona, pp. 64–71 (2004)
23. Oniani, A.: Georgian Idioms (in Georgian). Nakaduli, Tbilisi (1966)
24. Online Dictionary of Idioms. http://idioms.iliauni.edu.ge/. Accessed 20 Mar 2018
25. Ooi, V.: Computer Corpus Lexicography. Edinburgh University Press, Edinburgh (1998)
26. Rayfield, D.: A Comprehensive Georgian-English Dictionary. Garnett, London (2006)
27. Sag, I.A., Baldwin, T., Bond, F., Copestake, A., Flickinger, D.: Multiword expressions: a pain in the neck for NLP. In: Gelbukh, A. (ed.) CICLing 2002. LNCS, vol. 2276, pp. 1–15. Springer, Heidelberg (2002). https://doi.org/10.1007/3-540-45715-1_1
28. Sakhokia, T.: Georgian Figurative Expressions (in Georgian). Sakhelgami, Tbilisi (1950–1955)
29. Savary, A.: Computational inflection of multi-word units. a contrastive study of lexical approaches. Linguist. Issues Lang. Technol. CSLI Publ. 1(2), 1–53 (2008)
30. Shanidze, A.: The Basics of the Georgian Language Grammar (in Georgian). TSU, Tbilisi (1973)
31. Sinclair, J.: Corpus to corpus: a study of translation equivalence. Int. J. Lexicogr. 9(3), 171–178 (1996)
32. TLex Suite. Dictionary Compilation Software. https://tshwanedje.com/tshwanelex/
33. Tschenkeli, K.: Georgisch-Deutsches Wörterbuch. Amirani-Verlag, Zürich (1965)
34. TshwaneDJe Software and Consulting Homepage. http://tshwanedje.com/. Accessed 20 Mar 2018
35. Tsotsanidze, G., Loladze, A., Datukishvili, K.: Georgian Dictionary. BakurSulakauri, Tbilisi (2014)
36. Villavicencio, A., Baldwin, B., Waldron, B.:A Multilingual database of idioms. In: Lino, M. T., Xavier, M.F. (eds.) 2004 Proceedings of the Fourth International Conference on Language Resources and Evaluation, LREC, Lisbon, Portugal, pp. 1127–1130 (2004)
37. Villavicencio, A., Copestake, A., Waldron, B., Lambeau, F.: Lexical Encoding of MWEs. In: Tanaka, T., Villavicencio, A. (eds.) Second ACL Workshop on Multiword Expressions: Integrating Processing, Barcelona, Spain, pp. 80–87 (2004)
38. Zipf, G.: Selected Studies of the Principle of Relative Frequency in Language. Harvard University Press, Cambridge (1932)

Language as Mechanisms for Interaction: Towards an Evolutionary Tale

Ruth Kempson[1], Eleni Gregoromichelaki[1,2(✉)], and Christine Howes[3]

[1] King's College London, London, UK
`ruth.kempson@kcl.ac.uk`
[2] Heinrich Heine University, Düsseldorf, Germany
`elenigregor@gmail.com`
[3] University of Gothenburg, Gothenburg, Sweden
`christine.howes@gu.se`

Abstract. In this paper we present a view of natural language (NL) grammars compatible with enactive approaches to cognition. This perspective aims to directly model the group-forming properties of NL interactions. Firstly, NL communication is not taken as underpinned by convergence/common ground but modelled as the employment of flexible procedures enabling creative joint activities without overarching common goals. On this basis, we argue that a common non-individualistic pattern can be discerned across NL learning, individual and institutional NL change, and evolution. At all levels and stages, modelling of change relies on situated iteration leading to joint establishment and modification of practices. NL learning, change, and even NL emergence can all then be seen in gradualistic terms, with the higher-order organisation that incorporates NL grammars constituting an adaptive interactive system in continuity with the definition of living organisms as modelled in enactive approaches.

1 The Isolationist Background

Scientific modelling of high-level cognitive processes has recently turned towards the implementation of philosophical/psychological views advocating enactive approaches to cognition. Accordingly, for some time now, there has been a growing shift of emphasis in disciplines across cognitive science away from static representations of structure/content towards the dynamic, process-oriented modelling of skills and abilities that organisms employ, adjust, and perfect in dealing adequately with the ever-changing possibilities for action the environment affords [2,3,13,14,32,35,86]. In contrast, formal theorising about natural language (NL) has typically retained its characterisation as a code, an abstract system of rules and representations arbitrarily mapping forms to meanings. In this view, linguistic knowledge is codified as a 'grammar' mediating fixed mappings of phonological, syntactic and semantic representations. It is well-known that this characterisation is inadequate for any realistic application of this purported knowledge in a dynamically changing environment, issues pertaining to

© Springer-Verlag GmbH Germany, part of Springer Nature 2019
A. Silva et al. (Eds.): TbiLLC 2018, LNCS 11456, pp. 209–227, 2019.
https://doi.org/10.1007/978-3-662-59565-7_11

such applications being in principle precluded. As a result, such views are either presented as theories of an encapsulated cognitive capacity (as I-language or "competence": [10]) or are supplemented by invoking pragmatic competence underpinned by innate mechanisms of mindreading and altruism to bridge the gap [11,87]. Both solutions are undesirable if the aim is to account for basic NL properties. First, as we argue, the effects characterized as mindreading and altruism/cooperation are outcomes of the mechanisms that NLs instantiate, not themselves causes. Second, modelling of NLs as codes presupposes a synchronic and static view. This view has had a troubled ride in probing NL evolution. If a domain-specific, encapsulated capacity with arbitrary relations between levels of structure is assumed, gradualist accounts of NL evolution are precluded. Instead, the emergence of NL has been seen as a mutation, a so-called "sudden switch" (for a recent variant see [5]). In psychology, the code view precludes mechanisms that are subject to constant change and adaptation in response to events in the environment so that learning, plasticity, and cultural transmission are excluded in favour of biological determination (nativism) and prespecified unfolding of capacities ("maturation"). Accordingly, learning one's native language has been seen as requiring a Language Acquisition Device in which the child hypothesises a succession of grammars, increasingly approximating the adult grammar.

2   Interaction and Natural Language

But even a model of the most basic and mundane uses of NL, i.e. communication via conversational interaction, is beyond the static, isolationist accounts. In psycholinguistics, the code model methodology enriched with mind-reading capacities presuppose what has been characterized as the "cognitive sandwich" view [53]. According to this view, the mind is structured at three levels: perception and action are seen as separate from each other and peripheral; cognition, the locus of propositional thought, planning, and executive control, stands in between as the filling. Applied to the modelling of dialogue, this view postulates that low-level perception and action involve a series of independent coding/decoding modules which are separate from and coordinated by the higher processes of cognition. Communication then is explicated as the transfer or sharing of "meanings", conceived as propositions, from one individual mind to another. As a result, the perceiver/listener is modelled not as an interacting agent/actor but as a passive recipient decoding stimuli produced by a speaker and replicating the speaker's thought. This view contradicts empirical observations of dialogue data showing that production and comprehension in dialogue are as tightly interwoven as argued in current computational neuroscience models linking action, action perception, and joint action [2,13]. The most glaring case is data showing rapid exchange of speaker/hearer roles in conversation even within the building of a single structure:

(1) A: We're going to
   B: Marlborough
   C: Marlborough?
   B: to see Granny
   C: With the dogs?
   A: if you can keep them
     under control

(2) A: I need a a
   B: mattock. For breaking up clods
   of earth   [BNC]

(3) Jack: I just returned
   Kathy: from
   Jack: Finland   [data from 63]

The existence of such interactive constructing of utterances/meanings is problematic for all conventional grammars, since any dependencies are able to be split apart so that resolution is only possible across the turn divide (as in (2) and (3)). These interactions also show how the direction they may take is by no means confined to realising some over-arching intentionally held content anticipated in advance of the speech event. And, of course, constructions such as these may be uttered by a single interlocutor, for their own benefit or their interlocutors', refining and elaborating an initial sentence/thought, as in (4):

(4) Mary's back. Late last night. From the US. Tired and frustrated.

Not being a cognitively demanding task, even language-acquiring infants can join in, adding to a proffered frame, (5), or initiating a frame construction process, (6)–(7):

(5) A: Old MacDonald had a farm, E-I-E-I-O. On that farm he had a
   B: cow.
(6) *(2 year old on mum's bike waving at empty mooring over the canal)*
   Eliot: Daddy!
   Mother: That's right dear, you were here yesterday with Daddy
         clearing out the boat.   [direct observation]
(7) A: Bear.
   B: That's right dear, a panda.

The effect is one of rich potential for interactivity between participants, available from the earliest stages of NL development. Accordingly, in our view, what is needed to model such data is a grammar in which mechanisms of processing (actions) are modelled, rather than fixed mappings among form-contents. This involves a radical shift of assumptions, a shift that has been adopted in Dynamic Syntax – to which we now turn.

3 Language as Action

Dynamic Syntax (DS, [8,58]) is a grammar architecture whose core notion is incremental interpretation of word-sequences (comprehension) or linearisation of contents (production) relative to context. The DS syntactic engine, including the lexicon, is articulated in terms of goal-driven actions accomplished either by giving rise to expectations of further actions, by consuming contextual input, or by being abandoned as unviable in view of more competitive alternatives. Thus words, syntax, and morphology are all modelled as "affordances", opportunities for (inter-) action produced and recognised by interlocutors to perform step-by-step a coordinated mapping from perceivable stimuli (phonological strings) to conceptual mechanisms or vice-versa. To illustrate, we display below the (condensed) steps involved in the parsing of a standard long-distance dependency, *Who hugged Mary?*:[1]

$$(8) \quad \boxed{?Ty(t), \Diamond} \quad \overset{...who...}{\longrightarrow} \quad \boxed{\begin{array}{c} ?Ty(t) \\ \\ \mathbf{WH}:e, \Diamond \end{array}}$$

The task starts with a set of probabilistically-weighted predicted *interaction-control states* (ICSs) represented as a directed acyclic graph (DAG) keeping track of how alternatives unfold and are progressively abandoned (we show only one snapshot of an active DAG path above and only the syntactically-relevant part). Such ICSs include salient environmental information, means of coordination, e.g. "repair" [21], and the recent history of processing. On this basis, they induce triggering of goals to build/linearise conceptual mechanisms ('ad-hoc concepts') classified as belonging to ontological types (e for entities in general, e_s for events, $e \rightarrow (e_s \rightarrow t)$ for predicates, etc). In (9) above, the goal is realized as a prediction to process next a proposition of type t. This is shown as a one-node tree with the prediction $?Ty(t)$ and the ICS's current focus of attention, the pointer $\Diamond$. Such initiating predictions can be of any type since the model aims to integrate predictions generated through any multimodal means with linguistic "fragments" (see e.g. (6)–(7) earlier) seamlessly induced within such DAG frames [43]. Additionally, no extra discourse levels or machinery like QUD [36] or DRSs is needed since the predictions are generated by the totality of information available to the DAG path state (cf. [29,61,65]). With the pointer at a node including a predicted outcome, predictions of further affordances/subgoals are generated under the expectation of eventual satisfaction of the current goal either by the processing of (verbal) input (as a hearer) or by producing that input (as a speaker). For (8), one of the probabilistically-licensed next steps for English (executed by defined lexical and general computational *macros* of actions) is displayed in the second partial tree therein: a prediction that a structurally underspecified node (indicated by the dotted line) can be built and can accommodate the result of parsing/generating *who*. As illustrated here, temporary uncertainty about the eventual contribution of some element is implemented through *structural underspecification*. Initially "unfixed" tree-nodes model the retention of the contribution of the *wh*-element in a memory buffer until it can unify with some argument node in the upcoming local domain. Non-referential words like *who* and other semantically underspecified elements

---

[1] The detailed justification of this formalism as a grammar formalism is given elsewhere ([8,9,37,56,58], and others).

(e.g. pronominals, anaphors, auxiliaries, tenses) contribute underspecified place-holders in the form of so-called *metavariables* (indicated in bold font). Metavari-ables in turn trigger search for their eventual type-compatible substitution from among contextually-salient entities or predicates.

General computational and lexically-triggered macros always intersperse to develop a conceptual graph of available affordances, locally taking the form of a binary tree: in (9), the verb contributes both conceptual structure in the form of unfolding the tree further, fetching an ad-hoc concept (indicated as Hug') devel-oped according to contextual restrictions,[2] as well as placeholder metavariables for time and event entities to be supplied by the current ICS. Such conceptual structure is indefinitely extendible (see [18]) and "non-reconstructive" in the sense that it is not meant as an inner model of the world [43], (see also [14,15]). Instead, these structures function as 'interaction outcome indicators' [6] trig-gering possibilities of further (mental or physical) action, by either participant, extending or "repairing" the node elements, thus coordinating behaviour with selected aspects of the environment and each other.

Returning to the processing in (9), now NL-specific constraints kick in since the pointer $\Diamond$ is left at the argument node implementing the word-order restric-tion in English that the object needs to follow the verb:

(9)

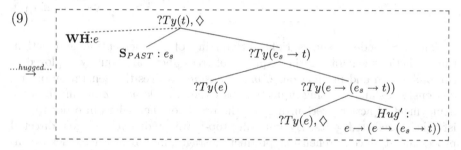

At this point, the word *Mary* can be processed to initiate the tracking of a contextually-identifiable individual ($Mary'$) at the argument node internal to the predicate (for the view that such concepts are skill adaptations allowing the accumulation of knowledge about individuals, see [67]). After this step, every-thing is in place for the structural underspecification to be resolved, namely, the node annotated by *who* can now unify with the subject node of the predicate, which results in an ICS that includes the minimal content of an utterance of

---

[2] In [22,25,27], this is modelled via a mapping onto a Type Theory with Records formulation, but we suppress these details here: see also [49,50,79].

Who hugged Mary? imposed as a goal ($?Q_{\mathbf{WH}}$) for the next action steps (either by the speaker or the hearer):

(10)

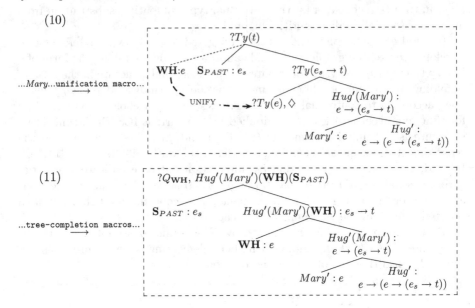

...*Mary*...unification macro...
$\longrightarrow$

(11)

...tree-completion macros...
$\longrightarrow$

The DS model assumes tight interlinking of NL perception and action: the predictions generating the sequence of trees above are equally deployed in comprehension and production. *Comprehension* involves the generation of predictions/goals and awaiting input to satisfy them, while *production* involves the deployment of action (verbalising) by the predictor themselves in order to satisfy their predicted goals. By imposing top-down predictive and goal-directed processing at all comprehension/production stages, interlocutor feedback is constantly anticipated and seamlessly integrated in the ICS [21,33,34,39,76]. Given that the total information state is modelled by the DAG as a single level, even "semantically empty" feedback such as backchannels (*mmm, uh-huh*), serves, through the same mechanisms that implement lexical and computational actions, the function of continually realigning interlocutors' processing contexts, whilst ensuring that the problem of maintaining coordination is computationally tractable [22,23,44,51]. Integration in the ICS can involve adding simple proposition-like structures such as (11) or locally linked structures of any type incrementally elaborating some node of a tree in the ICS. For this reason, maintaining even abandoned options is required to achieve the concurrent modelling of conversational phenomena like clarification, self/other-corrections etc, but also, quotation, code-switching, humorous effects and puns [39,49]. Given the modelling of word-by-word incrementality, at any point, either interlocutor can take over to realise the currently predicted goals in the ICS. This can be illustrated in the sharing of the dependency constrained by the locality definitive of reflexive anaphors:

(12) Mary: Did you burn
       Bob: myself? No.

As shown in (12), Mary starts a query involving an indexical metavariable contributed by *you* that is resolved by reference to the *Hearer'* contextual parameter currently occupied by *Bob'*:

(13)

$$\overset{Mary:Did\ you\ burn}{\longmapsto} \quad ?Ty(t), ?Q$$

$$S_{PAST} \qquad\qquad ?Ty(e_s \to t)$$

$$Ty(e), Bob' \qquad\qquad ?Ty(e \to (e_s \to t))$$

$$?Ty(e), \quad Ty(e \to (e \to (e_s \to t))),$$
$$\diamondsuit \qquad\qquad Burn'$$

With the ICS tracking the speaker/hearer roles as they shift subsententially, these roles are reset in the next step when it happens that Bob takes over the utterance. *Myself* is then uttered. Being a pronominal, it contributes a metavariable and, being a reflexive indexical, it imposes the restriction that the entity to substitute that metavariable needs to be a co-argument that bears the *Speaker'* role. At this point in time, the only such available entity in context is again *Bob'* which is duly selected as the substituent of the metavariable:

(14)

$$\overset{Bob:myself?}{\longmapsto} \qquad ?Ty(t), ?Q$$

$$S_{PAST} \qquad\qquad ?Ty(e_s \to t)$$

$$Ty(e), Bob' \qquad\qquad ?Ty(e \to (e_s \to t)), \diamondsuit$$

$$Ty(e), \quad Ty(e \to (e \to (e_s \to t))),$$
$$Bob' \qquad\qquad Burn'$$

As a result, binding of the reflexive is semantically appropriate, and locality is respected even though joining the string as a single sentence would be ungrammatical. This successful result relies on (a) the lack of a syntactic level of representation, and (b) the subsentential licensing of contextual dependencies. In combination, these design features render the fact that the utterance constitutes a joint action irrelevant for the wellformedness of the sequence of actions constituting the string production. This means that coordination among interlocutors here can be seen, not as propositional inferential activity, but as the outcome of the fact that the grammar consists of a set of licensed complementary actions that speakers/hearers perform in synchrony [38,40,42]. As DS models syntax as a process (actions), not the resulting product, semantically equivalent strings might result in identical trees but the record of the processes on the DAG will be distinct. Syntactic alignment [74, a.o.] and priming in experimental data is explained by the reuse of these actions [57] while the fact that all conversational processing phenomena are modelled in a single level explains the finding that such alignment occurs at levels below chance in general conversation [45]. Due to subsentential step-by-step licensing, speakers are not required to plan propositional units, so hearers can perform "repair" subsententially without need to reason about propositional intentions. Given that parsing/production are predictive activities [61,75], a current goal in the ICS may be satisfied by a current hearer, so that it yields the retrieval/provision of conceptual information that matches satisfactorily the original speaker's goals, as in (2)–(5), or be judged to require some adjustment that can be seamlessly and immediately provided by feedback extending/modifying the ensuing ICS:

(15) Ken: He said all the colored people uh walk- walk down the street
          and they may be all dressed up or somethin and these guys
          eh white- white guys'll come by with .hh
     Louise: mud.
     Ken: mud, ink or anything and throw it at 'em   [from 62]

The action dynamics in DS, and its emphasis on underspecification and update for both NL and context, reflect the formalism's fundamental cross-modal predictivity and integration of normative constraints from various sources, e.g. turn-taking conventions, within a single graph. This allows for parsimonious explanations of now standard psycholinguistic evidence of prediction from sentence processing studies [1,89, a.o.] without requiring internally structured predictive models (see [41] for comparison with [75]). The single-level assumption also allows for the fine-grained modelling of results of current turn-manipulation experiments showing that how people respond to truncated turns depends on how predictable the continuation is. Extremely predictable continuations do not even need to be articulated by either party in order to be taken as part of the interpretation, and continuations that are predictable in terms of structure but not content (such as those within a noun phrase) prompt dialogue participants to provide multi-functional utterances serving both as continuations and offering feedback as clarification requests ([52], cf. [29]). DS processing can model all these options since there is no notion of wellformedness defined over sentence-proposition mappings, only systematicity/productivity of procedures for incremental processing. Therefore, unlike non-incremental formalisms where explanation for these phenomena has to either be devolved to a parser external to the grammar or be relegated to performance "errors", fragmentary linguistic input/output and "repair" processes are not modelled as a problem for the interlocutors. Instead such processing is basic and constitutes the purpose of interaction which is to modify the interlocutors' cognitive and physical environments, a basic feature for learning and adaptation purposes.

4   Interaction and Language Learning

DS is a grammar modelling goal-directed coordination activity, with syntax transformed into a set of conditional update actions that induce or develop the processing environment of interacting agents. Depending on the moment-by-moment system-generated predictions of the next system state, processing strategies either attend to input verifying the predictions (as a hearer) or induce physical action realising the predictions (as a speaker), each agent manipulating the grammar mechanisms relative to their own capabilities, needs, desires, and goals but also as part of a system of coordinated processes with emergent properties [41,42]. This is unlike other frameworks [47,96] where these two apparently inverse activities are modelled as distinct, leaving mediating higher-order inference as the only means of modelling emergence and adaptation to the interlocutor's processing. DS, to the contrary, allows for variability without disruption in the affordances each agent perceives/pursues and, equally, divergence in what it is that they establish as the outcome of the interaction. Consequently, the DS update mechanisms have been shown to be learnable from child-directed semantically annotated data [25,26], where such asymmetries are crucial, and in the automatic induction of successful strategies serving task-specific dialogue games [24,27,54,81,94].

Given the embeddability of NL under domain-general skills and constraints as modelled by DS, learning an NL comes under one and the same domain of behavioural control, the establishment of sensorimotor contingencies, resulting from environmental and self-generated feedback [73,90]. In turn, given that the cultural/social environment is the main source of such contingencies for NL acquisition, the starting point for it is the now familiar observation (see [4]) that all utterance exchanges will necessarily involve moment by moment interaction between participants in communicative activities as they severally adjust jointly established action-control states to each other's desirable/undesirable affordances. Even from 4 months old, children enjoy interactive rituals like peek-aboo games in which there is no essential attribution of content. Moreover, [48] and [16] report on prelinguistic stages in which the caregiver characteristically uses their language fluently, invariably providing shaping feedback to the prelinguistic vocal behaviour of the child within their own conception of what this activity leads to (*mindshaping*: [95]). The result is an interactive effect between child and care-giver even in the absence of any expectation of mutual content duplication, the sole reward for both being the rich emotional bonding achieved by this interactive behaviour [88].

In the next phase of acquisition, the one-word utterance stage, successful communication again builds on the interaction between participants, despite, indeed riding upon, the asymmetry between them. This interactive behaviour rests on the reiteration of exchanges during which child and adult severally interpret what is offered them and engage in overt action to shared effect. As the child comes to isolate and so offer one-word utterances, there are notable structural patterns. On the one hand, the child may be offering some completion to a structured routine affordance just provided by the adult's utterance as in the nursery-rhyme exchange of (5). Or, given the coupling of the producer/comprehender systems, if the child is initiating some exchange with such a fragment, they can do so on the expectation that the adult will then develop it (as in (6) above). There are, furthermore, 'embedded correction' cases which can add elaboration, adjunct-like, to the child's offering, as in (7) [80]. In these earliest occurrences of NL, both producing and parsing such a construction would, following the anticipatory DS dynamic, involve predicting feedback and subsequently adjusting expectations (weightings), in the child's case, in favour of the carer's input stimulus. Learning an NL then is learning to exploit the affordances offered by interlocutors. In recurrent occurrences of such scenarios, sensorimotor contingencies will become entrenched so no matter how disjoint or asymmetric the construals of these events by the participants may be, at each stage, neither high-level inferences nor other-self mind-reading abductions need to be invoked (cf. [31,84]).

In fact, this asymmetry between participants in what they bring to bear in the conversational exchange continues across the lifespan, diagnostic of not just all expert/non-expert exchanges, but all dialogic encounters, as differences in experiences, cultural background, individual physiology and social communities all contribute to differences in people's language use, meaning that we never have the "same" language as anybody we are interacting with [17]. In consequence, variation and uncertainty lie at the heart of NL processing, and do not in general inhibit it. In any case, should such uncertainty be picked up on as problematic, NLs have tools specifically reifying the interaction and indicating need of clarification, correction, etc [19] so the pinpoint of uncertainty, if recognised as a hurdle, serves only to enrich the ongoing interaction by making explicit the implicit adjustment mechanisms of prediction-generation, which gradually becomes the basis for (explicit) inference and logical reasoning (instead of interaction relying on such capacities).

5   NL Evolution

The need for adjustment and change shapes all properties of NLs so that a dynamic perspective on NL abilities, rather than fixed form-content mappings, seems to us necessary. At all levels of interaction organisation, instead of high-level inferential processes deciphering hidden speaker intentions, it is domain-specific interaction patterns (*language games*) that allow for particular procedural conventions (i.e., in our view, words, syntax, semantics) to be modelled as emergent, learned, and adjusted during interaction [43,68,69]. This is possible because DS does not impose a single set of actions that must apply invariably to achieve each dialogue goal nor encoded speech act specifications modelling explicit propositional goals [40]. Instead, the composability of complex routines (*macros*) out of basic atomic actions can lead, through affordance competition [12], to plastic strategies that can be (re)deployed and refined at each instance to achieve results, with selection depending on the intersubjective processing environment structuring an interaction according to current needs. In this respect, computational work confirms the successful induction of domain-specific dialogue structures (language games) from very small amounts of unannotated data, with no dialogue act annotation but using instead a combination of DS and Reinforcement Learning [54,93]. Under such learning, multiple processing routes to the same dialogue goal are reinforced or inhibited by feedback depending on the situation and each individual's needs. Establishment of routinised processing also depends on efficiency in securing predictive success at minimal cost. For example, during reinforcement learning, the reward function upon reaching a goal penalises increasing dialogue time/length. This is a general constraint imposed by the organisation of the cognitive system itself, a property often invoked in linguistic pragmatics as determining communicative success [84], but here seen as the natural emergence of the way an organism has evolved to determine success in its task in manageable real time (the "lazy-brain hypothesis", [13]). This general view substitutes feedback-enhanced trial-and-error processing, selecting efficient routines through (inter)action, as the learning mechanism, in place of the need for internal world models and costly computation of others' internal mental states or common ground. Across multiple interactive situations this means that NL users can employ different strategies with different partners to reach the same outcome depending on their histories of interactions without local coordination failure. Long term, such tolerance of alternatives becomes a source of variation of the kind necessary for evolutionary selection.

Taking the fine details of such procedures to be the object of selection requires a view of evolution that is not confined to genetic modification [71,92]. Adjustments made possible by mechanisms loosely described as enculturation, niche construction, social learning, and cultural transmission [60] are involved here. Given that what is constructed during NL interactions are not world-mirroring models but repertoires of (inter)action-control states generating the next predicted inputs, DS follows the pattern assumed by enactive [72,86] and recent predictive coding models [14,15] for whole cognitive systems. In the latter, total brain-body organisms are described as instantiating predictive systems using previous experience at every step to anticipate with uncertainty the structure of the next incoming sensory array: perception, action (and imagination) all rely on probability distributions, rather than fixed decodings, over the incoming stimuli with different reliability weightings determining the ensuing adjustments as responses to error signals. Such weightings derive from (a) current attentional resources (as in [13]), and (b) the reinforcement history of the system, i.e., reward/punishment values assigned via personal and social exposure to cultural norms that dictate perspectives for understanding phenomena in context. In combination, these two factors determine that NL users exploit NL variability, not only for adjusting their understanding, but also for the purposes of attaching social/personal values to their actions, e.g., meanings signifying identity and group-memberships that go beyond denotational meaning (see e.g. [20]).

From the evolutionary point of view, variability is a constant property of a dynamically changing environment to which organisms have to actively respond controlling its influence moment by moment. To actively exert such control, an agent must store as part of its constitution predictable dependencies between its actions and the resulting sensory stimulation ('sensorimotor contingencies'). NL behaviour, as modelled by DS, is subsumed in this action resources organisation. Without imposing identity of strategies/outcomes at the level of individual agency, at the level of the social unit, co-construction of stimuli and meanings of the kind seen in split utterances (see (1)–(5) earlier) allows each interlocutor's processing to influence the other's actions by establishing feedback loops and thus lessening unpredictability leading towards temporary synergies of compatibility and coordination (see also [95]). For this fine-grained influence to take place, it is essential that what is sustaining the interactions is mechanisms flexibly shaping courses of actual/virtual actions (as in the DS DAG) rather than manipulation of stored fixed codes/intentions/goals/contents. Thus, on the DS perspective, the phenomenological phenomenon at some particular time in literate societies of a reified NL (code) can be seen as emerging from the high social values attached to stable (but, in fact, 'metastable' [55]) system states temporarily settling in short-term outcomes even though long-term the underlying basis is ephemeral ever-changing processes. Over the long term, by iterated interaction coordinations among groups of individuals, successful processing paths become progressively routinised and grammaticalised/lexicalised, i.e. easily activated as whole sequences of basic actions (macros). Cross-linguistic and diachronic analyses in DS show how the appearance of distinct NLs arises through the establishment of different such routinisations [7,9,59] invested with social value. This also provides the possibility of explanatory modelling of recent cognitive evidence that processing and interaction constraints affect directly the design of the grammar itself [28,77]. From this point of view, NLs can function adaptively, but also maladaptively in sedimenting prejudices and exclusions, because they comprise just mechanisms of storing and deploying reliable, systematic action-perception contingencies valid in particular human ecological niches, mainly the social environment. The perspective from which to develop the view of NLs as evolved systems is then broadly functionalist: narrowly defined, NL grammars store action-outcome contingencies for attracting and exploiting interaction with other humans (but also one's self in reasoning, planning, imagining) due to the reward values attached to interaction outcomes; in turn, such stored dispositions by prompting interaction and enabling it via the establishment of feedback loops and potential to aggregate as macros are shaped themselves to generalise efficiently due to iterative attempts at coordination with various partners and in various circumstances. Building on this basis, NL grammars can then become the underpinnings of systems of significance concerning moral/emotional considerations (e.g. altruism) and cultural group formations.

Though an adaptation-oriented view of NLs is advocated by some [11], the gradualist adaptation claim for NL grammars is disputed by others [30,64] invoking problems in reality caused by the code view of NLs. Despite marked differences between the various stand-points, two putative problems are assumed even in models taking the adaptionist perspective: (i) the problem of "signalling signalhood", i.e., identifying communicative intentions; and (ii) the assumption that successful communication requires establishing some fixed and shared signal-content correspondence ('compositionality') intended by the speaker to be recovered by the hearer. From the DS perspective, (i)–(ii) are artifacts of the reified code view of NLs. Regarding (i), "signalling signalhood" leads to definitional infinite regress and is only a consideration if a Gricean underpinning of communication is assumed under which the inadequacy of a code in this respect has to be supplemented by mindreading capacities. DS instead derives behaviour coordination via the NL mechanisms themselves, namely, the predictivity and adjustment of system resources. Any stimulus can be exploited not as an intentional 'signal' but as an affordance/conditional-action, depending on the current state of the agent. While retaining the assumptions of productivity and systematicity, DS rejects (ii), the standard compositionality requirement that imposes fixed NL form-content mappings. Instead, NLs are modelled as relatively reliable, but also fallible, processes, sets of domain-general basic procedures for licensing domain-specific action sequences that either assimilate input from the (social) environment (parsing) or induce behaviour to acquire that input (speech or other actions). Since it is not a reasonable assumption to impose duplication of needs and goals among individuals, variability of action-control mechanisms between conversational partners, or the same agent at different times, is expected. Moreover, fallibility is the main source of innovation, creativity, and increased efficiency [46] so this is not considered as an inherent problem that should have been eliminated by evolution. This is in line with evidence that in cross-generational acquisition, the evolution of underspecification/polysemy/ambiguity enhances rather than disadvantages language learning/change [60].

5.1   Language as an Adaptive Group-Creating Mechanism

In turning to evolutionary patterns, a distinctive feature of enactive approaches to the biology of living beings is that organisms are defined as autonomous systems that resist disorder and dissolution. Agents of various nested orders can be defined in this way, from cells, to individual brains, to dyads and groups, all repeatedly re-defining and configuring their own boundaries during interactions in ways that promote adaptive success and survival (see also [15,72]). NL coordination as modelled by DS is one such means of determining and maintaining boundaries ('group functional organisation', [92]), i.e., inducing temporary stabilities for joint action via social learning and cultural transmission.

Going beyond the level of the individual, the evolutionary concept of group selection has been resurrected by Sober and Wilson [82], who argue that groups must be treated as capable of constituting adaptive units in their own right, alongside individuals, for appropriate explanation of many evolutionary strands (the so-called *Multi-Level Selection Hypothesis*: [82,91]). Under this view, in most cases, group-level and individual-level pressures compete, with pressures for individual interests having to be outweighed by significant fitness enhancement at the level of competition among groups for group benefiting traits to be favoured. However, unlike many other macro-traits (e.g. moral behaviour), NL use is one of the few cases where both individual and group-level fitness seem to be affected given that NL abilities transcend the boundaries of individual vs social agency. Under the DS view, this is unsurprising since NLs are defined as means for interaction with interaction influencing each other's fitness being the very criterion that defines the concept of 'group' [82]. Instead of the standard notions of 'altruism' and 'cooperation' which omit the contribution of competition and individual vs social tensions as evolutionary forces, interaction in the DS sense can be seen as the basis of NL adaptivity (see also [70]). Moreover, given the abandonment of the code model, interaction and coordination between members of a group does not require that all members of the group share identical dispositions or intentions; as long as the propensities and goals of any interactants complement each other, they will be able to coordinate [66,69].

But we can also take a more fine-grained view, given that the concept of a 'group' is defined differentially relative to particular traits. Core NL features, namely, vagueness/ambiguity (i.e., open-endedness) and systematicity, have been shown to emerge solely from cultural transmission (iterative learning) without intentional design [60,78]. The challenge is to explain why this might be so. From this perspective, the relevance of individual- vs group-level distinctions in fitness enhancement arises. Firstly, a sufficiently loose concept of compositionality (*systematicity*) for affordance indicators allowing for variability in form and effects while nevertheless presuming on reliable predictable contributions is expected to arise at the individual level: individual memorization/learnability [60] needs a finite stock to systematic productive effect and, given that meaning is an emergent and relational feature of interactions, deterministic outcomes are not expected or needed. It is, however, cases where group adaptivity clearly outweighs individual adaptivity which provide the stronger evidence, solving what is otherwise puzzling, namely, that vagueness and ambiguity seem problematic at the individual level: open-endedness gives rise to the ever-present risk of misunderstanding between the interactive parties; and the related psychological correlate of non-determinism, uncertainty, phenomenologically seems hugely problematic for individuals. Nevertheless, the advantage of non-determinism at the group level is very striking. Open-endedness of action/perception outcomes via adaptable mechanisms is what enables group establishment. Individuals may interpret the world around them, including their interactions with other people, relative to their own needs and desires and still act collectively in coordination. But this can only be achieved if interactions between the parties do not demand identity of mechanisms or outcomes (see also [83]). Any such condition on achieving coordination would make NL communication appear impossible. In fact, it is the other way round, risk, uncertainty about even our own goals and resources, and vague offerings are the sine qua non of communication since meaning and innovation emerge through interaction, instead of being initially located in one mind and having to be transferred to another. Successful employment of only apparently shared terms in the service of variable purposes/meanings is then explainable if the assumption is not made that prior "common ground" and subsequent duplication of contents is a necessary presupposition for joint action. Here history provides numerous examples of how the unifying force of NL terms across disjoint communities can achieve striking social success through enabling otherwise conflicting sub-groups to cooperate under a single label; examples include *Solidarity* of Poland, *Coordinadora* in Bolivia, and the Zapatista rising in Mexico [85].

6   Conclusion

The DS perspective aims to directly model the group-forming properties of NL interactions. First, NL communication is not viewed as convergence/common ground but as the employment of procedures enabling creative joint activities without overarching common goals. Secondly, we have barely scratched the surface of a great number of issues here only perhaps to argue sufficiently that it is notable that a common non-individualistic pattern can be discerned across NL learning, individual and institutional NL change, and evolution. At all stages, modelling relies in situated iteration: the entrenching effect of assigning higher probability weightings to iterated processing paths (given DS assumptions) leading to routinisation; the setting up of shortcuts in response to cognitive pressures for economy; all being buttressed by the group functional organisation which the interactivity induces. NL learning, change and even NL emergence can all then be seen in gradualistic terms, hence the higher-order organisation that incorporates the NL system itself can be argued to constitute an adaptive interactive system in continuity with the definition of living organisms as modelled in enactive approaches.

References

1. Altmann, G., Kamide, Y.: Incremental interpretation at verbs: restricting the domain of subsequent reference. Cognition **73**(3), 247–264 (1999)
2. Anderson, M.L.: After Phrenology. Cambridge University Press, Cambridge (2014)
3. Anderson, M.L.: Of Bayes and bullets. In: Metzinger, T., Wiese, W. (eds.) Philosophy and Predictive Processing Frankfurt. MIND Group, Frankfurt am Main (2017)
4. Arnon, I., Casillas, M., Kurumada, C., Estigarribia, B.: Language in interaction: Studies in honor of Eve V Clark, vol. 12. John Benjamins, Amsterdam (2014)
5. Berwick, R.C., Chomsky, N.: Why Only Us: Language and Evolution. MIT, London (2015)
6. Bickhard, M.H., Richie, D.M.: On the Nature of Representation: A Case Study of James Gibson's Theory of Perception. Praeger, New York (1983)
7. Bouzouita, M.: The Diachronic Development of Spanish Clitic Placement. Ph.D. thesis, King's College London (2008)
8. Cann, R., Kempson, R., Marten, L.: The Dynamics of Language. Elsevier, Oxford (2005)
9. Chatzikyriakidis, S.: Clitics in Grecia Salentina Greek: a dynamic account. Lingua **119**(12), 1939–1968 (2009)
10. Chomsky, N.: Knowledge of Language: Its Nature, Origin, and Use. Praeger, Santa Barbara (1986)
11. Christiansen, M.H., Chater, N.: Language as shaped by the brain. Behav. Brain Sci. **31**(5), 489–509 (2008)
12. Cisek, P.: Cortical mechanisms of action selection: the affordance competition hypothesis. Phil. Trans. R. Soc. B: Bio. Sci. **362**(1485), 1585–1599 (2007)
13. Clark, A.: Surfing Uncertainty: Prediction, Action, and the Embodied Mind. Oxford University Press, Oxford (2016)

14. Clark, A.: Busting out: predictive brains, embodied minds, and the puzzle of the evidentiary veil. Noûs **51**(4), 727–753 (2017)
15. Clark, A.: How to knit your own Markov blanket. In: Metzinger, T., Wiese, W. (eds.) Philosophy and Predictive Processing. MIND Group, Frankfurt am Main (2017)
16. Clark, E.V., Casillas, M.: First language acquisition. In: Allan, K. (ed.) The Routledge Handbook of Linguistics, pp. 311–329. Routledge, London (2016)
17. Clark, H.H.: Communal lexicons. In: Malmkjær, K., Williams, J. (eds.) Context in Language Learning and Language Understanding, vol. 4, pp. 63–87. Cambridge University Press, Cambridge (1998)
18. Cooper, R.: Type theory and semantics in flux. In: Kempson, R., Asher, N., Fernando, T. (eds.) Handbook of the Philosophy of Science. Philosophy of Linguistics, vol. 14, pp. 271–323. North Holland, Oxford (2012)
19. Dingemanse, M., Roberts, S.G., Baranova, J., Blythe, J., Drew, P., et al.: Universal principles in the repair of communication problems. PLOS ONE **10**(9), e0136100 (2015)
20. Eckert, P.: Three waves of variation study: the emergence of meaning in the study of sociolinguistic variation. Annu. Rev. Anthropol. **41**, 87–100 (2012)
21. Eshghi, A., Howes, C., Gregoromichelaki, E., Hough, J., Purver, M.: Feedback in conversation as incremental semantic update. In: Proceedings of the 11th IWCS, pp. 261–271. ACL (2015)
22. Eshghi, A., Purver, M., Hough, J.: DyLan: Parser for Dynamic Syntax. Technical Report of Queen Mary University, London (2011)
23. Eshghi, A.: DS-TTR: an incremental, semantic, contextual parser for dialogue. In: Proceedings of the 19th SemDial (2015)
24. Eshghi, A., Lemon, O.: How domain-general can we be? Learning incremental dialogue systems without dialogue acts. In: Proceedings of the 18th SemDial (2014)
25. Eshghi, A., Purver, M., Hough, J.: Probabilistic induction for an incremental semantic grammar. In: Proceedings of the 10th IWCS, pp. 107–118. Potsdam (2013)
26. Eshghi, A., Purver, M., Hough, J., Sato, Y.: Inducing lexical entries in an incremental semantic grammar. In: Proceedings of CSLP2012 (2012)
27. Eshghi, A., Shalyminov, I., Lemon, O.: Interactional dynamics and the emergence of language games. In: Proceedings of FADLI Workshop, ESSLLI (2017)
28. Fedzechkina, M., Chu, B., Florian Jaeger, T.: Human information processing shapes language change. Psychol. Sci. **29**(1), 72–82 (2017). https://doi.org/10.1177/0956797617728726
29. Ferreira, F., Chantavarin, S.: Integration and prediction in language processing: a synthesis of old and new. Curr. Dir. Psychol. Sci. **27**(6), 443–448 (2018)
30. Fitch, W.T., Hauser, M.D., Chomsky, N.: The evolution of the language faculty: clarifications and implications. Cognition **97**(2), 179–210 (2005)
31. Friston, K., Frith, C.: A duet for one. Conscious. Cogn. **36**, 390–405 (2015)
32. Gallagher, S.: Enactivist Interventions: Rethinking the Mind. Oxford University Press, Oxford (2017)
33. Gargett, A., Gregoromichelaki, E., Howes, C., Sato, Y.: Dialogue-grammar correspondence in Dynamic Syntax. In: Proceedings of the 12th SemDial (2008)
34. Gargett, A., Gregoromichelaki, E., Kempson, R., Purver, M., Sato, Y.: Grammar resources for modelling dialogue dynamically. Cogn. Neurodyn. **3**(4), 347–363 (2009)
35. Gibson, J.J.: The Ecological Approach to Visual Perception. Psychology Press, Oxford (1979)

36. Ginzburg, J.: The Interactive Stance. OUP, Oxford (2012)
37. Gregoromichelaki, E.: Conditionals: a Dynamic Syntax account. Ph.D. thesis, King's College London (2006)
38. Gregoromichelaki, E.: Grammar as action in language and music. In: Orwin, M., Kempson, R., Howes, C. (eds.) Language, Music and Interaction, pp. 93–134. College Publications, London (2013)
39. Gregoromichelaki, E.: Quotation in dialogue. In: Saka, P., Johnson, M. (eds.) The Semantics and Pragmatics of Quotation. Perspectives in Pragmatics, Philosophy & Psychology, vol. 15. Springer, Cham (2017). https://doi.org/10.1007/978-3-319-68747-6_8
40. Gregoromichelaki, E., Kempson, R.: Joint utterances and the (split-) turn taking puzzle. In: Mey, J.L., Capone, A. (eds.) Interdisciplinary Studies in Pragmatics, Culture and Society. Perspectives in Pragmatics, Philosophy & Psychology, vol. 4. Springer, Heidelberg (2016). https://doi.org/10.1007/978-3-319-12616-6_28
41. Gregoromichelaki, E., Kempson, R., Howes, C., Eshghi, A.: On making syntax dynamic. In: Wachsmuth, I., de Ruiter, J.P., Jaecks, P., Kopp, S. (eds.) Alignment in Communication, pp. 57–85. John Benjamins, Amsterdam (2013)
42. Gregoromichelaki, E., Kempson, R., Purver, M., Mills, G.J., Cann, R., Meyer-Viol, W., Healey, P.G.T.: Incrementality and intention-recognition in utterance processing. Dialogue and Discourse 2(1), 199–233 (2011)
43. Gregoromichelaki, E., et al.: Completability vs (in)completeness. Acta Linguistica Hafniensia (forthcoming)
44. Gregoromichelaki, E., Sato, Y., Kempson, R., Gargett, A., Howes, C.: Dialogue modelling and the remit of core grammar. In: Proceedings of IWCS (2009)
45. Healey, P.G.T., Purver, M., Howes, C.: Divergence in dialogue. PLOS ONE 9(6), 98598 (2014)
46. Healey, P.G.T., de Ruiter, J.P., Mills, G.J.: Editors' introduction: miscommunication. Top. Cogn. Sci. 10(2), 264–278 (2018)
47. Hendriks, P.: Asymmetries Between Language Production and Comprehension. Studies in Theoretical Psycholinguistics, vol. 42, 1st edn. Springer, Dordrecht (2014). https://doi.org/10.1007/978-94-007-6901-4
48. Hilbrink, E.E., Gattis, M., Levinson, S.C.: Early developmental changes in the timing of turn-taking: a longitudinal study of mother-infant interaction. Front. Psychol. 6, 1492–1492 (2015)
49. Hough, J.: Modelling incremental self-repair processing in dialogue. Ph.D. thesis, Queen Mary University of London (2014)
50. Hough, J., Purver, M.: Probabilistic type theory for incremental dialogue processing. In: 2014 Proceedings of the EACL, pp. 80–88. ACL, April 2014
51. Howes, C., Eshghi, A.: Feedback relevance spaces. In: 2017 Proceedings of IWCS (2017)
52. Howes, C., Healey, P.G.T., Purver, M., Eshghi, A.: Finishing each other's ... responding to incomplete contributions in dialogue. In: Proceedings of the 34th Annual Meeting of the Cognitive Science Society, pp. 479–484 (2012)
53. Hurley, S.: The shared circuits model. Behav. Brain Sci. 31(1), 1–22 (2008)
54. Kalatzis, D., Eshghi, A., Lemon, O.: Bootstrapping incremental dialogue systems. In: Proceedings of Conference on Neural Information Processing Systems (2016)
55. Kelso, J.S.: Multistability and metastability: understanding dynamic coordination in the brain. Phil. Trans. R. Soc. B 367(1591), 906–918 (2012)
56. Kempson, R., Cann, R., Gregoromichelaki, E., Chatzikyriakidis, S.: Language as mechanisms for interaction. Theor. Linguist. 42(3-4), 203–276 (2016)

57. Kempson, R., Gregoromichelaki, E.: Action sequences instead of representational-levels. Behav. Brain Sci. **40**, 296 (2017)

58. Kempson, R., Meyer-Viol, W., Gabbay, D.: Dynamic Syntax: The Flow of Language Understanding. Blackwell, Oxford (2001)

59. Kiaer, J.: Processing and interfaces in syntactic theory: the case of Korean. Ph.D. thesis, King's College London (2007)

60. Kirby, S., Smith, K., Cornish, H.: Language, learning and cultural evolution. In: Cooper, R., Kempson, R. (eds.) Language in Flux, pp. 81–108. College Publications, London (2008)

61. Kuperberg, G.R., Jaeger, T.F.: What do we mean by prediction in language comprehension? Lang. Cogn. Neurosci. **31**(1), 32–59 (2016)

62. Lerner, G.H.: Collaborative turn sequences. In: Conversation Analysis: Studies from the First Generation, pp. 225–256. John Benjamins, Amsterdam (2004)

63. Lerner, G.H.: On the place of linguistic resources in the organization of talk-in-interaction: grammar as action in prompting a speaker to elaborate. Res. Lang. Soc. Interact. **37**(2), 151–184 (2004)

64. Lightfoot, D.: The Development of Language. Blackwell, Oxford (1999)

65. Lowder, M.W., Ferreira, F.: Prediction in the processing of repair disfluencies. J. Exp. Psychol.: Learn. Mem. Cogn. **42**(9), 1400–1416 (2016)

66. Millikan, R.: The Varieties of Meaning: The Jean-Nicod Lectures. MIT Press, Cambridge (2004)

67. Millikan, R.G.: On Clear and Confused Ideas. Cambridge University Press, Cambridge (2000)

68. Mills, G., Gregoromichelaki, E.: Establishing coherence in dialogue: sequentiality, intentions and negotiation. In: Proceedings of the 14th SemDial (2010)

69. Mills, G.J.: Dialogue in joint activity. New Ideas Psychol. **32**, 158–173 (2014)

70. Nettle, D., Dunbar, R.I.M.: Social markers and the evolution of reciprocal exchange. Curr. Anthropol. **38**(1), 93–99 (1997)

71. Odling-Smee, F.J., Laland, K.N., Feldman, M.W.: Niche Construction: The Neglected Process in Evolution, vol. 37. Princeton University Press, Princeton (2003)

72. Paolo, E.A.D., Cuffari, E.C., Jaegher, H.D.: Linguistic Bodies. MIT Press, Cambridge (2018)

73. Piaget, J.: The Construction of Reality in the Child. Basic Books, New York (1954)

74. Pickering, M.J., Ferreira, V.S.: Structural priming: a critical review. Psychol. Bull. **134**(3), 427 (2008)

75. Pickering, M.J., Garrod, S.: An integrated theory of language production and comprehension. Behav. Brain Sci. **36**, 329–347 (2013)

76. Purver, M., Gregoromichelaki, E., Meyer-Viol, W., Cann, R.: Splitting the 'I's and crossing the 'You's. In: Proceedings of the 14th SemDial, pp. 43–50 (2010)

77. Roberts, S.G., Levinson, S.C.: Conversation, cognition and cultural evolution. Interact. Stud. **18**(3), 402–429 (2017)

78. Santana, C.: Ambiguity in cooperative signaling. Philos. Sci. **81**(3), 398–422 (2014)

79. Sato, Y.: Local ambiguity, search strategies and parsing in dynamic syntax. In: Gregoromichelaki, E., Kempson, R., Howes, C. (eds.) The Dynamics of Lexical Interfaces. CSLI Publications, Stanford (2011)

80. Saxton, M.: Child Language. Sage, Los Angeles (2017)

81. Shalyminov, I., Eshghi, A., Lemon, O.: Challenging neural dialogue models with natural data. In: Proceedings of the 21st SemDial. pp. 98–106 (2017)

82. Sober, E., Wilson, D.S.: Unto Others, vol. 218. Harvard University Press, Cambridge (1998)

83. Sowa, J.F.: Language games. In: Pietarinen, A.V. (ed.) Game Theory and Linguistic Meaning, pp. 17–38. Elsevier, Amsterdam (2007)
84. Sperber, D., Wilson, D.: Relevance, 2nd edn. Blackwell, Oxford (1995)
85. Tarrow, S.: Power in Movement. Cambridge University Press, Cambridge (2011)
86. Thompson, E.: Mind in Life. Harvard University Press, Cambridge (2007)
87. Tomasello, M.: Origins of Human Communication. MIT Press, Cambridge (2008)
88. Trevarthen, C.: The child's need to learn a culture. Child. Soc. **9**(1), 5–19 (1995)
89. Trueswell, J.C., Tanenhaus, M.K.: Approaches to Studying World-Situated Language Use. MIT Press, Cambridge (2005)
90. Vygotski, L.S.: Thought and Language. MIT Press, Cambridge (2012)
91. Wilson, D.S.: Darwin's Cathedral. University of Chicago Press, Chicago (2002)
92. Wilson, D.S.: Does Altruism Exist?: Culture, Genes, and the Welfare of Others. Yale University Press, New Haven (2015)
93. Yu, Y., Eshghi, A., Lemon, O.: Incremental generation of visually grounded language in situated dialogue. In: Proceedings of INLG, Los Angeles (2016)
94. Yu, Y., Eshghi, A., Lemon, O.: Learning how to learn: an adaptive dialogue agent for incrementally learning visually grounded word meanings. In: RoboNLP (2017)
95. Zawidzki, T.W.: Mindshaping. MIT Press, Cambridge (2013)
96. Zeevat, H.: Language Production and Interpretation. Brill, Frankfurt (2014)

Bridging Inferences in a Dynamic Frame Theory

Ralf Naumann[✉] and Wiebke Petersen

Institut für Sprache und Information, Universität Düsseldorf, Düsseldorf, Germany
`naumann@phil-fak.uni-duesseldorf.de`

Abstract. In this article we develop a theory of bridging inferences in a dynamic frame theory that is an extension of Incremental Dynamics. In contrast to previous approaches bridging is seen as based on predictions/expectations that are triggered by discourse referents in a particular context where predictions are (more specific) instances of Questions under Discussion. In our frame theory each discourse referent is associated with a frame f that contains the information known about it in the current context. Predictions/QuDs are modelled as sets F of extensions of this frame relative to a (possibly complex) attribute about whose value no information is given so far. A continuation of the current context answers a question if it introduces a frame f' that contains information about the value of the attribute corresponding to the question. The set F is constrained by a probability distribution on the domain of frames. Only those extensions are considered whose conditional probability in the current context is high. The relation between f and f' can be restricted in several ways. Bridging inferences correspond to those restrictions in which (i) the frames belong to the semantic representations of two clauses and (ii) the relation is established by a separate update operation (The research was supported by the German Science Foundation (DFG) funding the Collaborative Research Center 991. We would like to thank the two reviewers as well as the editors for helpful comments and suggestions).

1 Introduction

It is by now a well-known fact that the semantic processing of an utterance usually involves different sources of information which are used in parallel to arrive at a coherent interpretation of this utterance in the given context. Four principle sources must be distinguished: (i) the (linguistic) meaning of the lexical items; (ii) (non-linguistic) world and situational knowledge, (iii) the prior linguistic context and (iv) the information structure of the text, i.e. the way sentences are related by coherence relations and questions under discussion. A prime example of this interplay between different sources of information are bridging inferences. [AL98, 83p.] take bridging to be 'an inference that two objects or events that are introduced in a text are related in a particular way that isn't explicitly stated, and yet the relation is an essential part of the content of the

© Springer-Verlag GmbH Germany, part of Springer Nature 2019
A. Silva et al. (Eds.): TbiLLC 2018, LNCS 11456, pp. 228–252, 2019.
https://doi.org/10.1007/978-3-662-59565-7_12

text in the sense that without this information, the lack of connection between the sentences would make the text incoherent.' Examples of bridging inferences are given in (1) and (2).

(1) a. Lizzy met a dog yesterday. The dog was very friendly. [AL98, 86p.]
     b. John unpacked the picnic. The beer was warm. [CH77]
     c. I was at a wedding last week. The mock turtle soup was a dream.
         [Geu11]
     d. I've just arrived. The camel is outside and needs water. [AL98, 86p.]

(2) a. In the group there was one person missing. It was Mary who left.
     b. John partied all night. He's going to get drunk again today.
     c. Jack was going to commit suicide. He bought a rope. [Cha83]

Bridging inferences are most prominently related to definite descriptions, witness the examples in (1). This is, however, not the only possibility. They can also be triggered by 'it'-clefts, (2-a), temporal adverbials like 'again', (2-b), and indefinites like 'a rope' in (2-c), as shown by the examples in (2).

Common to all bridging inferences is (i) a new discourse referent is introduced (see [Bur06] for neurophysiological evidence) and (ii) a dependency (bridging) relation between this discourse referent (corresponding to the bridged expression) and a discourse referent that has already been introduced in the linguistic context (denoting an antecedent object) is established. The difference between (1-a) on the one hand and (1-b), (1-c) on the other is that in the former case the newly introduced discourse referent must be identical to an already introduced discourse referent (familiarity condition) whereas in the latter two cases the newly introduced discourse referent is required to be distinct from any discourse referent introduced so far (novelty condition).[1]

Bridging inferences are often related to a presupposition. For example, the definite description 'the dog' in (1-a) triggers the presupposition that there is a unique dog in the context. The bridging inference consists in establishing a link between 'a dog' and 'the dog'. In this case the dependency relation is the identity relation. The dog introduced in the first sentence by the indefinite is identical to the dog denoted by the definite description in the second sentence. In (2-b) 'again' can be used felicitously only if John got drunk before today. The bridging inference is the inference that the previous occasion of John getting drunk was concurrent with his partying all night the day before. As noted in [Cla77] and [AL98], bridging inferences may occur in the absence of presupposition triggers as well. An example is (2-c) with the bridged expressing 'a rope' and the bridging inference that the rope is related to the planned suicide. It was the instrument to be used by Jack in his plan. An important further aspect of bridging inferences is that they provide additional information about the antecedent object. For example, in (1-d) the additional information is about the arriving event. The means of transport used in this event, or the presupposed moving event

---

[1] This difference will be reflected in the formal theory by having two bridging conditions. See below, Sects. 6 and 7 for details.

leading to the arrival, was a camel. In (2-c) the rope is the instrument used in the planned suicide and in (1-a) a comprehender gets to know that the dog introduced in the first sentence was very friendly. (1-c) shows that the dependency can be indirect. The turtle soup is directly related as a part (starter) of the meal which was served at the wedding. The examples in (1) and (2) in addition show that the dependency relation can be instantiated by various forms of relations: (a) identity (1-a), (b) constituent part-of (e.g. (1-b)) or concurrency, (2-b) (cf. [Cla77] for a comprehensive taxonomy). Due to lack of space, we will restrict bridging inferences to those cases involving NPs of the form 'the N' and 'an N' and hence to examples like those in (1) and (2-c).

The rest of the article is organized as follows. In Sect. 2 we provide the formal definition of the frame theory in which bridging examples are analyzed. In Sect. 3 this theory is embedded into Incremental Dynamics, [vE07]. Section 4 outlines the approach of [AL98] which is the starting point for our analysis of bridging presented in the subsequent sections. We start by relating bridging inferences to 'Questions under Discussion' in Sect. 5. In the following section this relation is informally extended to our frame theory. Finally, in Sect. 7 the ideas developed in the preceding two sections are formally made precise. In particular, we define two bridging conditions that relate a bridged expressions to its antecedent and two corresponding update operations.

2   Frame Theory

At their core frames are attribute-value structures. Their strength for an analysis of bridging inferences lies in the fact that they allow for a fine-grained analysis of individuals and events.

Consider the frame in Fig. 1.

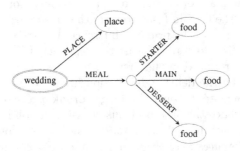

Fig. 1. Wedding frame

This frame can be taken as a partial description of a wedding.[2] This wedding takes place at a particular location and the meal served had three parts: a starter,

---

[2] Alternatively, it can be taken as a frame scheme or a frame type. In this case it refers to the set of weddings which have a location and in which the meal is made up by a starter, a main course and a dessert. In the text, a frame depicted is always meant as an instantiated frame in the sense that each node has a particular object as value.

a main course and a dessert. This example shows that a frame contains two different kinds of information: relational information which links two objects in a frame via a chain of attributes and sortal information which classifies an object in the frame as belonging to a particular class (or sort) of objects. Relational information is represented by labeled arcs where the label indicates the arc. Sortal information is represented by circles with the sort being indicated by the label inside the circle. For example, the wedding is mapped to its location by the attribute PLACE and to the dessert served by the chain of attributes MEAL and DESSERT in that order. All three components of the meal are classified as being of sort **food**. Relational and sortal information are linked in a particular way. For each attribute, there is a source sort and a target sort. For example, the attribute PLACE has as source sort **physical object** including both individuals (human beings, engines, dogs etc.) and events (weddings, hittings, buying, eatings etc.) and as target sort objects of sort **place**. The target sort of the attributes STARTER, MAIN and DESSERT are all **food**. The same holds for the chains made up by the attributes MEAL and STARTER, MEAL and MAIN as well as MEAL and DESSERT.

Next, we will make the informal characterization given above more precise. One way of looking at the above figure is in terms of a relational model $\mathcal{M}$. Each chain of attributes is satisfied in a corresponding model relative to two objects and each sort formula is satisfied relative to a single object. This perspective on frames makes them similar to possible worlds which, too, are taken as relational models according to one formal representation. We will follow the lead of possible world semantics and two-sorted type theory in which possible worlds are objects of a domain D_w (and not relational models) and take frames as elements of a domain D_f of frames (and not as relational models). The link between a frame and the relational structure associated with it is defined indirectly, again similar to two-sorted type theory. Instead of interpreting attributes as functional relations on $D_o \times D_o$ with D_o the domain of objects comprising both individuals (human beings, chair, dogs etc.) and events (writings, pushings etc.), they are interpreted as ternary relations on $D_f \times D_o \times D_o$. For example, for ATTR an atomic attribute symbol like MEAL or STARTER, $[\![\text{ATTR}]\!]$ is a function that assigns to a frame f a binary relation on D_o s.t. $[\![\text{ATTR}]\!](f)(o)(o')$ is true if o and o' are related by ATTR in f.[3] Similarly, sort formulas are interpreted as (boolean combinations of) binary relations on $D_f \times D_o$. As already mentioned above, this way of relativizing the interpretation of expressions is similar to the way information is made world-dependent in two-sorted type theory. The formal definitions are given next. Let $\Sigma = \langle Sort, Attr \rangle$ be a frame signature of (atomic) sort and attribute symbols, respectively, with $Sort \cap Attr = \emptyset$. The frame language $\mathcal{L}$ based on Σ is defined in (3) and its interpretation is given in (4). The interpretation function $[\![\,]\!]$ assigns to each $\sigma \in Sort$ a binary relation on $D_f \times D_o$ and to each ATTR $\in Attr$ a ternary relation on $D_f \times D_o \times D_o$.

---

[3] Strictly speaking, it assigns to a frame f a 1-place function as attributes are required to be functional.

(3)   a. $\phi ::= \sigma \,|\, \neg\phi \,|\, \phi_1 \wedge \phi_2$

      b. $\pi ::= \Delta \,|\, \text{ATTR} \,|\, \pi_1 \cap \pi_2 \,|\, \pi_1 \bullet \pi_2 \,|\, \uparrow\phi \,|\, \downarrow\phi$

(4)   a. $[\![\sigma]\!](f)(o)=1$ iff $\langle f, o \rangle \in I(\sigma)$.

      b. $[\![\neg\phi]\!](f)(o)=1$ iff $[\![\phi]\!](f)(o)=0$.

      c. $[\![\phi \wedge \psi]\!](f)(o)=1$ iff $[\![\phi]\!](f)(o)=1$ and $[\![\psi]\!](f)(o)=1$.

      d. $[\![\text{ATTR}]\!](f)(o)(o')=1$ iff $\langle f, o, o' \rangle \in I(\text{ATTR})$.

      e. $[\![\pi \cap \pi']\!](f)(o)(o')=1$ iff $[\![\pi]\!](f)(o)(o')=1$ and $[\![\pi']\!](f)(o)(o')=1$.

      f. $[\![\pi \bullet \pi']\!](f)(o)(o')=1$ iff $\exists o'' : [\![\pi]\!](f)(o)(o'')=1$ and $[\![\pi']\!](f)(o'')(o')=1$.

The clauses for sort formulas are self-evident. Boolean operations besides $\neg$ and $\wedge$ are defined in the usual way. At the level of relational information, $\bullet$ is sequencing. It is used to built chains of attributes and is defined only if the target sort of the first attribute is a subsort of the source sort of the second attribute (details follow below). The intersection $\cap$ operator is similar to (boolean) conjunction at the level of sortal formulas. It requires that two objects in a frame satisfy both the relation formulas π and π'. Its main use in our frame theory is explained below.

So far, sortal and relational information are not connected with each other. However, as was said above, each attribute has both a source and a target sort. Therefore, one wants to say that the object at the end of a chain π satisfies the sortal information expressed by the sortal formula ϕ (3-a). Similarly, this information should also be expressible for the source sort. It is therefore necessary to go from the relational to the sortal level. This is achieved by two operators $\uparrow$ and $\downarrow$. Formula $\uparrow\phi$ is true at a triple $\langle f, o, o' \rangle$ if o' satisfies the sortal information ϕ in f, i.e. one has ϕ is true for $\langle f, o' \rangle$. Hence, $\uparrow$ 'projects' a relation in a frame to the second object in this relation and classifies it by the sortal information expressed by its argument. By contrast, $\downarrow$ projects to the first object. The satisfaction clauses are given in (5).

(5)   a. $[\![\uparrow \phi]\!](f)(o)(o')=1$ iff $[\![\phi]\!](f)(o')=1$.

      b. $[\![\downarrow \phi]\!](f)(o)(o')=1$ iff $[\![\phi]\!](f)(o)=1$.

Having $\uparrow$ and $\downarrow$ together with $\cap$ allows to express the information that at the end (beginning) of a chain sortal information ϕ holds. This is achieved by relation formulas of the form $\pi \cap \uparrow\phi$ and $\pi \cap \downarrow\phi$. For example, in the wedding frame one has MEAL $\bullet$ STARTER $\cap$ $\uparrow$**food**. This formula expresses that the wedding is related to an object of sort **food** by the chain MEAL $\bullet$ STARTER. Of course, this information can be made more specific by requiring that the starter is a subsort of **food**, e.g. a (mock turtle) soup: MEAL $\bullet$ STARTER $\cap$ $\uparrow$**soup**. As it stands, we also need to say that the object at the root of a frame satisfies the sortal information ϕ without any additional relational information. This case arises for instance for minimal frames which only contain sortal but no relational information. This kind of information is expressed by means of the null-ary operator Δ. Δ holds of a triple $\langle f, o, o' \rangle$ if one has $o = o'$. Hence, $\Delta \cap \downarrow\phi$ is true in a frame f and objects o and o' if ϕ is true in f for o and o' is identical to o. For example, in the wedding frame $\Delta \cap \downarrow$**wedding** is true at the root of the frame. Note that this relation formula does not contain any (chain of) attributes. The satisfaction clause for Δ is given below.

(6) $[\![\Delta]\!](f)(o)(o')=1$ iff $o=o'$.

The three domains D_f (frames), D_o (objects) and D_w (possible worlds) are related in the following way. First, for each frame f, there is a unique object $o \in D_o$ about which f contains information. This relation is captured by a function $root$ which assigns to each $f \in D_f$ the object $root(f) \in D_o$. If $root(f) = o$, f is called a frame associated with o. Second, each frame belongs to a unique possible world $w \in D_w$. This relation is captured by a function IN that maps each $f \in D_f$ to the world $IN(f) \in D_w$ to which it belongs.[4] Given these functions, a frame can be taken as a partial description of its root in the world to which the frame belongs.

The relation between a frame and a particular relational structure is defined in terms of a function θ that maps a frame f to the set of relations about which it contains information relative to its referent $root(f)$. Elements of $\theta(f)$ are based on relation formulas $\text{ATTR}_1 \bullet \cdots \bullet \text{ATTR}_n \cap \uparrow\sigma$ for chains of length greater 0 and $\Delta \cap \downarrow\sigma$ for sortal information at the root of the frame. Hence, $\theta(f)$ contains for each chain π in the frame this chain together with sortal information at the end of the chain and sortal information about its root. For example, for the wedding frame above one has $\theta(f_{wedding}) = \{\Delta \cap \downarrow\textbf{wedding},\ \text{PLACE} \cap \uparrow\textbf{place},\ \text{MEAL} \cap \uparrow\textbf{meal},\ \text{MEAL} \bullet$ $\text{STARTER} \cap \uparrow\textbf{food},\ \text{MEAL} \bullet \text{MAIN} \cap \uparrow\textbf{food},\ \text{MEAL} \bullet \text{DESSERT} \cap \uparrow\textbf{food}\}$.[5] Due to the use of $\cap$, Δ, $\uparrow$ and $\downarrow$ all elements of θ for a frame f are relation formulas and, hence, interpreted as functional relations on $D_f \times D_o \times D_o$. To underline that θ is based on chains in a frame, we write $\pi \in \theta(f)$ whenever there is a $\sigma \in Sort$ such that $\pi \cap \uparrow\sigma \in \theta(f)$. θ is closed both under prefixes of attribute chains and supersorts. For closure under prefixes of attribute chains, one has: if $\text{ATTR}_1 \bullet \cdots \bullet \text{ATTR}_n \cap \uparrow\sigma \in$ $\theta(f)$ then $\text{ATTR}_1 \bullet \cdots \bullet \text{ATTR}_{n-1} \cap \uparrow\sigma' \in \theta(f)$ for σ' the target sort (or one of its subsorts) of the attribute ATTR_{n-1}.[6] Closure under supersorts says that if $\pi \cap \uparrow\sigma \in \theta(f)$ and σ' is a supersort of σ, then $\pi \cap \uparrow\sigma' \in \theta(f)$. In the sequel the

---

[4] If possible worlds and frames are taken as relational models, the relation between them can be made precise in the following way. Each frame is a particular submodel $\mathcal{M}$ of a possible world $\mathcal{M}_w$. $\mathcal{M}$ is constructed from $\mathcal{M}_w$ as follows. In a first step one forms the reduct $\mathcal{M}'$ of $\mathcal{M}_w$ to the language $\mathcal{L}$ on which the frame is based. In a second step, one considers the set S of submodels $\mathcal{N}$ of $\mathcal{M}'$ that satisfy the axioms imposed on the frame. A frame is then any minimal model in S. See [NP17] for details.

[5] Though the elements are relations, we write for example $\pi \cap \uparrow\sigma$ instead of $[\![\pi \cap \uparrow\sigma]\!]$ to ease readability.

[6] This definition of closure under prefixes assumes a function that maps each ATTR to its target sort σ. In [NP17] this functional relation between an attribute and a sort is defined in an extension of the current theory that is based on an order-sorted logic for attributes and sorts. In particular, one has in the signature of this logic a $Sort \times Sort$-indexed family of sets of (attribute) function symbols $(\mathcal{A}_{\sigma,\sigma'})_{\sigma,\sigma' \in Sort}$. For $\text{ATTR} \in \mathcal{A}_{\sigma,\sigma'}$, one writes $\text{ATTR} : \sigma \to \sigma'$. σ is called the *source sort* and σ' the *target sort* of ATTR. See [NP17] for details.

chains with a supersort of a given sort will not be included if the value of θ is given for a frame f. Frames with the same referent (root) can be ordered according to the information they contain about their common referent. This is captured by the relation $\sqsubseteq$ on the domain of frames.

(7) $f \sqsubseteq f'$ iff $root(f) = root(f')$ and $IN(f) = IN(f')$ and $\forall \pi.(\pi \in \theta(f) \rightarrow \pi \in \theta(f'))$ and if $\pi \cap \uparrow \sigma \in \theta(f)$, and hence $\pi \in \theta(f')$, then there is some σ' with $\pi \cap \uparrow \sigma' \in \theta(f')$ such that $\forall o.\forall o'.[\![\pi \cap \uparrow \sigma']\!](f')(o)(o') \rightarrow [\![\pi \cap \uparrow \sigma]\!](f)(o)(o')))$.

$f \sqsubseteq f'$ holds if f and f' have the same root and belong to the same world. In addition, the information contained in f is a subset of the information contained in f'. This is the case if all chains belonging to $\theta(f)$ also belong to $\theta(f')$ and whenever a pair of objects satisfies a chain of f' that already belongs to f, the pair satisfies the same chain in f as well. The latter condition is necessary to account for the fact that the more specific frame f' may differ from the subsumed frame f by (a) the set of chains and (b) the specificity of the sortal restrictions added to the chains. Implicit in $\sqsubseteq$ is the fact that a frame is a partial description of an object. For example, the wedding frame $f_{wedding}$ at the beginning of this section is a particular element of the hierarchy for frames of sort **wedding**. It does not contain information about the bride or the broom. Adding this information yields a frame $f'_{wedding}$ with more information about the concept 'wedding'. This latter frame is higher in the frame hierarchy since one has $f_{wedding} \sqsubseteq f'_{wedding}$. $\theta(f'_{wedding})$ is $\theta(f_{wedding})$ augmented with chains for the bride and the broom. By contrast, leaving out the chain PLACE results in a less informative frame $f^*_{wedding}$ for which one has $\theta(f^*_{wedding}) = \theta(f_{wedding}) - \{PLACE \cap \uparrow \textbf{place}\}$. The minimal element in the 'wedding' frame hierarchy has $\theta(f^{min}_{wedding}) = \{\Delta \cap \downarrow \textbf{wedding}\}$. This information $(\Delta \cap \downarrow \textbf{wedding})$ can be further generalized to $\Delta \cap \downarrow \textbf{object}$. Though this information no longer classifies the wedding as a wedding and therefore does not, when taken in isolation, correspond to a frame in the 'wedding' frame hierarchy, it is the minimal frame in the 'object' frame hierarchy.

A second relation between two frames is that of one frame being a subframe of another. Let us illustrate this notion by some examples from the wedding frame. First, the whole wedding frame is a subframe of itself. The frame starting at the MEAL attribute is a subframe of the wedding frame. Let this subframe be f_{meal}. Its information is given by $\theta(f_{meal}) = \{\Delta \cap \downarrow \textbf{meal}, STARTER \cap \uparrow \textbf{food}, MAIN \cap \uparrow \textbf{food}, DESSERT \cap \uparrow \textbf{food}\}$. f_{meal} is the maximal subframe starting at the MEAL attribute. This subframe is a partial description of the meal that was served at the wedding. One of the subframes of this frame is the frame whose only attribute is STARTER which partially describes a meal by saying that it has a starter of sort **food**. These examples show that a subframe is always defined relative to a chain of attributes π corresponding to a relation formula $\pi \cap \uparrow \sigma$. For subframes starting at the root, one has $\pi = \Delta$ and for other subframes π always is of the form $\text{ATTR}_1 \bullet \cdots \bullet \text{ATTR}_n$. This notion, denoted by $\preceq_\pi$, is defined in (8).

(8) $f' \preceq_\pi f$ iff $\pi \in \theta(f)$ and f' satisfies the following conditions with respect to f and π: (a) $IN(f') = IN(f)$. (b) $root(f') = \iota o'.[\![\pi]\!](f)(root(f))(o')$. (c) Let $\theta_\pi(f) = \{\pi' \cap \uparrow\sigma' \mid \pi \bullet \pi' \cap \uparrow\sigma' \in \theta(f)\}$ and let S be a prefix-closed subset of $\theta_\pi(f)$ (i.e., $S \subseteq \theta_\pi(f)$ with if $\pi \bullet \text{ATTR} \in S$, then $\pi \in S$). Then $\theta(f') = S \cup \{\Delta \cap \downarrow\sigma \mid \pi \cap \uparrow\sigma \in \theta(f)\}$.

A frame f' can only be a subframe of f with respect to a chain π if π is a chain in f: $\pi \in \theta(f)$. In order to determine a subframe f' it is sufficient to specify the value of θ for f' and its root and the world it belongs to. A subframe is always required to be in the same world as its superframe: $IN(f') = IN(f)$. The root of f' is the object at the end of the chain π in f: $root(f') = \iota o'.[\![\pi]\!](f)(root(f))(o')$. The object o' is related to a set of o'-rooted frames f'. f' is a subframe of f only if f' contains a subset of the information about o' that f contains about o'. The information contained in f about o' is given by the suffixes π' with sortal information σ' of all chains $\pi \bullet \pi'$ such that $\pi \bullet \pi' \cap \uparrow\sigma' \in \theta(f)$. Let this set be $\theta_\pi(f) = \{\pi' \cap \uparrow\sigma' \mid \pi \bullet \pi' \cap \uparrow\sigma' \in \theta(f)\}$. The information about o' contained in f' is then a subset of $\theta_\pi(f)$ together with the sortal information at the root which is given by $\Delta \cap \downarrow\sigma$. The fact that $\theta(f')$ is required to be only a subset of the corresponding set in f accounts for the fact that the subframe relative to π and f is in general not unique as shown by the example of the two subframes starting at the MEAL attribute above. Since θ is required to be closed under chain-prefixes, it follows that if a chain $\pi' = \text{ATTR}_1 \bullet \cdots \bullet \text{ATTR}_n$ is in $\theta(f')$, then $\pi'' = \text{ATTR}_1 \bullet \cdots \bullet \text{ATTR}_{n-1}$ is in $\theta(f')$ too. The relation $\preceq$ is the union of the $\preceq_\pi$ for π an admissible chain for frames of the sort in question.

The relations $\sqsubseteq$ and $\preceq$ will play a central role in the analysis of bridging inferences below in Sects. 6 and 7. The interpretation of a bridged expression, for example 'the mock turtle soup' in a context in which a wedding was introduced previously, provides a frame f^* that is required to be a subframe of an extension of the frame for the wedding.

3   Combining Incremental Dynamics and Frame Theory

Our frame theory is integrated in *Incremental Dynamics*, [vE07, Nou03]. In this framework information states are defined as sets of stacks (also called 'contexts') and not as sets of (partial) variable assignments, as it is standardly done in model-theoretic semantics. A stack can be thought of as a function from an initial segment $\{0, \ldots, n-1\}$ of the natural numbers $\mathbb{N}$ to entities of a domain D_o that are stored in the stack. Hence, a stack can equivalently be taken as a sequence of objects $\{\langle 0, d_0 \rangle, \ldots, \langle n-1, d_{n-1} \rangle\}$ of length n. If c is a stack, $|c|$ is the length of c. The objects stored in a stack are the *discourse objects*. By $c(i)$ we denote the object at position i at stack c. A link between stack positions and discourse objects that are stored at a position is established by two operations. First, there is a pushing operation:

(9) $c^\sqcap d := c \cup \{\langle |c|, d \rangle\}$.

Pushing an object d on the stack extends the stack by this element at position $|c|$. The pushing operation will be used in the interpretation of the domain extension operation that introduces a new object on the stack, which, in turn is part of the interpretation of determiners (see below for details). The second operation retrieves an object from the stack.

(10) $ret := \lambda i. \lambda c. \iota d. c(i) = d$.

We write $c[i]$ for $ret(i)(c)$. The retrieval operation will become part of the interpretation of common nouns and verbs. For details on Incremental Dynamics without frames see [Nou03, vE07].

In the remainder of this section, we are going to incorporate frames into Incremental Dynamics. The first, and most important, modification is related to the type of objects that are stored in a stack. Storing only objects is insufficient to account for bridging inferences. Recall that a bridging inference basically consists in relating information about two objects in the discourse with each other. For example, a mock turtle soup becomes related to a wedding. Viewed from the perspective of our frame theory, in order to establish such an inference it is necessary to look both at the information got in a discourse about an object and, in addition, at possible ways of how this information can be extended. Consider again the following example from the introduction involving a wedding.

(11)   I was at a wedding last week. The mock turtle soup was a dream.

After processing the first sentence, a new object o has been introduced in the stack about which one has got the information that it is a wedding. The frame f that contains this information is given by the conditions $\theta(f) = \{\Delta \cap \downarrow \boldsymbol{wedding}\}$ and $root(f) = o$. It is a minimal frame of sort **wedding** because there is no relational information linking the wedding to other objects. In order to relate the mock turtle soup to the wedding by a bridging inference, one uses both the information that it is a wedding, got from bottom-up processing, and the conceptual knowledge that weddings can be related to objects of sort **mock turtle soup** by the chain MEAL • STARTER. The former information is given by $\theta(f)$ whereas the latter is given by the frame hierarchy for objects of sort **wedding**. For example, given that a comprehender knows that an object is related to a frame f with $\theta(f) = \{\Delta \cap \downarrow \mathbf{wedding}\}$ containing bottom-up information, he applies top-down conceptual knowledge about weddings to infer that the wedding can be related to a mock turtle soup in the way described above. In Sect. 7 we will model these two kinds of information by assigning to a stack position pairs $\langle o, \langle f_o, F_o \rangle \rangle$ consisting of an object o and a pair consisting of a frame f_o containing the information about o got from bottom-up processing, and a set F_o, which is a set of frames each element of which extends f_o along $\sqsubseteq$ in a particular way. For the moment, we will stick to the simpler modelling and take a stack position to be a pair $\langle o, f_o \rangle$ consisting of an object, called the *object* *component* and an associated frame, called the *frame component*. Such pairs are called *discourse objects*.

Second, the notions of possibility and information state from Incremental Dynamics have to be adapted. We assume that an information state models the epistemic state of a comprehender, i.e. both his (factual) beliefs (knowledge) and his discourse information. This distinction will be represented by defining a possibility as a pair $\langle c, w \rangle$ consisting of a stack c (discourse component) and a world w (factual component). Possible worlds model epistemic uncertainty. An information state is a set of possibilities.

Finally, the lexicon has to be adapted. Since information expressed by common nouns and verbs is always sortal or relational, it is related to frames in our frame theory. This kind of information expresses either that in a frame two objects are related by a chain of attributes or that an object satisfies some sortal information. Hence, frames must become part of the interpretation of these lexical items. This is achieved in the following way.

Common nouns are translated as (atomic) sort expressions whereas the translation of verbs is based on a neo-Davidsonian decompositional analysis. Unary event predicate expressions are translated as (atomic) sort expressions and thematic relation expressions are translated as (atomic) attribute expressions, i.e. as elements of *Attr*. The interpretation of n-ary predicative expressions is lifted in a way similar to Incremental Dynamics without frames (see [vE07] for details). In particular, the type e is replaced by the type of indices ι with variables $i, j, \ldots$. In (12) the interpretations of $\exists$ and common nouns in terms of our discourse and factual components (i.e., stacks and possible worlds) are given. Let $A_w = \{\langle o, f_o \rangle \mid o \in D_o \wedge root(f_o) = o \wedge IN(f_o) = w \wedge \theta(f_o) = \{\Delta \cap \downarrow \mathbf{object}\}\}$ for $w \in D_w$. That is A_w consists for each world $w \in D_w$ of all pairs $\langle o, f_o \rangle$ with $o \in D_o$ and f_o is the most general frame of o expressing only that o is of sort **object**.

(12)   a. $\exists := \lambda s.\lambda s'.\exists \alpha (s = \langle c, w \rangle \wedge s' = \langle c', w \rangle \wedge c' = c^\frown \alpha \wedge \alpha \in A_w)$

        b. $\lambda i.\lambda s.\lambda s'.\exists f'(s = \langle c, w \rangle \wedge s' = \langle c', w \rangle \wedge |c'| = |c| \wedge \forall j \, (0 \leq j < |c| \wedge$
          $j \neq i \rightarrow c'[j] = c[j]) \wedge c[i] = \langle o, f_o \rangle \wedge f_o \sqsubseteq f' \wedge [\![cn]\!](f')(o) \wedge \theta(f') =$
          $\theta(f_o) \cup \{\Delta \cap \downarrow cn\} \wedge c'[i] = \langle o, f' \rangle).$

In (12-a), $\exists$ introduces a new discourse object on the stack. The information associated with this object is the most general one since it is only required to be of sort **object**, which is true of all elements in D_o. This information is subsumed by any further information that is eventually added about the newly introduced object, for example by a head noun. According to (12-b), common nouns are not interpreted as pure tests, which would be their typical analysis in [vE07] but as operations on possibilities. The input and the output possibilities s and s' differ only with respect to position i of their respective discourse components. The semantic contribution of a common noun is to add sortal information. This is modelled by requiring that there is a frame f' that extends f_o ($f_o \sqsubseteq f'$) s.t. o satisfies the sortal information in f': $[\![cn]\!](f')(o)$, and by adding this sortal information to $\theta(f_o)$ to yield $\theta(f')$: $\theta(f') = \theta(f_o) \cup \{\Delta \cap \downarrow cn\}$. Finally, f_o is replaced by f' in the output possibility: $c'[i] = \langle o, f' \rangle$. Thus, there is both a test and an update operation associated with the interpretation of a common noun.

4 The Approach of Asher and Lascarides 1998

Since our approach is similar in spirit to that of Asher and Lascarides [AL98], we will begin by sketching their approach. Their analysis is based on Chierchia's [Chi95] analysis of definite descriptions as anaphoric. One way of analyzing 'the N' is given in (13-a). Chierchia enriches this meaning by adding a free $n + 1$-ary relational constant that is functional in its last argument, (13-b). R links the argument x to an n-tuple of objects $y_1 \ldots y_n$. Functionality requires that given $y_1 \ldots y_n$ x is uniquely determined by R: $R(y_1, \ldots, y_n, x_1) \wedge R(y_1, \ldots, y_n, x_2) \rightarrow x_1 = x_2$. On this analysis, 'the N' denotes a (unique) N which is related by some dependency relation to an n-tuple $y_1 \ldots y_n$.

(13)   a. $\iota x.N(x)$.
        b. $\iota x.[R(y_1, \ldots, y_n, x) \wedge N(x)]$.

[AL98] claims that in the case of a bridging inference, R is a binary functional relation B that has to hold between the antecedent object y and the denotation of the definite description x: $B(y, x)$. Hence, lexical semantics provides an under-specified relation B which functions as the bridge or the dependency relation and which must be determined by finding an appropriate value by connecting it to an object in the present discourse context.

How are B and y specified (resolved)? A common strategy is based on coherence relations. Consider e.g. (14).

(14)   a.   John took engine E1 from Avon to Dansville.
        b.   He picked up the boxcar (and took it to Broxburn).

Let K_α be a semantic representation of the first sentence, K_β a semantic representation of the second sentence and K_τ a semantic representation of the context in which the second sentence is interpreted so that K_α is a part (subrepresentation) of K_τ. K_α and K_β introduce two events of taking and picking up, respectively. This information is sufficient to defeasibly infer that the two sentences are related by the coherence relation *Narration*. Coherence relations are associated with (non-defeasible) rules that allow to infer additional information about the discourse referents introduced in the three semantic representations. For *Narration* one has: (i) e_α precedes e_β and (ii) if e_α and e_β have the same actor, the location of this actor at the end of e_α is the same as his location at the beginning of e_β. In addition, one has (iii): lexical semantic information about 'pick up' allows the inference that the theme of e_β is located at this location too. Together, (ii) and (iii) yield $In(boxcar, Dansville)$ since Dansville is the location of the actor (i.e. John) at the end of the taking event ($=e_\alpha$) and the boxcar is the theme of the picking-up event ($=e_\beta$). The condition $In(boxcar, Dansville)$ is added to K_β. Resolving B to the function of containment In and assigning y the value *Dansville*, (which is part of an update operation) yields the required bridging inference because a relation between a discourse referent introduced in the first and a discourse referent introduced in the second sentence has been established.

In the above example the derivation of a coherence relation between the two sentences yielded the required bridging inference. However, as noted in [AL98, p.104], often there is not enough information in K_β to infer a particular coherence relation between it and the previous context because K_β contains non-resolved material (B and y) and is therefore underspecified. As a result, B and y must be resolved *before* a coherence relation can be established between the two sentences. The coherence relation is then used as a constraint on the resolution used. The resolution should be such that discourse coherence is maximized (principle 'Maximize Discourse Coherence'). Consider the following variant of example (1-d) from the introduction.

(15)   a.   John arrived yesterday at 3pm.
        b.   The camel was outside and needed water.

A possible coherence relation linking the two sentences is *Background*. However, this relation also applies if the second sentence is replaced by its present tense variant 'The camel is outside and needs water'. However, due to the tense shift a bridging inference should not be possible. Thus, a different strategy is needed. First, one uses lexical semantics to infer that 'arrive', being a motion verb, defines a thematic relation 'mode of transport' (besides the theme-relation that has already been introduced during processing the first sentence). Second, one uses world knowledge to infer that camels can be used as such a mode of transport. Both pieces of information are not yet elements of the semantic representations.

(16)   a. $\forall e(arrive(e) \rightarrow \exists z.Means\text{-}of\text{-}Transport(e, z))$.
        b. $\forall x(camel(x) \rightarrow can\text{-}be\text{-}used\text{-}as\text{-}Means\text{-}of\text{-}Transport(x))$.

Using the additional information in (16), a possible resolution is given by $y = e_{arrive}$ and $B = Means\text{-}of\text{-}Transport$. so that one has $Means\text{-}of\text{-}Transport(e_{arrive}, x)$ with x the camel. This is the required bridging inference because the referent of the definite description is linked to an object introduced in the first sentence. In addition, $Means\text{-}of\text{-}Transport(e_{arrive}, x)$ can be used to infer that the two sentences are related by the coherence relation *Result*. The state of the camel needing water described in the second sentence is the result (or was caused) by the arrival, or, more precisely, by the motion event presupposed by the arriving event.[7] When taken together, one gets a coherent interpretation of (15) because the two sentences are connected by the coherence relation *Result* so that the principle 'Maximize Discourse Coherence' is satisfied.

Let us make the following observations about the second strategy proposed by [AL98]: (a) B is part of the semantic representation of a definite

---

[7] See [Pn97] for a formal analysis in which achievement verbs like 'arrive' are analyzed as boundary events of other, non-boundary events.

description due to the familiarity constraint imposed by the determiner 'the'. It is therefore independent of any constraints that are imposed related to coherence considerations though it is used to establish a coherence relation in the above example. This strategy fails if the bridged expression is an indefinite like 'a rope' in (2-c) in the introduction since for indefinites a novelty condition rather than a familiarity condition applies. This raises the question where in the semantic representation B and y come from if bridging inferences are not triggered by definite descriptions. (b) A distinction is made between (lexical) semantic properties of an expression that are part of its current semantic representation and properties for which this does not hold. And (c) B is resolved to a property of the latter kind of properties. Observations (b) and (c) already contain one possible answer to the problem raised in the first observation. If B is ultimately (resolved to) a semantic property associated with a (candidate) antecedent object, it should be related to the semantic representation of this object instead of with the semantic representation of the bridged expression. The semantic contribution of the bridged expression to a bridging inference is to provide a value for this property relative to the antecedent object: $B(x) = y$. Definite descriptions are then the special case in which the existence of an appropriate B is required by the semantics of 'the'. On this perspective a bridging inference is triggered, in effect, by the antecedent object: there is a semantic property associated with this object the value of which is unknown for this object. Besides being directly applicable to bridged expressions that are not definite descriptions, a second advantage of this perspective is that it does not directly rely on the use of coherence relations and can therefore be applied across the board to bridging inferences. In the remainder of this article, we are going to work out this perspective on bridging inferences.

5   Bridging Inferences and 'Questions Under Discussion'

If bridging inferences are triggered by the antecedent object, a question to be raised is what happens if B is not already part of the semantic representation of a (candidate) antecedent object. As we have seen in the previous section, a link to coherence relations cannot be the answer because often a bridging inference needs to be done without relying on information provided by these relations. A second strategy to establish coherence between a context and its continuation is based on the notion of a *Question under Discussion* (QuD). According to [KR17], 'in QuD-models of discourse interpretation, clauses cohere with the preceding context by virtue of providing answers to (usually implicit) questions that are situated within a speaker's goal-driven strategy of inquiry.' If an object is introduced into a discourse, this introduction is in general not bare in the sense that no sortal and relational information is associated with it. Initial additional information is given by common nouns for individuals (e.g. it is a wedding) and verbs for events (e.g. 'it is a hitting'). This information can be extended in various ways in the subsequent discourse. However, such extensions are in general not arbitrary but are related to particular questions that are raised in relation to these objects and which depend on the context in which the object is introduced. More generally, one has: In a QuD-model of discourse every newly introduced object raises a set of questions (cf. [RR16]). For objects, i.e. individuals and events introduced by common nouns like 'suicide', these questions are related to possibly complex properties these objects have and, therefore, to sortal and relational information about them. The corresponding rhetorical relation is called *Entity-Elaboration*. If o is the object 'under discussion', the canonical form of an Entity-Elaboration is 'What about o?'. Events that are introduced in the interpretation of verbs raise questions that are related to a particular coherence relation. Examples of questions are 'And then?', 'Why?', 'So what?' and 'How were things like then?', (cf. [RR16]). The relation to coherence between sentences is the following. At each stage τ of a discourse there is a set of active questions related to the objects that have already been introduced into the discourse. An extension of τ with a sentence ϕ is coherent only if this continuation implicitly contains at least one answer to at least one active question raised in τ and thereby automatically links an object already introduced to information provided in the continuation ϕ.

Let's illustrate this with one example from the introduction. In the first sentence of (2-c) Jack and a (planned) suicide are introduced. One therefore gets QuDs that are related to Entity-Elaboration: *What about Jack?* and *What about the (planned) suicide?*. Possible answers are: $\exists y.\exists e.buy(e,jack,y) \land rope(y)$, $\exists y.instrument(e_s,y) \land rope(y) \land depressed(jack)$. Note that the free variables in the answers refer to objects that are introduced in the first sentence whereas the existentially bound variables are objects that are introduced in the second sentence.

6 Bridging in Our Frame Theory: Bridging Relations and QuDs

Let us relate the results of the preceding section to our theory of frames. Information about an object o achieved by bottom-up processing is stored in the frame component f_o of the discourse object $\langle o, f_o \rangle$. The frame f_o contains at least sortal information which classifies o, e.g. as a wedding or a car. In our frame theory knowing the sort of an object is directly related to knowledge about the frame hierarchy for objects of this sort. The frame f_o is an element of this hierarchy, usually the minimal element (if only sortal information is known). Frames f for which $f_o \sqsubseteq f$ holds are extensions of f_o in which additional information about o is provided. A frame f that extends f_o by a chain π will be called a π-extension of f_o. This notion is defined in (17).

(17)   A frame f^π is a π-extension of a frame f if (a) $\pi \notin \theta(f)$, (b) $f \sqsubseteq f^\pi$, (c) $\pi \in \theta(f^\pi)$ and (d) for all π' that are not a prefix of π: if $\pi' \in \theta(f^\pi)$ then $\pi' \in \theta(f)$.

Two kinds of π-extensions must be distinguished. Let f^π be a π-extension of f with $\pi \cap \uparrow\sigma \in \theta(f^\pi)$, that is σ is the sortal information given in f^π at the end of chain π: (a) If σ is the target sort of π, this information already follows from conceptual knowledge. For example, knowing that o is a wedding, one knows that it took place at a particular location which is of sort **place**. This information is implied by knowledge of the frame hierarchy because if π is admissible for frames associated with objects of a particular sort, then its values are restricted by a particular sortal constraint expressed by its target sort. (b) If σ is not the target sort or if additional information beyond that sort is provided, the information contained in f_o is properly extended in the sense that it is neither implied by f_o nor does it follow from conceptual knowledge. Let us make this distinction between the two kinds of π-extensions explicit by defining π-extensions that do not introduce information already implied by knowledge of the frame hierarchy as *non-factual* π-extensions. Such extensions are similar to presupposed information in the sense that the information associated with them is not asserted and part of any extension of the current information state. Any particular extension will involve at least one factual π-extension. For example, getting to know that someone attended a wedding, a non-factual π-extension involves a meal served at this wedding. A particular extension of this information state containing this information can provide information about the particular dishes served at that meal, e.g. a mock turtle soup as starter.

(18)   A non-factual π-extension f^π of a frame f is a π-extension of f with $\pi \cap \uparrow\sigma \in \theta(f^\pi)$ for which σ is the target sort of π and for each prefix π^p of π, one has $\pi^p \cap \uparrow\sigma' \in \theta(f^\pi)$ only if σ' is the target sort of π^p. This latter condition ensures that for prefixes too, no factual information is introduced.

This relationship between bottom-up information and top-down conceptual knowledge suggests the following strategy to model QuDs with frames. A QuD is always related (i) to a discourse object $\langle o, f_o \rangle \in c$ that is already on the stack and (ii) a non-factual π-extension f^π of f_o. Non-factual π-extensions with $\pi \cap \uparrow\sigma \in \theta(f^\pi)$ for which the sort σ is the target sort of π are *underspecified* answers to QuDs. A proper (or non-underspecified) answer related to π must provide additional information about this value and is therefore related to a (factual) π-extension f'^π in which new factual information about the value of π is provided so that one has $f^\pi \sqsubseteq f'^\pi$. How is f'^π related to the linguistic context? f'^π must be related to the bridged expression and therefore to a part of the semantic representation of a sentence ϕ that is a continuation of the stage τ of the current discourse. However, f'^π is in general not the frame that is introduced with the semantic representation of the bridged expression as a constituent of ϕ. Rather, this expression introduces a frame f^*. Since f^* provides information about the value of π, it follows that the relation has to be defined in terms of the subframe relation $\preceq$: $f^* \preceq f'^\pi$. When taken together, one gets (19).

(19) $\exists f'^\pi . f^\pi \sqsubseteq f'^\pi \wedge f^* \preceq f'^\pi$.

According to (17), a π-extension adds to a frame a chain π (together with its prefixes due to the definition of θ). This accounts for the fact that for a bridging relation of length ≥ 1 the antecedent object is related to a second object.

A direct consequence of this definition is that it does not account for cases in which the bridging relation is identity. In this case the information used in the bridging inference is already used when the antecedent object was introduced in the first place. Hence, the bridged expression by itself does, at least in general, not provide new information about the antecedent object. For example, in (1-a) ('Lizzy met a dog yesterday. The dog was very friendly.') the sortal information **dog**, given by $\Delta \cap \downarrow$**dog**, is already an element of $\theta(f)$, i.e. the frame information associated with the antecedent object 'a dog' from the first sentence. The information provided by the bridged expression is *given* information relative to the antecedent object so that the frame associated with the bridged expression does not give rise to a proper extension of the frame associated with the antecedent object. Identity has therefore to be treated in a different way. There are at least the following arguments for such a separate treatment. First, if the dependency relation is identity, the antecedent object is always related to itself. Second, the identity relation is possible only with bridged expressions of the form 'the N'. For example, in 'Lizzy met a dog yesterday. A dog was friendly' the two occurrences of 'a dog' cannot refer to the same dog. Third, and most importantly, there is empirical evidence that bridged expressions of the form 'the N' with the bridging relation being identity are processed differently in the brain. [Bur06] found a difference in the P600 effect, an ERP-component, during online semantic processing between the identity relation on the one hand and bridged DPs like 'the engine' and new DPs like 'a rope' on the other hand. Related to these arguments is the following observation. The additional information about the antecedent object is provided by a verbal expression, e.g. 'is friendly' in (1-a). Since the information is related to the same object, the relation between f and f^* can be defined by $\sqsubseteq$ alone (a reference to $\preceq_\pi$ is not needed):

(20) $\exists f' : f \sqsubseteq f' \wedge f^* \sqsubseteq f'.$

The above discussion has shown that a distinction has to be made between the bridging relation and a QuD. A QuD always involves a π-extension of the frame associated with the antecedent object. This is the case because new information about this object is provided. The case of a bridging relation is more complex. For establishing coherence, a π-extension is not necessary as shown by bridging inferences based on identity. This difference shows up in the way new information is added. In the case of identity (20) applies which does not require the $\preceq$ relation because no anaphoric relation to a second object is established. We are now ready to define QuDs in our frame theory. QuDs are represented as underspecified answers while answers to QuDs as (specified) answers.

(21)   a.   A QuD raised by a discourse object $\langle o, f_o \rangle$ is a set of frames F such that each element of F is a non-factual π-extension of f_o for some chain π.

         b.   An answer to a QuD raised by a discourse object $\langle o, f_o \rangle$ is a frame f^* which stands to f_o in the bridging relation given in (19).

A necessary condition for discourse coherence is defined in (22).

(22)   A continuation ϕ of a stage τ of a discourse is coherent relative to τ only
if semantic processing ϕ introduces a frame f^* that is an answer to a
QuD raised by a discourse object belonging to the discourse component
at stage τ.

From the above discussion it follows that in our frame theory the bridging relation B can be defined either at the level of objects or at the level of frames. Let $\langle o', f_{o'} \rangle$ be a discourse object introduced in the continuation ϕ. At the level of objects one gets $[\![\pi]\!](f_o)(o)(o')$; the antecedent object o and the 'bridged' object o' are related by π. At the level of frames, B is defined by a relation between f_o and $f_{o'}$; this relation $Rel_B(f_o, f_{o'})$ holds if there is a π-extension of f_o of which $f_{o'}$ is a subframe. In contrast to defining B at the level of objects, the definition of Rel_B at the frame level is not functional. For a given frame f_o, there are many frames that satisfy the definition. $f_{o'}$ depends on the information that is given about o' in the continuation.

Relating B to a QuD and modelling the latter as underspecified answers has the effect that the dependency relation itself is not underspecified in the sense that it is represented as a free variable as in [AL98]. Underspecification comes in because the value of the chain is constraint only by its target sort, information that is given to a comprehender independently of any discourse information. Having different π-extensions accounts for the fact that answers can involve various relations linking an antecedent object to an object to which it stands in a dependency relation. An example is given in (23).

(23)   I took my car for a test drive. The engine/brakes/tyres made a weird
noise.

This change of perspective on bridging inferences is made possible due to the shift in location of the bridging (dependency) relation. It is no longer related to the bridged expression but to the semantic representation of a (candidate) antecedent object.

Let us illustrate the above discussion by the following example and the two associated frames for the antecedent object denoted by 'my car' and the bridged object denoted by 'the engine'.

(24)   I took my car for a test drive. The engine made a weird noise.

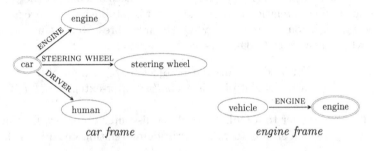

car frame   *engine frame*

When the car is introduced in the first sentence, only sortal information is provided.[8] Hence, one has $\theta(f_{car}) = \{\Delta \cap \downarrow\textbf{car}\}$.[9] However, using his knowledge about the frame hierarchy, a comprehender knows that there are (frame-) extensions of f_{car} that provide additional information about the car. Three possible extensions are related to the attributes ENGINE, STEERING_WHEEL and DRIVER. One has: $f_{car} \sqsubseteq f_{car}^{\text{ENGINE}}$, $f_{car} \sqsubseteq f_{car}^{\text{STEERING\_WHEEL}}$ and $f_{car} \sqsubseteq f_{car}^{\text{DRIVER}}$. At the level of the function θ one has: $\theta(f_{car}^{\text{ENGINE}}) = \theta(f_{car}) \cup \{\text{ENGINE} \cap \uparrow\textbf{engine}\}$, $\theta(f_{car}^{\text{STEERING\_WHEEL}}) = \theta(f_{car}) \cup \{\text{STEERING\_WHEEL} \cap \uparrow\textbf{steering\_wheel}\}$ and $\theta(f_{car}^{\text{DRIVER}}) = \theta(f_{car}) \cup \{\text{DRIVER} \cap \uparrow\textbf{person}\}$. Each extension corresponds to one possible underspecified answer to the QuD related to the Entity-Elaboration 'What about the car?'.

The frames f_{car}^{ATTR} only contain minimal information about the object o' to which the car is related by ATTR $\cap \uparrow\sigma$. The only information about o' is $\Delta \cap \downarrow\sigma$ where σ is the target sort of ATTR. However, the information provided about o' in a continuation will in general be richer because it contains all information got about it in this continuation. For instance, in the example at hand one gets to know that the car emitted a weird noise. Let the discourse object related to 'the engine' in the second sentence be $\langle o', f_{o'}\rangle$. A possible value for $\theta(f_{o'})$ is $\{\Delta \cap \downarrow\textbf{engine}, \text{EMISSION} \cap \uparrow\textbf{sound}, \text{EMISSION} \bullet \text{PITCH} \cap \uparrow\textbf{weird}\}$. As a result, $f_{o'}$ contains more information about the engine of the car than f_{car}^{ENGINE}. $f_{o'}$ is depicted in the figure below.

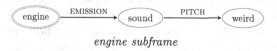

engine subframe

$f_{o'}$ is not a subframe of f_{car}^{ENGINE} because it contains more information about the engine than f_{car}^{ENGINE}. However, there is an extension $f_{car}^{\text{ENGINE}'}$ of f_{car}^{ENGINE} for which $f_{o'} \preceq f_{car}^{\text{ENGINE}'}$ holds: $\theta(f_{car}^{\text{ENGINE}'}) = \theta(f_{car}^{\text{ENGINE}}) \cup \{\text{ENGINE} \bullet \text{EMISSION} \cap \uparrow\textbf{sound}, \text{ENGINE} \bullet \text{EMISSION} \bullet \text{PITCH} \cap \uparrow\textbf{weird}\}$.

7 The Formal Account: Extending Frame-Based Incremental Dynamic with QuDs

In order to implement the strategy for bridging inferences developed in the preceding section we first have to extend the theory from Sect. 2 with QuDs. Recall that QuDs are always related to a particular discourse object $\langle o, f_o\rangle$ and are modelled as a set F_o of non-factual π-extensions of f_o. One way of integrating F_o into the theory is at the level of stack positions. Instead of storing pairs of the form $\langle o, f_o\rangle$ one stores pairs of the form $\langle o, \langle f_o, F_o\rangle\rangle$. The frame f_o will be called the factual frame component and the set F_o the set of non-factual

---

[8] We leave out the information related to the verb 'take'.

[9] Recall that $\Delta \cap \downarrow\phi$ restricts the relation expressed by $\downarrow \phi$ to those elements $\langle o, o'\rangle$ from $D_o \times D_o$ for which one has $o = o'$, i.e. $\downarrow \phi$ is restricted to the diagonal of $D_o \times D_o$.

π-extensions. There are at least two arguments in favour of this option. First, as will be shown below, the bridging constraint is integrated as a part of the normal process of semantic composition and not, say, as part of a more global pragmatic component that operates on given semantic structures. As an effect, F_o must be available locally during the process of semantic composition. The second argument is the context-sensitive character of QuDs. Though they are always raised relative to discourse objects of a particular kind, what counts as an answer to them depends on the information available in the current context. Hence, F_o cannot be an element of the model but must be a component of an information state.

Adding F_o to the information stored about an object o on a stack c is only a first step in modelling the bridging constraint. The second step is to incorporate the bridging inference. Relative to this step two related questions have to be answered: (i) What kind of operation is associated with this step?, and (ii) Where in the process of semantic composition should this operation be placed? Let's begin with the first question for a definite description 'the N'. Recall that a bridging inference operates at two different levels. First, and foremost, it is related to making a text coherent by linking information in a continuation to an object that has already been introduced. Second, a comprehender gets additional information about that object by relating the new information to it. Hence, a bridging inference consists of two operations: establishing a dependency relation (bridging condition) and, if successful, an update operation on the frame component of the antecedent object.

Let's turn to the second question. Establishing the bridging condition (BC) amounts to testing either (19) or (20). This can be done as soon as the head noun has been semantically processed. By contrast, the update operation has to be executed after the verbal element Q has been processed. When taken together, one gets that the test of the bridging condition is done after processing the nominal element P and the update operation is executed after processing Q. In (25) the two test conditions are defined. (26) contains the corresponding update operations and (27) the interpretation of 'the'. The revised interpretation of $\exists$ is given in (27-d). One has: $A_w = \{\langle o, \langle f_o, F_o \rangle\rangle \mid o \in D_o \wedge root(f) = o \wedge IN(f) = w \wedge \theta(f_o) = \{\Delta \cap \downarrow\mathbf{object}\} \wedge F_o \subseteq \{f \mid f \text{ is a non-factual } \pi\text{-extension of } f_o\}\}$ for $w \in D_w$.[10] π^1 and π^2 are the first and second projection functions for pairs.[11]

---

[10] Requiring that F_o be a subset of the non-factual π-extensions of f_o raises the question of how this set can be further restricted. In general we taken the determination of the initial F_o to be context-specific, based on probabilities. We will come back to this question at the end of this section.

[11] Recall from Sects. 2 and 3 that an information state is a set of possibilities and that a possibility is a pair $\langle c, w \rangle$ consisting of a stack c and a world w. In contrast to Sect. 3 our stack elements are now pairs $\langle o, \langle f_o, F_o \rangle\rangle$ with object o, its frame f_o and a set of π-extensions F_o. In (25)–(27) o is used for objects, f for frames, F for sets of frames, c for stacks, i, j for stack indices and s for possibilities. Note that while π is used for chains of attributes, π^1 and π^2 denote the projection function.

(25)   a. $BC_1 := \lambda f.\lambda j.\lambda i.\lambda s.\lambda s'.\exists c.\exists o.\exists o'.\exists f_o.\exists F_o.\exists f_{o'}.\exists F_{o'}.$
$(s=s' \wedge \pi^1(s)=c \wedge |\pi^1(s)| = i \wedge c[i]=\langle o', \langle f_{o'}, F_{o'}\rangle\rangle \wedge j < i \wedge$
$c[j]=\langle o, \langle f_o, F_o\rangle\rangle \wedge f_o \sqsubseteq f \wedge f_{o'} \sqsubseteq f).$

        b. $BC_2 := \lambda f.\lambda j.\lambda i.\lambda s.\lambda s'.\exists c.\exists o.\exists o'.\exists f_o.\exists F_o.\exists f_{o'}.\exists F_{o'}.\exists f_o^\pi.$
$(s=s' \wedge \pi^1(s)=c \wedge c[i]=\langle o', \langle f_{o'}, F_{o'}\rangle\rangle \wedge j < i \wedge c[j]=\langle o, \langle f_o, F_o\rangle\rangle \wedge f_o^\pi \in$
$F_o \wedge f_o^\pi \sqsubseteq f \wedge f_{o'} \preceq_\pi f).$

(26)   a. $Update_1 := \lambda f.\lambda j.\lambda i.\lambda s.\lambda s'.\exists c, c'.\exists o, o'.\exists f_o.\exists f_o^\pi.\exists F_o.\exists f_{o'}.\exists F_{o'}.\exists F_o'.$
$(\pi^2(s)=\pi^2(s') \wedge \pi^1(s)=c \wedge \pi^1(s')=c' \wedge c \approx_j c' \wedge c[i]=\langle o', \langle f_{o'}, F_{o'}\rangle\rangle \wedge$
$j < i \wedge c[j]=\langle o, \langle f_o, F_o\rangle\rangle \wedge f_o^\pi \in F_o \wedge f_o \sqsubseteq f_o^\pi \wedge f_o \sqsubseteq f \wedge f_{o'} \sqsubseteq f \wedge$
$c'[j]=\langle o, \langle f, F_o'\rangle\rangle \wedge \theta(f)=\theta(f_o) \cup \theta(f_{o'}) \wedge F_o'=F_o - \{f' \mid f' \in F_o \wedge f' \sqsubseteq f\}).$

        b. $Update_2 := \lambda f.\lambda j.\lambda i.\lambda s.\lambda s'.\exists c, c'.\exists o, o'.\exists f_o.\exists f_o^\pi.\exists F_o.\exists f_{o'}.\exists F_{o'}.\exists F_o'.$
$(\pi^2(s)=\pi^2(s') \wedge \pi^1(s)=c \wedge \pi^1(s')=c' \wedge c \approx_j c' \wedge c[i]=\langle o', \langle f_{o'}, F_{o'}\rangle\rangle \wedge$
$j < i \wedge c[j]=\langle o, \langle f_o, F_o\rangle\rangle \wedge f_o^\pi \in F_o \wedge f_o \sqsubseteq f_o^\pi \wedge f_o \sqsubseteq f \wedge f_{o'} \preceq_\pi$
$f \wedge c'[j]=\langle o, \langle f, F_o'\rangle\rangle \wedge \theta(f)=\theta(f_o) \cup \{\pi \bullet \pi' \mid \pi' \in \theta(f_{o'})\} \wedge F_o'=F_o -$
$\{f' \mid f' \in F_o \wedge f' \sqsubseteq f\}).$

        c. $c \approx_i c' := |c|=|c'| \wedge \forall j (0 \leq j < |c| \wedge j \neq i \rightarrow c'[j]=c[j]).$[12]

(27)   a. $[\![the]\!] := \lambda P.\lambda Q.\lambda s.\exists f.\exists j.(\exists \cdot P(|\pi^1(s)|) \cdot [BC_1(f)(j)(|\pi^1(s)|) \cdot$
$Q(|\pi^1(s)|) \cdot Update_1(f)(j)(|\pi^1(s)|) \cup BC_2(f)(j)(|\pi^1(s)|) \cdot Q(|\pi^1(s)|) \cdot$
$Update_2(f)(j)(|\pi^1(s)|)])(s).$

        b. $\phi \cdot \psi := \lambda s.\lambda s'.\exists s''(s'' \in \phi(s) \wedge s' \in \psi(s'')).$[13]

        c. $\phi \cup \psi := \lambda s.\lambda s'.s' \in \phi(s) \vee s' \in \psi(s).$

        d. $\exists := \lambda s.\lambda s'.\exists \alpha(s = \langle c, w\rangle \wedge s' = \langle c', w\rangle \wedge c' = c^\frown \alpha \wedge \alpha \in A_w)$

The tests of the bridging conditions $BC_{1/2}$ in (25) are part of the interpretation of the determiners 'the' and 'a'. They therefore introduce the 'bridged' object if they are a constituent of a bridged expression. Their semantic function is to test for the bridging relations (19) and (20). BC_1 corresponds to (20) and therefore tests on the identity (or a sort subsumption) relation whereas BC_2 corresponds to (19) and is thus related to bridging involving a relation other than identity. For BC_1 the 'bridged' object o' is stored at position $i = |\pi^1(s)|$, i.e. at the last position of the stack since it has just been introduced. Its associated factual frame is $f_{o'}$ and the set of non-factual π-extensions is $F_{o'}$. In order for (20) to be satisfied, there has to be an antecedent object o that has already been introduced so that it is stored at a position j preceding $i = |\pi^1(s)|$. Recall that for (20) no non-factual π-extensions are used. Hence, the set F_o of non-factual π-extensions associated with o at j does play no role. All that is required is that there is a frame f that extends both the factual frame component associated with o' and that is associated with o: $f_o \sqsubseteq f \wedge f_{o'} \sqsubseteq f$. The constraint that there is a factual π-extension providing new factual information is built into the update operation in (26). The difference between BC_1 and BC_2 consists in the

---

[12] $c \approx_i c'$ says that the stacks c and c' differ at most w.r.t. the value assigned to position i.

[13]   In the definitions of $\cdot$ and $\cup$, ϕ and ψ map possibilities (i.e. pairs $\langle c, w\rangle$ consisting of a stack and a world) to sets of possibilities.

bridging relation. Since for BC_2 this relation is (19), establishing this relation always involves a non-factual π-extension since the antecedent object o is related to another object o' by a chain of attributes π. Hence, one has that a non-factual π-extension f_o^π belonging to F_o must be extended by f and the factual frame component $f_{o'}$ associated with the 'bridged' object has to be a subframe of f : $f_o^\pi \sqsubseteq f \wedge f_{o'} \preceq_\pi f$.

The two update operations $Update_{1/2}$ in (26) operate on the frame component of the antecedent object because this component has to be updated due to the new information provided by the 'bridged' object. $Update_1$ is used for bridging inferences involving the identity relation. This operation therefore corresponds to BC_1. By contrast, $Update_2$ applies to bridging inferences where the antecedent object is related to a second object and, hence, BC_2 is used. In $Update_1$ the updated frame f for the antecedent object has to satisfy three conditions: (i) it extends the 'old' frame f_o: $f_o \sqsubseteq f$; (ii) it extends the frame $f_{o'}$ associated with the bridged DP: $f_{o'} \sqsubseteq f$; and it extends a non-factual π-extension: $f_o^\pi \sqsubseteq f$.

Let us illustrate these conditions with the example of Lizzy and the dog: 'Lizzy met a dog yesterday. The dog was very friendly.' One has: $\theta(f_o) = \{\Delta \cap \downarrow\textbf{dog}\}$, $\theta(f_{o'}) = \{\Delta \cap \downarrow\textbf{dog}, \text{BEHAVIOUR} \cap \uparrow\textbf{friendly}\}^{14}$ and $\theta(f_o^{\text{BEHAVIOUR}}) = \{\Delta \cap \downarrow\textbf{dog}, \text{BEHAVIOUR} \cap \uparrow\textbf{behaviour}\}$. $f_{o'}$ is the frame that results after processing the VP. For the final updated frame f we have that $\theta(f)$ is the union of $\theta(f_o)$ and $\theta(f_{o'})$: $\theta(f) = \{\Delta \cap \downarrow\textbf{dog}\} \cup \{\Delta \cap \downarrow\textbf{dog}, \text{BEHAVIOUR} \cap \uparrow\textbf{friendly}\} = \{\Delta \cap \downarrow\textbf{dog}, \text{BEHAVIOUR} \cap \uparrow\textbf{friendly}\}$. The way $\theta(f)$ is construed ensures that it is the minimal frame satisfying the three conditions. In this particular case the (updated) frame f is identical to f_o: This follows from the fact that the original information about the dog got in the first sentence is minimal, only sortal information is provided, and from the fact that the bridging relation is identity. As a result, the information got in the second sentence is still about o and repeats the sortal information from the first sentence. Furthermore, this example shows that the update operation does not simply take the frame f that passed the bridging constraints $BC_{1/2}$ as the new factual frame component of the antecedent object. This frame has in addition to comprise the information got about the dependent object o' from processing the verbal element. The new non-factual π-extensions component F_o' is the old one minus those non-factual π-extensions in this set that are subsumed by f because the corresponding QuDs have been answered.

$Update_2$ differs from $Update_1$ in the way the updated frame f is related to the new information provided by $f_{o'}$. Since the antecedent object and the object denoted by the bridged DP are not identical and are therefore related by a chain of attributes with length greater 0, $f_{o'}$ cannot be extended by f but has to be a subframe of f. Let us illustrate this with the car example: 'I took my car for a test drive. The engine made a weird noise'. One has: $\theta(f_o) = \{\Delta \cap \downarrow\textbf{car}\}$, $\theta(f_{o'}) = \{\Delta \cap \downarrow\textbf{engine}, \text{EMISSION} \cap \uparrow\textbf{sound}, \text{EMISSION} \bullet \text{PITCH} \cap \uparrow\textbf{weird}\}$ and $\theta(f_o^{\text{ENGINE}}) = \{\Delta \cap \downarrow\textbf{car}, \text{ENGINE} \cap \uparrow\textbf{engine}\}$. The frame $f_{o'}$ is the frame for the

---

[14] Recall that the value of θ for a frame f is closed under supersorts. Hence, $\theta(f_{o'})$ is, in effect the set $\{\Delta \cap \downarrow\textbf{dog}, \text{BEHAVIOUR} \cap \uparrow\textbf{friendly}, \text{BEHAVIOUR} \cap \uparrow\textbf{behaviour}\}$. This set is a superset of the set $\theta(f_o^{\text{BEHAVIOUR}})$ given next.

engine after processing the VP in the second sentence. For the updated frame f, $\theta(f)$ is construed as follows. The set $\theta(f_o) = \{\Delta \cap \downarrow \mathbf{car}\}$ is extended by chains $\pi \bullet \pi'$ where π is given by f_o^π: ENGINE and π' is an element from $\theta(f_{o'})$. Since there are three elements, one gets the chains ENGINE $\bullet$ ($\Delta \cap \downarrow \mathbf{engine}$), ENGINE $\bullet$ EMISSION $\cap \uparrow \mathbf{sound}$ and ENGINE $\bullet$ EMISSION $\bullet$ PITCH $\cap \uparrow \mathbf{weird}$. Since ENGINE $\bullet$ ($\Delta \cap \downarrow \mathbf{engine}$) has the same satisfaction conditions as ENGINE $\cap \uparrow \mathbf{engine}$), one gets $\theta(f) = \{\Delta \cap \downarrow \mathbf{car}$, ENGINE $\cap \uparrow \mathbf{engine}$, ENGINE $\bullet$ EMISSION $\cap \uparrow \mathbf{sound}$, ENGINE $\bullet$ EMISSION $\bullet$ PITCH $\cap \uparrow \mathbf{weird}\}$. Similar to the way f is construed in the case of an identity relation, the construction of f ensures that f is the minimal frame satisfying the three conditions. The new non-factual π-extensions component is the old one minus those non-factual π-extensions in this set that are subsumed by f because the corresponding QuDs have been answered. This is again similar to the case of the first update operation.

Finally, we turn to the interpretation of the definite and indefinite determiner. Processing the definite determiner 'the' consists of two branches (using the choice operation) after processing the head noun P (27). In the first branch BC_1 succeeds followed by the interpretation of the verbal element Q and the update operation $Update_1$. In the second branch BC_2 succeeds followed by the interpretation of the verbal element and the update operation $Update_2$. On this interpretation of 'the', a definite description always is a bridging expression for which the bridging constraint has to be satisfied. One can therefore say that it 'signals' that there is a relation to the previous context.

For the determiner 'a', only bridging condition BC_2 applies, as shown above in Sect. 6 by the example 'Jack was going to commit suicide. He bought a rope'. The reason is that the frame associated with the discourse referent introduced by 'a rope' is not a subframe of the frame associated with the discourse referent introduced by 'a suicide' since the roots of the frames are different. Furthermore, this condition is only a *sufficient* condition to ensure discourse coherence. The bridging condition can equally be satisfied by another frame $f_{o''}$ introduced in the continuation. Hence, for 'a', both BC_2 and the update operation $Update_2$ must be optional. We model this by replacing the first branch in the interpretation of 'a' by a branch that only executes P and Q without any test or update operation.

(28) $\quad [\![a]\!] := \lambda P. \lambda Q. \lambda s. \exists f. \exists j. (\exists \cdot P(|\pi^1(s)|) \cdot [Q(|\pi^1(s)|) \cup BC_2(f)(j)(|\pi^1(s)|) \cdot$
$\quad Q(|\pi^1(s)|) \cdot Update_2(f)(j)(|\pi^1(s)|)])(s).$

If the update operation $Update_2$ is not obligatory for indefinites, the following problem can arise. Processing a continuation can be successful, i.e. there is a (non-empty) output information state, without successfully checking the BC for at least one possibility. It is therefore necessary to explicitly test for this satisfaction. One way of doing this is during the combination of two sentences. There are at least two ways of how this testing can be done: at the level of the first (factual) frame component or on the second, QuD-related frame component. We will choose the first option. If a continuation contains information about a discourse object α that is already on the stack at some position i in some possibility s of the output of the first sentence, then the factual frame component f'_o at position i in a successor possibility s' of s must be a proper frame extension of the frame component f_o at position i in s: $f_o \sqsubset f'_o$. The notion of a successor possibility is defined in (29).

(29)   a. $s \trianglelefteq s' := \exists c.\exists w.\exists c'.\exists w'.s=\langle c, w\rangle \wedge s'=\langle c', w'\rangle \wedge w=w' \wedge \exists c'' : c'=c'' c'' \wedge$
$\forall i : 0 \le i < |c| \to c[i] <_i c'[i]$.

b. $c[i] <_i c'[i] := \exists o.\exists o'.\exists f.\exists f'.\exists F.\exists F'.c[i]=\langle o, \langle f, F\rangle\rangle \wedge c'[i]=\langle o', \langle f',$
$F'\rangle\rangle \wedge o=o' \wedge f \sqsubseteq f' \wedge F \supseteq F')$.

A possibility s' is a successor of a possibility s if they share the same world component and, therefore, contain information about objects and frames in the same world. Furthermore, s' possibly extends the discourse information of s in the following respects. First, it can contain information about more objects: $\exists c'' : c'=c'' c''$. Second, w.r.t. to the discourse objects in the discourse component c in s one has: the same objects are stored in the respected positions. For the frame components, one has that s' contains at least the information that s contains about the stored objects. The factual frame components are related by $\sqsubseteq$ and the QuD-component by $\supseteq$. It is not necessary that a successor possibility s' of s properly extends the information in s about a discourse objects. In this case one has $f = f'$ and $F = F'$. The bridging condition BC_test, defined in (30), captures the constraint of a proper extension for the factual frame component.

(30) $BC\_test(s, s') := s \trianglelefteq s' \wedge \exists i : 0 \le i < |\pi^1(s)| \wedge \pi^1(\pi^2([\pi^1(s)](i))) \sqsubseteq$
$\pi^1(\pi^2([\pi^1(s')](i)))$.

Each successor s' of a possibility s in the input information state must properly extend the information associated with at least one discourse object that is an element of the stack in s. The requirement that *each* successor possibility has to satisfy the bridging constraint (for at least one position) is necessary because a comprehender does not know in advance which of these successors will eventually be eliminated. If BC_test is added to the definition of combining two sentences, one gets the required global test of discourse coherence.

(31) $\phi \cdot_D \psi := \lambda s.\lambda s'.\exists s''(\phi(s)(s'') \wedge \psi(s'')(s') \wedge BC\_test(s'', s'))$.

Note that $\cdot_D$ is different from $\cdot$ defined in (27-b) above. $\cdot$ is used at the lexical level to combine constituents of sentences that are built from $\exists$, dynamic properties, $BC_{1/2}$ and $Update_{1/2}$. By contrast, $\cdot_D$ is used at the discourse level to combine sentences.

A final question we need to address, is 'How is the set F determined?' Simply assuming that F consists of all (non-factual) π-extensions of the current factual frame component yields a set that is likely to be infinite. Relating F to the notion of prediction provides a possible way of analysing how F can be restricted to a proper subset of all possible non-factual π-extensions. In [NP17] we present an account that bases predictions on probabilities. The key idea is to define for each position i on a stack a probability measure Pr_i on subsets of the range of θ, i.e. relations on $D_f \times D_o \times D_o$. Expectations are ranked in such a way that pre-activation is restricted to those extensions whose probability exceeds a particular value.[15]

8   Conclusion

In this article we have developed a theory of bridging inference in frame theory. Using frames, bridging inferences can be modelled as update operations involving frames. In contrast to previous approaches, no 'incompleteness' in form of free variables is needed. Rather, incompleteness is replaced by underspecification. Following models of QuDs and results of neurophysiological research on predictions during semantic processing in the brain, each discourse object is related to a set of possible ways of how information about this object can be extended by a continuation of the discourse. Extensions are based on a particular chain of attributes and on knowledge of the frame hierarchy associated with objects of a particular sort. These extensions are underspecified in the sense that except for the constraint imposed by the target sort nothing is known in the discourse about the value of the chain. The bridging inference consists in relating one extension with a frame that is introduced in a continuation of the discourse. The implicit character of bridging inferences shows up in the fact that establishing and testing for them is modelled by separate update operations that, by themselves, are not needed in the process of semantically combining the constituents of a sentence and/or a discourse. The difference between definite descriptions and indefinites lies in the way they are related to QuDs/predictions. Whereas definite descriptions *always* discharge a bridging constraint and therefore ensure discourse coherence, this is only a possibility for indefinites. They can, but need not, be related to a previously introduced discourse object by a bridging (dependency) relation.

References

[AL98]   Asher, N., Lascarides, A.: Bridging. J. Semant. **15**, 83–113 (1998)

[Bur06] Burkhardt, P.: Inferential bridging relations reveal distinct neural mechanisms: evidence from event-related brain potentials. Brain Lang. **98**(2), 159–168 (2006)

[CH77]   Clark, H.H., Haviland, S.E.: Comprehension and the given-new contract. In: Freedle, R.O. (ed.) Discourse Production and Comprehension, pp. 1–40. Ablex Publishing, Hillsdale (1977)

[Cha83] Charniak, E.: Passing markers: a theory of contextual influence in language comprehension. Cogn. Sci. **7**, 171–190 (1983)

[Chi95] Chierchia, G.: Dynamics of Meaning: Anaphora, Presupposition and the Theory of Grammar. University of Chicago Press, Chicago (1995)

---

[15] Therefore, the use of default logic in [AL98] and weighted abduction in [HSAM93] is replaced by probability measures on frame hierarchies.

[Cla77] Clark, H.: Bridging. In: Johnson-Laird, P., Wason, P. (eds.) Thinking: Readings in Cognitive Science, pp. 411–420. Cambridge University Press, Cambridge (1977)

[Geu11] Geurts, B.: Accessibility and anaphora. In: von Heusinger, K., Maienborn, C., Portner, P. (eds.) Semantics. Handbooks of Linguistics and Communication Science, vol. 2, pp. 1988–2011. DeGruyter (2011). Chapter 75

[HSAM93] Hobbs, J.R., Stickel, M.E., Appelt, D.E., Martin, P.A.: Interpretation as abduction. Artif. Intell. **63**(1–2), 69–142 (1993)

[KR17] Kehler, A., Rohde, H.: Evaluating an expectation-driven question-under-discussion model of discourse interpretation. Discourse Process. **54**(3), 219–238 (2017)

[Nou03] Nouwen, R.: Plural pronominal anaphora in context. Ph.D. thesis, Netherlands Graduate School of Linguistics Dissertations, LOT, Utrecht (2003)

[NP17] Naumann, R., Petersen, W., Thomas, G.: Underspecified changes: a dynamic, probabilistic frame theory for verbs. In: Sauerland, U., Solt, S. (eds.) Proceedings of Sinn und Bedeutung 22, vol. 2 (2018)

[Pn97] Piñón, C.: Achievements in an event semantics. In: Lawson, A. (ed.) Proceedings SALT VII, pp. 276–293. Cornell University, Ithaca (1997)

[RR16] Reyle, U., Riester, A.: Joint information structure and discourse structure analysis in an underspecified DRT framework. In: Hunter, J., Simons, M., Stone, M. (eds.) Proceedings of the 20th Workshop on the Semantics and Pragmatics of Dialogue (JerSem), New Brunswick, pp. 15–24. Rutgers University (2016)

[vE07] van Eijck, J.: Context and the composition of meaning. In: Bunt, H., Muskens, R. (eds.) Computing Meaning, vol. 83, pp. 173–193. Springer, Dordrecht (2007). https://doi.org/10.1007/978-1-4020-5958-2_8

Misfits: On Unexpected German *Ob*-Predicates

Kerstin Schwabe[(⊠)]

Leibniz-Zentrum für Allgemeine Sprachwissenschaft Berlin, Schützenstraße 18,
10117 Berlin, Germany
schwabe@leibniz-zas.de

Abstract. German subjectively veridical *sicher sein* 'be certain' can embed *ob*-clauses in negative contexts, while subjectively veridical *glauben* 'believe' and nonveridical *möglich sein* 'be possible' cannot. The Logical Form of *F isn't certain if M is in Rome* is regarded as the negated disjunction of two sentences $\neg(c_f\ \sigma \lor c_f\ \neg\sigma)$ or $\neg c_f\ \sigma \land \neg c_f\ \neg\sigma$. *Be certain* can have this LF because $\neg c_f\ \sigma$ and $\neg c_f\ \neg\sigma$ are compatible and nonveridical. *Believe* excludes this LF because $\neg b_f\ \sigma$ and $\neg b_f\ \neg\sigma$ are incompatible in a question-under-discussion context. It follows from this incompatibility and from the incompatibility of $b_f\ \sigma$ and $b_f\ \neg\sigma$ that $b_f\ \neg\sigma$ and $\neg b_f\ \sigma$ are equivalent. Therefore *believe* cannot be nonveridical. *Be possible* doesn't allow the LF either. Similar to *believe*, $\neg p_f\ \sigma$ and $\neg p_f\ \neg\sigma$ are incompatible. But unlike *believe*, $p_f\ \sigma$ and $p_f\ \neg\sigma$ are compatible.

Keywords: German interrogative embedding predicates ·
Contrary and complementary opposites · Neg-raising

1 Introduction

A glance into the ZAS data base of German clause embedding predicates shows that 666 out of 1795 clause embedding predicates embed *ob*-clauses 'whether/if-clauses' – cf. Stiebels et al. [19]. You find not only *fragen* 'ask', *wissen* 'know' and *bedenken* 'consider', which are more or less omnipresent when issues of interrogative embedding are discussed, but also unexpected verbs such as *sicher sein* 'be certain' – cf. (1) to (3). Such predicates account for nine percent of the *ob*-predicates. They only combine with an *ob*-clause if they are in the scope of a nonveridical operator – cf. (1) to (3). They are, so to speak, misfits among the *ob*-clause embedding predicates.

The author thanks the reviewers for their valuable support as well as Antonios Tsinakis, Peter Öhl and Robert Fittler for several intensive discussions.

© Springer-Verlag GmbH Germany, part of Springer Nature 2019
A. Silva et al. (Eds.): TbiLLC 2018, LNCS 11456, pp. 253–274, 2019.
https://doi.org/10.1007/978-3-662-59565-7_13

(1)   a.   Frank fragt, ob Maria in Rom ist.
             'Frank asks if Maria is in Rome.'
       b.   Frank fragt nicht, ob Maria in Rom ist.
             'Frank doesn't ask if Maria is in Rome.'

(2)   a.   Frank weiß, ob Maria in Rom ist.
             'Frank knows if Maria is in Rome.'
       b.   Frank weiß nicht, ob Maria in Rom ist.
             'Frank doesn't know if Maria is in Rome.'

(3)   a.   #Frank ist sicher, ob Maria in Rom ist.
             'Frank is certain if Maria is in Rome.'
       b.   Frank ist nicht sicher, ob Maria in Rom ist.
             'Frank isn't certain if Maria is in Rome'.

The outsiders can be intuitively divided into four groups:

i. *be certain*-predicates. *ausgehen* 'expect', *ausmachen* 'realize', *ausschließen* 'exclude', *begreifen* 'comprehend', *bekannt sein* 'be known', *beschwören* 'conjure', *bestätigen* 'confirm', *bewusst sein* 'be aware', *dementieren* 'deny', *einleuchten* 'be clear', *entsinnen* 'recall', *gewahr werden* 'become aware', *sicher sein* 'be certain', *verbergen* 'conceal', *verstehen* 'understand', *wahrnehmen* 'perceive', *übereinstimmen* 'agree', *überzeugt sein* 'be convinced', *vergessen* 'forget', *widerlegen* 'refute', ...

(4)   a.   Er ist "nicht sicher", ob der damals erzielte Überschuß von 307 Millionen wieder erreicht wird. ZDB 788: TIGER
             'He is "not sure" whether the surplus of 307 million achieved at that time will be achieved again.'
       b.   Inzwischen habe der Staatsschutz die Ermittlungen aufgenommen, da nicht auszuschließen sei, ob die Anschläge mit Aktionen autonomer Gruppen zusammenhängen. DWDS 1284: DWDS BZ 1997
             'In the meantime, the state security authorities have started investigations, since it cannot be ruled out whether the attacks are connected with actions by autonomous groups.'

ii. *forsee*-predicates. *absehen* 'foresee', *ahnen* 'guess', *hellsehen* 'predict', *voraussehen* 'foresee', *vorausahnen* 'anticipate', ...

(5)   Technik kann nicht vorausahnen, ob ein Kind auf dem Fahrrad gleich auf die Straße fährt – das kann nur der Mensch. ZDB 22906: DWDS nun 2012
       'Technology cannot predict whether a child on a bicycle will ride straight onto the road – only humans can do that.'

iii. *determine*-predicates. *einigen* 'reach an agreement', *garantieren* 'guarantee', *verantworten* 'be accountable', *versprechen* 'promise', ...

(6)   Es will keine Angela Merkel, die sagt, ich kann nicht versprechen, ob ich es
        besser kann, aber ich will es versuchen. ZDB 8574: DWDS BZ 2005
        'It doesn't want Angela Merkel saying I can't promise if I can do better, but I
        want to try.'

iv. *concern*-predicates. anfechten 'bother', auffallen 'notice', ausmachen 'care', interessieren 'interest', jucken 'care', kümmern 'care', stören 'bother', tangieren 'concern', …

(7)   Den Urlaubern macht es nichts aus, ob sie in Antalya oder Alicante am Strand
        liegen, … ZDB 25700: DWDS Zeit 1999
        'It doesn't matter to the tourists whether they are on the beach in Antalya or
        Alicante.'

There are inherently negative predicates that can embed *ob*-clauses. They are partly
opposites of *be certain-*, *determine-* or *concern*-predicates – cf. (8) and Sect. 4.5.

v. Inherently negative predicates. ausstehen 'be pending', entfallen 'slip so.'s mind',
entgehen 'not recognize', ignorieren 'disregard', unklar sein 'be unclear', unsicher sein
'be uncertain', unterschlagen 'suppress', verbergen 'mask', vergessen 'forget', ver-
heimlichen 'conceal', vernachlässigen 'disregard'.

(8)   Die Klage hat argumentiert, dass das Hinrichtungsprotokoll verfassungswidrig
        ist, weil das Lähmungsmittel für die Neuromuskulatur *verbirgt*, ob das Schlaf-
        mittel funktioniert … ZDB 22008: DeWaC-5 P 257450634
        '… because the suppressant conceals whether the barbiturate is working.'

All these examples show that the matrix predicate is in the scope of a nonveridical
operator. This does not always have to be a negation element. Nonveridical contexts
are for example also modal verbs and polarity questions – cf. (9) and (10).

(9)   Beide Partner sollten sich also ganz sicher sein, ob sie die gemeinsame Geburt
        wollen. ZDB 25375: DWDS BZ 1995
        'Both partners should therefore be quite sure whether they want the joint birth.'

(10)   Aber ist sie sicher, ob sie das wirklich will? DWDS 25678: DWDS Zeit 1998
        'But is she sure she really wants this?'

As to the use of the predicates of the classes *i* to *iv* and *v* imagine a path that begins
with the question state of an individual α and ends at best with α's knowledge state
regarding the question – cf. Schwabe [16]. For α's question state it is characteristic that
there is a question $\{\sigma, \neg\sigma\}$ and that α wants that α knows that σ or α knows that $\neg\sigma$.
Question states can be related to by predicates like *sich fragen* 'wonder' or *argwöhnen*
'suspect'. A question state can be followed by α's question act which is addressed to β or
by some mental activity of α. A question act can be related to by predicates like *fragen*
'ask', *nachhaken* 'ask further questions', *bitten* 'ask' and *betteln* 'beg'. A mental activity
can be denoted by *bedenken* 'ponder' or *beobachten* 'observe'. If β knows the answer
and asserts it, β performs a proper response act. This act can be denoted by predicates
like *ankündigen* 'announce' and *bestimmen* 'determine'. α's finding out the answer by

some mental activity can or be denoted by *herausfinden* 'find out' or *merken* 'notice'. If α believes β's true answer or what α found out and α is aware of the truth of it, α knows the answer.

However, as reality shows, β often does not react in the intended way. The reason for this may be that β does not know the answer or simply does not want to give it. While β's ignorance can be denoted by the negated predicates from the classes *i* to *iv* as well as from class *v*, β's lack of interest in the answer can be expressed by negating predicates like *interessieren* 'interest' and by inherently negative predicates like *egal sein* 'not care'. Predicates like *ankündigen* 'announce', *bestimmen* 'determine', *herausfinden* 'find out' or *merken* 'notice' as well as the predicates of *i* to *v* belong to the class of responsive predicates, that is, to predicates that relate to the answers of a question – cf. Lahiri [13] and Spector and Égré [18]. If predicates like *herausfinden* 'find out' embed an *ob*-clause, they relate to the true answer to a question, that is, either σ or ¬σ. They embed question extensions in terms of Groenendijk and Stokhof [10] or they are objective-veridical in terms of Schwabe and Fittler [17]. Predicates like *nicht sicher sein* 'not be certain' refer to the possible answers to the question, that is, to both σ or ¬σ. According to Schwabe [16] *find out*-predicates are proper responsive predicates and *be certain*-predicates are improper ones. The latter have in common that they are not objective-veridical like *herausfinden* 'find out' or *wissen* 'know' and they are not potentially factive like *bedauern* 'regret'. They also share the ability to express an epistemic attitude of the matrix subject towards the embedded proposition and to occur in question-response contexts. But this also applies to predicates such as *glauben* 'believe' and *möglich sein* 'be possible'. That is, β can answer α's question (11a) with (11b) and also with (11c).

(11)   a.   Is Maria going to Rome?

   b.   I think she's going to Rome.

   c.   It's possible for me she's going to Rome.

This raises the intriguing question of why predicates like *sicher sein* 'be certain' can embed *ob*-clauses in negative contexts, while predicates like *glauben* 'believe' and *möglich sein* 'be possible' cannot.

In Sect. 2, two approaches to this issue are briefly presented. The conclusion will be that Öhl's [15] suggestion that the embedding behavior of *be certain* is due to its characteristic of being subjective veridical is not sufficient. To better understand the semantic properties of the *be certain-*, *believe-* and *be possible*-predicates which are discussed in Sect. 4, Sect. 3 introduces the Logical Form of constructions with embedded *ob*-clauses.

2   Subjective Veridicality

Adger and Quer [1:109] distinguish between Q(uestion selecting)-predicates like *ask* and P(roposition selecting)-predicates like *tell*. The latter class includes the set of TF (true-false) predicates discussed by Ginzburg [9] like *assume*, *claim* and *maintain*. These predicates, according to Adger and Quer, indicate the subject's epistemic commitment to the truth or falsity of the embedded proposition. Their semantics is incompatible with that of an *if*-clause [1:125]. According to Adger and Quer, the class of P-predicates also includes predicates like *admit, hear, say, be obvious* and *be clear,* a predicate class the predicates of which embed *if*-clauses in negative but not in affirmative contexts. They are suggested to lack the lexical specification to be incompatible with questions.

With Adger and Quer the following is not clear: there's no reason why *be certain* predicates shouldn't be TF predicates in affirmative contexts. Why then can they embed *ob*-questions in negative contexts? Öhl [15] encounters a similar problem. He adapts Giannakidou's [8] concept of subjecttively veridical predicates, which is briefly summarized here for the purposes of the paper.

(12)   *Veridicality* and *Nonveridically*   (Definition 1 in Giannakidou [7])

    a.   A propositional operator F is veridical iff Fp entails or presupposes that p is true in some individual's model $M(i)$. P is true in $M(i)$ iff $M(i) \subseteq p$, i.e. if all worlds in $M(i)$ are p-worlds.

    b.   Otherwise, F is nonveridical.

Epistemic model of an individual i
An epistemic model $M(i) \in M$ is a set of worlds associated with an individual i representing worlds compatible with what i believes or knows.

Truth in an epistemic model (= full commitment)
A proposition p is true in an epistemic model $M(i)$ iff $M(i) \subseteq p$: $\forall w \, [w \in M(i) \rightarrow w \in \{w' \mid p(w')\}]$.

By replacing "epistemic model $M(i)$" by "information state $W(i)$", Giannakidou can distinguish between *veridical, antiveridical* and *nonveridical* information states and ultimately between *veridical, antiveridical* and *nonveridical* propositional operators. Unbiased questions, so Giannakidou, convey typical nonveridical information states.

(13)   *(Non)veridicality* and *(Non)homogeneity*   (Definition 3 in Giannakidou [8])

    a.   An information state (a set of worlds) $W(i)$ relative to an epistemic agent i is *veridical* with respect to a proposition p iff all worlds in $W(i)$ are p-worlds. (*Positively homogeneous state*).

    b.   An information state $W(i)$ relative to an epistemic agent i is *antiveridical* with respect to a proposition p iff all worlds in $W(i)$ are $\neg p$-worlds. (*Negatively homogeneous state*).

    c.   An information state $W(i)$ relative to an epistemic agent i is *nonveridical* with respect to a proposition p iff $W(i)$ is partitioned into p and $\neg p$-worlds. (*Nonhomogeneous state*).

Öhl makes the property of subjective veridicality responsible for the ability of predicates like *be clear* or *be certain* to embed *if*-clauses. However, if one takes into account predicates such as *accept* and *believe* that are subjectively veridical according to (12) and that do not allow questions in negative contexts, one quickly sees that this property is not sufficient. The following will show that predicates such as *be clear* and *be certain*, which are subjectively veridical in affirmative contexts, are able to convey nonveridical information states in negative contexts. Predicates like *be possible* are not able to do this. Although they are nonveridical in affirmative contexts, they cannot connect with questions. The reason for this results from the Logical Form of constructions with responsive predicates and *ob*-clauses. It will also be shown why predicates like *believe*, when they occur in negative contexts, are not able to convey nonveridical states.

3 Logical Form of Responsive *Ob*-Constructions

With regard to the predicate *sicher sein* 'be certain', the initial situation is as follows: On the one hand there is its argument structure as shown in (14) and the embedded question (15). The latter is represented in the manner of Adger and Quer [1] or Hamblin [11], respectively. On the other hand there is the embedding construction (16). How do the predicate and the question come together?

(14) *sicher sein* 'be certain'
$\lambda p \, \lambda x \, \lambda e$ [certain (p, x, e)]

(15) $\lambda p \, [p = \sigma \vee p = \neg\sigma]$
$\lambda p \, [p = $ Maria is in Rome $\vee p = \neg$ Maria is in Rome]

(16) Frank ist nicht sicher, ob Maria in Rom ist.
'Frank isn't certain whether M is in Rome.'
F is not certain that M is in R and F is not certain that she is not in R.
$\forall p \, \forall e \, [\neg \,(\text{certain } (p, f, e) \wedge (mr = p)) \wedge \neg \,(\text{certain } (p, f, e) \wedge (\neg mr = p))]$
abbreviated: $\neg c_f \, \sigma \wedge \neg c_f \neg\sigma$

Adapting Adger and Quer's [1] Logical Form of constructions with unselected *if*-clauses, we suggest a polarity sensitive operator (17i) that takes the *ob*-clause (15), thus yielding the Operator Phrase (17ii). The latter, on its part, applies to an objective or polarity sensitive matrix predicate (14) and creates (17iii). If (17iii) is combined with the subject and then with the negation operator, the final Logical Form [[NegP]] (17v, vi) obtains.

(17) Frank ist nicht sicher, ob Maria in Rom ist.
'Frank isn't certain whether M is in Rome.'
[… not … [$_{vP}$ [$_{vP}$ Frank … [$_{v'}$ [$_{OP}$ O t_{CP}] sicher ist]]] [$_{CP}$ ob M in R ist]]]

i. $[\![O]\!]$ $= \lambda R \, \lambda P_{P \in OVP \cup PSP} \, \lambda x \, \exists p \, \exists e \, [(P \,(p, x, e)) \wedge R \,(p)]$

ii. $[\![OP]\!]$ $= \lambda P_{P \in OVP \cup PSP} \, \lambda x \, \exists p \, \exists e \, [(P \,(p, x, e)) \wedge ((mr = p) \vee (\neg mr = p))]$

iii. $[\![v']\!]$ $= \lambda x \, \exists p \, \exists e \, [c \,(p, x, e) \wedge ((mr = p) \vee (\neg mr = p))]$

iv. $[\![VP]\!]$ $= \exists p \, \exists e \, [(c \,(p, f, e) \wedge (mr = p)) \vee (c \,(p, f, e) \wedge (\neg mr = p))]$

v. $[\![NegP]\!]$ $= \neg \, \exists p \, \exists e \, [(c \,(p, f, e) \wedge (mr = p)) \vee (c \,(p, f, e) \wedge (\neg mr = p))]$

vi. $= \forall p \, \forall e \, [\neg \,(c \,(p, f, e) \wedge (mr = p)) \wedge \neg \,(c \,(p, f, e) \wedge (\neg mr = p))]$

'F is not certain that M is in Rome and he is not certain that she is not in Rome.'

The equivalent representations (17v, vi) show that there is a disjunction or conjunction of two sentences, both sharing the matrix predicate but differinh with respect to their complementary propositions. This disjunction or conjunction can be regarded as the reduction of the structure 'Frank is not certain if Maria is in Rome'. Additionally, this operator existentially binds the variable p contributed by the question and the predicate as well as the eventuality variable e provided by the verb.

4 Intranegative Opposites

As shown in 2, Öhl [15] sees the subjective veridicality of predicates like *sicher sein* 'be certain' as the only reason for their ability to embed *ob*-clauses in a polarity environment. It was also pointed out that this condition cannot be sufficient since *glauben* 'believe' is subjectively veridical too but is always incompatible with *ob*-clauses. This section examines why among the subjectively epistemic predicates like *sicher sein* 'be certain' and *glauben* 'believe' only predicates like *sicher sein* embed *ob*-clauses in polarity contexts. Predicates like *sicher sein* will be defined as contrary positive intranegative opposites. It will be suggested that these predicates can embed *ob*-clauses in a polarity context because this context turns them into subjectively nonveridical predicates. It will be shown that *glauben* 'believe' in a question context does not become subjectively nonveridical when negated. It will be defined as a complementary intranegative opposite. Finally, it will be questioned why predicates like *möglich sein* 'be possible', which are originally subjectively nonveridical, cannot embed *ob*-clauses. It will turn out that a subjectively epistemic predicate embeds an *ob*-clause only if it is subjectively nonveridical in a polarity context. This condition is fulfilled for *sicher sein* 'be certain' but not for *glauben* and *möglich sein*.

4.1 Positive Contrary Intranegative Opposites: The *Be Certain* Case

If an individual α addresses a question like *Is Maria in Rome?* {mr, ¬mr} to an individual β, then α believes that β knows that Maria is in Rome or that β knows that Maria is not in Rome. If β doesn't know if Maria is in Rome but is willing to react to the question, he or she can do so – depending on his or her epistemic attitude – by means of a predicate like *sicher sein* 'be certain'. This can be reported by expressions as given in (18a-e):

(18) a. Frank ist sicher, dass Maria in Rom ist. c_f mr
          'Frank is certain that Maria is in Rome.'

      b. Frank ist sicher, dass Maria nicht in Rom ist. c_f ¬mr

      c. Frank ist nicht sicher, dass Maria in Rom ist.   ¬c_f mr

      d. Frank ist nicht sicher, dass Maria nicht in Rom ist.   ¬c_f ¬mr

      e. Frank ist nicht sicher, ob Maria in Rom ist.   ¬c_f mr ∧ ¬c_f ¬mr
          'Frank isn't certain whether Maria is in Rome.'

The Logical Form for (18e) is given in Sect. 3 – cf. (17v, vi). It consists of the disjunction of two complex propositions, both sharing the matrix predicate and differing in their complementary embedded propositions. The complementarity of the embedded propositions contributes to the fact that *sicher sein* 'be certain' is intranegative when it embeds an *ob*-clause. All complex propositions as given with (18a-e) represent epistemic states of Frank. The relationships between all these states can be visualized with the help of the Figs. 1 and 2. Figure 1 shows two inverse epistemic scales, one hosting Frank's epistemic evaluation grades for σ and the other one with his epistemic evaluation grades for ¬σ. At the rightmost pole of the σ-scale, there is 'Frank knows that Maria is in Rome' or kn_f σ, just followed by c_f σ. And at the leftmost pole of the ¬σ-scale, there is kn_f ¬σ with c_f ¬σ in tow. On the σ-scale, the degrees kn_f σ and c_f σ are followed by degrees that are not kn_f σ and c_f σ. The converse is also true: the degrees kn_f ¬σ and c_f ¬σ are also followed by their negations.

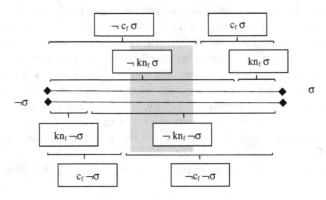

Fig. 1. Epistemic scales: *sicher sein* 'be certain'

Figure 1 illustrates that the negation phases ¬c_f σ and ¬c_f ¬σ overlap. It is important to underline that the overlaps involve other epistemic attitudes of the epistemic subject towards σ and ¬σ. This is just the case that is denoted by expressions like *Frank ist nicht sicher, ob Maria in Rom ist* 'Frank isn't certain whether Maria is in Rome'. Such an expression is compatible, as we will see below, with *Frank hält es für möglich, dass Maria in Rom ist.* 'Frank considers it possible that Maria is in Rome'.

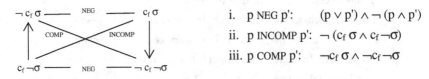

i.   p NEG p': $(p \lor p') \land \neg (p \land p')$

ii. p INCOMP p': $\neg (c_f \, \sigma \land c_f \, \neg \sigma)$

iii. p COMP p': $\neg c_f \, \sigma \land \neg c_f \, \neg \sigma$

Fig. 2. Epistemic square of *sicher sein* 'be certain'

Figure 2 shows *i.* that $c_f \, \sigma$ and $\neg c_f \, \sigma$, on the one hand, and $c_f \, \neg\sigma$ and $\neg c_f \, \neg\sigma$, on the other, are negatives of each other,[1] *ii.* that $c_f \, \neg\sigma$ and $c_f \, \sigma$ are incompatible, and *iii.* that $\neg c_f \, \sigma$ and $\neg c_f \, \neg\sigma$ are compatible. It follows from the incompatibility of $c_f \, \sigma$ and $c_f \, \neg\sigma$ that $c_f \, \sigma$ entails $\neg c_f \, \neg\sigma$ and that $c_f \, \neg\sigma$ entails $\neg c_f \, \sigma$. With these terminological clarifications it is now possible to define the notion positive contrary intranegative opposite.[2]

Definition 1: positive contrary intranegative opposites. Let q and q' be operators with a predicate type argument. Let the predicate domains of q and q' be such that q yields a truth value for a predicate expression σ iff q' yields a truth value for the negative opposite of σ. q and q' are *positive contrary intranegative opposites* to each other iff: for any predicate expressions σ or σ' eligible as operands of q and q': if σ NEG σ', then qσ INCOMP q'σ' and $\neg$qσ COMP $\neg$q'σ'.

The property to be a contrary positive intranegative opposite that becomes non-veridical in the scope of a nonveridical operator enables predicates like *sicher sein* 'be certain' to embed *ob*-clauses. Something similar can be shown for objective-veridical predicates like *wissen* 'know'. And it's certainly not hard to show even for *forsee*- and *determine*-predicates that they are positive intranegative opposites.

As we will see in the subsequent sections, *glauben* 'believe' and *möglich sein* 'be possible' do not meet the condition of being a contrary positive intranegative opposite.

4.2 Complementary Intranegative Opposites: The *Believe* Case

This paragraph seeks to examine why a predicate like *glauben* 'believe' does not accept an *ob*-clause if it is in the scope of a nonveridical operator. This is a quite pertinent question because *glauben* is subjectively veridical just as *sicher sein* 'be certain' is. The reason why *glauben* fails is that it is a complementary intranegative opposite and not a contrary one like *sicher sein* 'be certain'. What does this mean and entail?

Similar to *sicher sein* 'be certain', *glauben* 'believe' is related to two epistemic states: $b_f \, \sigma$ and $b_f \, \neg\sigma$ as well as their negations $\neg b_f \, \sigma$ and $\neg b_f \, \neg\sigma$.

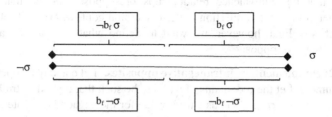

Fig. 3. Epistemic scales: *glauben* 'believe'

---

[1] $c_f \, \sigma$ and $\neg c_f \, \sigma$ are negatives of each other because they yield opposite truth values – cf. Löbner [14:485].

[2] The formulation of the following definition is based on that of Löbner [14:486] for dual operators.

A comparison between Figs. 1 and 3 reveals that $\neg b_f\,\sigma$ and $\neg b_f\,\neg\sigma$ do not overlap unlike $\neg c_f\,\sigma$ and $\neg c_f\,\neg\sigma$ did. The reason for this is that in the context of the question $\{\sigma, \neg\sigma\}$, the epistemic subject either believes σ or $\neg\sigma$, which means that $\neg b_f\,\sigma$ and $\neg b_f\,\neg\sigma$ are mutually exclusive. Similar to the negative *be certain*-states $\neg c_f\,\sigma$, $\neg c_f\,\neg\sigma$ and $\neg c_f\,\sigma \wedge \neg c_f\,\neg\sigma$, $\neg b_f\,\sigma$ and $\neg b_f\,\neg\sigma$ involve other epistemic attitudes of Frank towards σ and $\neg\sigma$.

As shown in Fig. 4, $b_f\,\sigma$ and $b_f\,\neg\sigma$ on the one hand, and $\neg b_f\,\sigma$ and $\neg b_f\,\neg\sigma$, on the other, are incompatible or yield opposite truth values. In terms of Löbner [14:485], they are in a negative relation NEG to each other. As for *sicher sein* 'be certain', recall that $c_f\,\sigma$ and $c_f\,\neg\sigma$ were incompatible and $\neg c_f\,\sigma$ and $\neg c_f\,\neg\sigma$ were compatible.

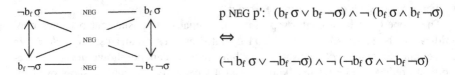

$$p \text{ NEG } p': (b_f\,\sigma \vee b_f\,\neg\sigma) \wedge \neg (b_f\,\sigma \wedge b_f\,\neg\sigma)$$
$$\Leftrightarrow$$
$$(\neg b_f\,\sigma \vee \neg b_f\,\neg\sigma) \wedge \neg (\neg b_f\,\sigma \wedge \neg b_f\,\neg\sigma)$$

Fig. 4. Epistemic square: *glauben* 'believe'

Figure 4 and (19) show that the negativity relation of $b_f\,\sigma$ and $b_f\,\neg\sigma$ entails that $\neg b_f\,\sigma$ and $b_f\,\neg\sigma$ are equivalent. The same applies for $b_f\,\sigma$ and $\neg b_f\,\neg\sigma$.

(19) $[(b_f\,\sigma \vee b_f\,\neg\sigma) \wedge \neg (b_f\,\sigma \wedge b_f\,\neg\sigma)]$ $\Leftrightarrow$

$[(b_f\,\sigma \vee b_f\,\neg\sigma) \wedge (\neg b_f\,\sigma \vee \neg b_f\,\neg\sigma)]$ $\Leftrightarrow$

$[(\neg b_f\,\sigma \Rightarrow b_f\,\neg\sigma) \wedge (b_f\,\sigma \Rightarrow \neg b_f\,\neg\sigma)]$ $\Leftrightarrow$

$[(\neg b_f\,\neg\sigma \Rightarrow b_f\,\sigma) \wedge (b_f\,\neg\sigma \Rightarrow \neg b_f\,\sigma)]$ $\Leftrightarrow$

$[(\neg b_f\,\sigma \Leftrightarrow b_f\,\neg\sigma) \wedge (b_f\,\sigma \Leftrightarrow \neg b_f\,\neg\sigma)]$

Provided that $\neg b_f\,\sigma$ and $b_f\,\neg\sigma$ are equivalent, the disjunction '$b_f\,\sigma \vee b_f\,\neg\sigma$', which would result if *glauben* 'believe' embedded an *ob*-clause, is tautological. It should be emphasized that the equivalence relationships presuppose that the matrix subject believes either σ or $\neg\sigma$ in a question-under-discussion context. We will return to this issue in Sect. 4.6. First, however, we want to define what is meant by a "complementary intranegative opposite".

Definition 2: complementary intranegative opposites. Let q and q' be operators with a p-type argument. Let the p-domains of q and q' be such that q yields a truth value for a predicate expression σ iff q' yields a truth value for the negative opposite of σ. q and q' are *complementary intranegative opposites* or *intranegative negatives* iff: for any p-expressions σ or σ' eligible as operands of q and q': if σ NEG σ', then qσ NEG q'σ' and $\neg$qσ NEG $\neg$q'σ'.[3]

This complementarity property applies, for instance, for *denken* 'think', *erwarten* 'expect', *hoffen* 'hope', *meinen* 'think', *wollen* 'want' and *wahrscheinlich sein*

---

[3] This definition corresponds to Löbner's [14:486] definition of dual opposites.

'be likely'. Whereas *wahrscheinlich sein* is always a complementary intranegative opposite, predicates like *glauben* 'believe', *hoffen* 'hope', *meinen* 'think', *denken* 'think' and *wollen* 'want' have this property only in question-under-discussion contexts. This property prevents the just mentioned predicates from embedding *ob*-clauses in the scope of a nonveridical operator. Let us remember, for a subjectively veridical predicate it is necessary to be subjectively nonveridical when taking an *ob*-clause. Predicates like *sicher sein* 'be certain' become nonveridical if they are in the scope of a nonveridical operator. The negated disjunction '$\neg$ (c_f σ $\lor$ c_f $\neg\sigma$)' they form when combined with a question and a nonveridical operator is well-formed because $\neg c_f$ σ and $\neg c_f$ $\neg\sigma$ are compatible. Imagine a predicate like *glauben* 'believe' combined with a question and being negated. The resulting negated disjunction $\neg$ (b_f σ $\lor$ b_f $\neg\sigma$) would be inadmissible. The reason for this is that the negated disjunction would contradict the condition that b_f σ and b_f $\neg\sigma$ are disjoint and incompatible or in a NEG-relationship or complementary, respectively – see Fig. 4. Another reason is that, assuming that $\neg b_f$ σ and b_f $\neg\sigma$ are equivalent, the contradiction '$\neg b_f$ σ $\land$ b_f σ' would result. In the next section, we will see why a nonverdical predicate like *möglich sein* 'be possible' cannot combine with a question and a nonveridical operator.

4.3   Negative Contrary Intranegative Opposites: The *Be Possible* Case

As already mentioned above, a predicate like *möglich sein* 'be possible' never embeds *ob*-clauses. The predicates $poss_f$ σ and $poss_f$ $\neg\sigma$ are subjectively nonveridical while their negatives are antiveridical in terms of Giannakidou [8]. The subjectively nonveridical $poss_f$ σ and $poss_f$ $\neg\sigma$ enable an overlap as illustrated in Fig. 5.

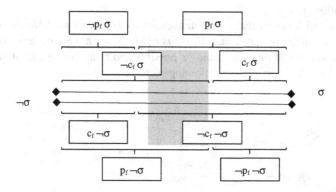

Fig. 5. Epistemic scales: *sicher sein* 'be certain', *möglich sein* 'be possible'

Figure 6 below shows the compatibility properties of *möglich sein* 'be possible'. It illustrates that the positive intranegatives $poss_f$ σ and $poss_f$ $\neg\sigma$ are compatible, unlike the positive intranegatives c_f σ and c_f $\neg\sigma$ and that the negative intranegatives $\neg poss_f$ σ and $\neg poss_f$ $\neg\sigma$ are incompatible.

It follows from the incompatibility of $\neg poss_f$ σ and $\neg poss_f$ $\neg\sigma$ that $\neg poss_f$ σ entails $poss_f$ $\neg\sigma$ and that $\neg poss_f$ $\neg\sigma$ entails $poss_f$ σ. The definition of a contrary negative intranegative opposite is as follows:

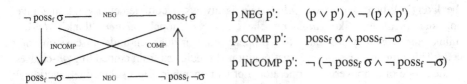

$$p \text{ NEG } p': \quad (p \vee p') \wedge \neg (p \wedge p')$$

$$p \text{ COMP } p': \quad \text{poss}_f \, \sigma \wedge \text{poss}_f \, \neg\sigma$$

$$p \text{ INCOMP } p': \quad \neg (\neg \text{poss}_f \, \sigma \wedge \neg \text{poss}_f \, \neg\sigma)$$

Fig. 6. Epistemic square: *möglich sein* 'be possible'

Definition 3: contrary negative intranegative opposites. Let q and q' be operators with a predicate type argument. Let the predicate domains of q and q' be such that q yields a truth value for a predicate expression σ iff q' yields a truth value for the negative opposite of σ. q and q' are *contrary negative intranegative opposites* to each other iff: for any predicate expressions σ or σ' eligible as operands of q and q': if σ NEG σ', then qσ COMP q'σ' and ¬qσ INCOMP ¬q'σ'.

The compatibility properties of *sicher sein* 'be certain', *glauben* 'believe' and *möglich sein* 'be possible' specified in the individual definitions can be formulated as the following syntactically relevant compatibility restrictions.

(20) a. *sicher sein* 'be certain': $\neg (c_\alpha \, \sigma \wedge c_\alpha \neg\sigma)$

  b. *glauben* 'believe': $(b_\alpha \, \sigma \vee b_\alpha \, \neg\sigma) \wedge \neg (b_\alpha \, \sigma \wedge b_\alpha \, \sigma \neg\sigma)$

  c. *möglich sein* 'be possible': $(p_\alpha \, \sigma \vee p_\alpha \neg\sigma)$

The compatibility restriction of *sicher sein* 'be certain' does not exclude the combination $\neg c_\alpha \, \sigma \wedge \neg c_\alpha \, \neg\sigma$ which is the Logical Form of *Frank is not certain if Maria is in Rome*. And the compatibility restriction of *möglich sein* 'be possible' does not exclude the combination $p_\alpha \, \sigma \wedge p_\alpha \, \neg\sigma$.

Figures 5 and 7 reveal that the sentences *Frank is certain that Maria is in Rome* and *It is not possible for Frank that Maria is not in Rome* are equivalent.[4] If you take the epistemic squares of *be certain* and *be possible* and rotate the square of *be possible* horizontally by 180°, you get the following picture:

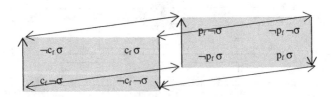

Fig. 7. Epistemic squares: *sicher sein* 'be certain', *möglich sein* 'be possible'

The equivalence of '$\neg c_f \, \sigma \wedge \neg c_f \, \neg\sigma$' and 'poss$_f \, \sigma \wedge$ poss$_f \, \neg\sigma$' raises the question of why *be certain* allows question embedding in a polarity context and *be possible* does

---

[4] Löbner [14:494] demonstrates that *be certain* is the dual of *be possible*. The duality relationship implies that $c_f \, \sigma$ and $\neg p_f \, \neg\sigma$ are equivalent. Horn [12:325) points out that $\neg c_f \, \sigma$ and $p_f \, \neg\sigma$ are equivalent.

not. The reason for this can be found by combining the question *if Maria is in Rome* with the matrix predicates *sicher sein* 'be certain' and *möglich sein* 'be possible' – cf. the derivation (17) of *sicher sein* in Sect. 3.

(21) $[\![VP]\!]$ = $\exists p\, \exists e\, [((\text{cert}\,(p, f, e)) \wedge (mr = p)) \vee ((\text{cert}\,(p, f, e)) \wedge (\neg mr = p))]$

$c_f\, \sigma \vee c_f\, \neg\sigma$

(22) $[\![VP]\!]$ = $\exists p\, \exists e\, [((\text{poss}\,(p, f, e)) \wedge (mr = p)) \vee ((\text{poss}\,(p, f, e)) \wedge (\neg mr = p))]$

$\text{poss}_f\, \sigma \vee \text{poss}_f\, \neg\sigma$

Sentences with Logical Forms like (21) and (22) prohibit themselves for pragmatic reasons. A question embedding would simply be trivial. However, while (21) can be saved by negation as we have seen in Sect. 3, this is not possible with (22). The reason for this is is simply the compatibility restriction of *be possible*, which cannot be negated. That is, $\neg\text{poss}_f\, \sigma$ and $\neg\text{poss}_f\, \neg\sigma$ is always excluded, which implies that *möglich sein* 'be possible' cannot be polarity sensitive and thus, unlike *sicher sein* 'be certain', not copatible with questions – cf. (17).

4.4 Implications and Horn-Scales

Figure 8 summarizes the relationship of *wissen* 'know', *sicher sein* 'be certain', *glauben* 'believe' and *möglich* sein 'be possible' to the epistemic scales. It illustrates the possible equivalences and implications and the strength of the implicative potential.

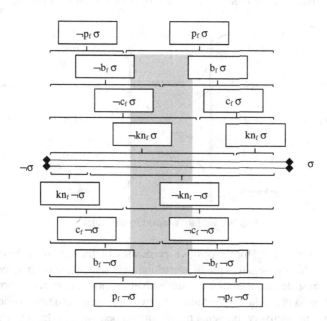

Fig. 8. Epistemic scales: *wissen* 'know', *sicher sein* 'be certain', *glauben* 'believe', *möglich sein* 'be possible'

Both $\neg kn_f\ \sigma$ and $\neg kn_f\ \neg\sigma$ as well as $\neg c_f\ \sigma$ and $\neg c_f\ \neg\sigma$ overlap. The same is true for $p_f\ \sigma$ and $p_f\ \neg\sigma$. The predicates involved are subjectively nonveridical. *Glauben* 'believe' doesn't allow such overlaps provided it is used in a question under discussion context. From the previous section, we know that only the overlaps formed by negated predicates, that is, by subjectively nonveridical predicates, are compatible with *ob*-clauses.

The subjectively nonveridical predicates $\neg c_f\ \sigma$ and $\neg c_f\ \neg\sigma$ as well as their equivalent pendants $p_f\ \neg\sigma$ and $p_f\ \sigma$ allow only one implication each. That is, both $\neg c_f\ \sigma$ and $p_f\ \neg\sigma$ imply $\neg kn_f\ \sigma$. And $\neg c_f\ \neg\sigma$ and $p_f\ \sigma$ entail $\neg kn_f\ \neg\sigma$. These predicates can be described as small implication triggers. In comparison, the complementary subjectively veridical *glauben* 'believe' is a medium implication trigger because $b_f\ \sigma$ implies $\neg c_f\ \neg\sigma$, which in turn implies $\neg kn_f\ \neg\sigma$. And $b_f\ \neg\sigma$ implies $\neg c_f\ \sigma$ which implies $\neg kn_f\ \sigma$. The subjectively veridical predicates $c_f\ \sigma$ and $c_f\ \neg\sigma$ and their equivalent antiveridical pendants $\neg p_f\ \neg\sigma$ and $\neg p_f\ \sigma$ are large implication triggers since each of them allows three implications. Thus, $c_f\ \sigma$ or $\neg p_f\neg\sigma$ entails $b_f\ \sigma$, which in turn entails $p_f\ \sigma$ and $\neg kn_f\ \neg\sigma$. And $c_f\ \neg\sigma$ or $\neg p_f\ \sigma$, respectively, implies $\neg b_f\ \sigma$ which implies $\neg c_f\ \sigma$ and $\neg kn_f\ \sigma$.

The division into small, medium and large predicates corresponds to that of Horn [12:325] into weak, middle and strong predicates. He places these predicates on two vertical scales. One scale is for positive verbs like *be certain, be likely* and *be possible*, the other is for their negatives. The weak verbs are at the bottom of each scale. Instead of *believe* Horn has *be likely* as a medium verb. The reason for this is that he, contrary to what is assumed here, does not consider $b_f\ \sigma$ and $b_f\ \neg\sigma$ to be disjunctive – cf. (20b). Figure 9 shows a version of Horn's [12] scales slightly adapted to this paper. Horn's predicates are indicated by indices, as are the ones discussed by Löbner.

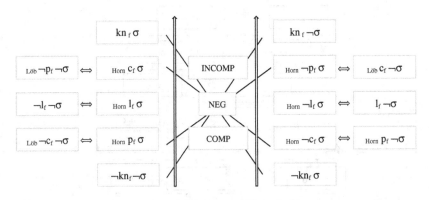

Fig. 9. Slightly adapted Horn Scales

It is easy to see that Horn's positive predicates of the left scale have negative equivalents and that his negative predicates of the right scale have positive equivalents. At the top of the scales are the contrary, subjectively veridical $c_f\ \sigma$ and $c_f\ \neg\sigma$ as well as their equivalent antiveridical pendants $\neg p_f\ \neg\sigma$ and $\neg p_f\ \sigma$. Their veridicality or antiveridicality, respectively, prevents them from embedding *ob*-clauses. The medium predicates are in the middle of each scale. They include $l_f\ \sigma$ as well as $\neg l_f\ \sigma$ according to Horn but also b_f σ as well as $\neg b_f$ σ according to Löbner and this paper. The weak predicates *möglich sein* 'be possible' and *nicht sicher sein* 'not be certain' are at the bottom of the scales. Both are, as we have seen in the previous sections, nonveridical. That is, the truth values of their embedded propositions σ or $\neg\sigma$ are not decided. This indecisiveness is a prerequisite for a question. However, we know that only *nicht sicher sein* can embed an *ob*-clause. *Möglich sein* 'be possible' fails because a sentence with it and an *ob*-clause is uninformative and it becomes antiveridical if it is negated.

4.5 Lexical Opposites and Their Intranegative Opposites

So far, we have been dealing with contrary and complementary intranegative opposites like, *be certain, be possible* and *believe* that have the syntactically formed negatives *not be certain, not be possible* and *not believe*. Apart from these, there are also lexical opposites. So we distinguish between lexically complementary opposites (LC) like *certain* and *uncertain*, on the one hand, and lexically contrary opposites like *deny* and *confirm* (Lc). As far as the LC-opposites are concerned, there are those who have complementary intranegative opposites (LCC) and those whose intranegative opposites are contrary (LCc). Lc-opposites always have contrary intranegative opposites (Lcc).

LCC. Lexical complementary opposites with complementary intranegative opposites are *wahr sein* 'be true' and *falsch sein* 'be false' as well as *wahrscheinlich sein* 'be likely' and *unwahrscheinlich sein* 'be unlikely'. They are not interesting for our purposes, as they do not allow the embedding of questions. Why don't they do this? Let's look at the opposites *wahr sein* 'be true' and *falsch sein* 'be false'. Both are in a NEG-relation to each other because false$_f$ σ is equivalent to $\neg$true$_f$ σ. Just as with *believe*, $\neg$true$_f$ σ and $\neg$true$_f$ $\neg\sigma$ as well as false$_f$ σ and false$_f$ $\neg\sigma$ are in a NEG-relation to each other. However, as we have seen in Sect. 4.1, the embedding of an *ob*-clause requires predicates that are contrary positive intranegative opposites. It is obvious that LCC verbs do not meet this condition.

LCc. Lexical complementary opposites with contrary intranegative opposites include, for instance, *erinnern* 'recall' and *vergessen* 'forget', *sicher sein* 'be certain' and *unsicher sein* 'be uncertain' as well as *zeigen* 'show' and *verbergen* 'conceal'. Predicates like *vergessen* 'forget', *unsicher sein* 'be uncertain', and *verbergen* 'conceal' belong to the predicate class we called inherently negative predicates in Sect. 1. They all embed *ob*-clauses like their negated opposites *nicht erinnern* 'not recall' and *nicht zeigen* 'not show', and *nicht sicher sein* 'not be certain' cf. (23a, b).

(23) a. Frank hat vergessen, ob er einen Pass besitzt. ($f_f\sigma$)
'Frank forgot if he had a passport.'

   b. Frank kann sich nicht erinnern, ob er einen Pass besitzt. ($r_f\sigma$)
'Frank can't remember if he has a passport.'

The compatibility restrictions of *erinnern* 'recall' and *vergessen* 'forget' in (24a, b) as well as Fig. 10 illustrate that *erinnern* is a positive contrary intranegative opposite like *sicher sein* 'be certain' and that *vergessen* is a contrary negative intranegative opposite like *möglich sein* 'be possible' – cf. (20a, c).

(24)   a.   *erinnern* 'recall': $\neg\,(r_f\,\sigma \wedge r_f\neg\sigma)$
              sicher sein 'be certain': $\neg\,(c_\alpha\,\sigma \wedge c_\alpha\neg\sigma)$
       b.   *vergessen* 'forget': $(f_f\,\sigma \vee f_f\neg\sigma)$
              möglich sein 'be possible': $(p_\alpha\,\sigma \vee p_\alpha\neg\sigma)$

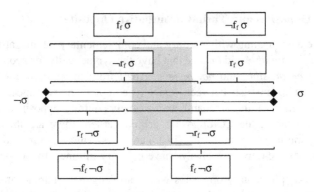

Fig. 10. epistemic squares: *sich erinnern* 'remember' and *vergessen* 'forget'

As for the epistemic square of *forget* in Fig. 11, its original rectangle is rotated 180° vertically.

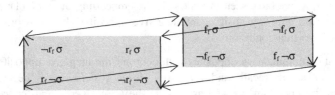

Fig. 11. Epistemic squares: *sich erinnern* 'remember' and *vergessen* 'forget'

Unlike *möglich sein* 'be possible', *vergessen* can embed an *ob*-clause – cf. (23a). The reason for this is that it is inherently negative.

Lcc. There are a few lexical contrary opposites with contrary intranegative opposites in German. They include the epistemic predicates *bestätigen* 'confirm' and *dementieren* 'deny' or *widerlegen* 'deny' – see (25a, b). And we also find these contrary predicates with respect to *concern*-predicates like *helfen* 'help' and *schaden* 'harm' or *begeistern* 'impress' or *stören* 'bother'. As shown in (25a, b), lexical contrary opposites can be conjoined in a nonveridical context.

(25) a. $\neg\text{conf}_f\,\sigma \wedge \neg\text{deny}_f\,\sigma$

Die Bankgesellschaft wollte am Sonntag weder bestätigen noch dementieren, dass der Vorstand Feddersen ... mit einem Mandat betraut habe. DWDS BZ 2001

'On Sunday, the bank company neither wanted to confirm nor deny that the board had entrusted Feddersen with a mandate ....'

b. $\neg\text{conf}_f\,\neg\sigma \wedge \neg\text{deny}_f\,\neg\sigma$

Doch wollte man bei dem Unternehmen weder bestätigen noch dementieren, dass die Kreditlinien ... nicht verlängert werden sollen. DWDS TS 2002

'However, the company neither wanted to confirm nor deny that the credit lines ... were not to be extended.'

(26a, b) illustrate that the verbs, taken alone, are exactly like *sicher sein* 'be certain' positive contrary intranegative opposites. Therefore, they can embed *ob*-clauses.

(26) a. $\neg\text{conf}_f\,\sigma \wedge \neg\text{conf}_f\,\neg\sigma$

Er muss dann bestätigen, ob er diese Dienste weiter nutzen oder diese Nummern sperren lassen will. ZDB 24083: DWDS TS 2003

'He must then confirm whether he wants to continue using these services or have these numbers blocked.'

b. $\neg\text{deny}_f\,\sigma \wedge \neg\text{deny}_f\,\neg\sigma$

Als sogar Helmut Kohl nicht ganz eindeutig dementierte, ob er ... im Waldspaziergang nun ein "akzeptables Ergebnis" erkenne, war höchste Alarmstufe erreicht: ... ZDB 25511: DWDS Zeit 1983

'When even Helmut Kohl did not quite clearly deny whether ... he recognized an "acceptable result" during the walk in the woods, the highest alarm level was reached: ...

As shown in (27a, b) to (29) and Figs. 12 and 13 below, lexical contrary predicates can embed *ob*-clauses when they seem coordinated.

(27) a.   Bracht wollte nicht *bestätigen*, aber auch nicht *dementieren*, ob die Bürgschaft .... beim DFB angekommen sei. ZDB 3004: DWDS BZ 1994

'B neither wanted to confirm nor to deny whether the DFB received the security.'

b.   ..., weder begeistert noch stört es mich, ob Palin aus religiösen Gründen eine Abtreibung ablehnt ... ZDB 24121: DWDS Zeit 2008

'It neither thrills nor does it disturb me whether Palin opposes abortion for religious reasons.'

At this point, we only discuss structures like (27a). Its rough syntactic representation (28) shows the conjunction of two complex sentences where the embedded *ob*-clauses are raised to the right and the subject in the second conjunct is elided.

(28)   Frank hat nicht bestätigt, und nicht dementiert, ob Maria in Rom ist.
'Frank didn't confirm, and didn't deny, whether Maria is in Rome.'

[ConjP [ConjP [CP1 Frank ... [NegP ... nicht [VP tFrank ... [v' [ORP tCP] bestätigt]]]]]

[Conj' und [CP2 ~~Frank~~ ... [NegP ... nicht [VP tFrank ... [v' [ORP tCP] dementiert]]]]]]]

[CP ob Maria in Rom ist]]

The derivation of the Logical Form of (28) takes place in a similar way as the derivation of the Logical Form of 'Frank is not certain whether Maria is in Rome.' – see (17).

(29)   ∀p ∀e [[¬ (confirm (p, f, e)) ∧ (mr = p)] ∧ [¬ (confirm (p, f, e)) ∧ (¬mr = p)]

∧ [¬ (deny (p, f, e)) ∧ (mr = p)] ∧ [¬ (deny (p, f, e)) ∧ (¬mr = p)]]

((F has not confirmed that mr) and (F has not confirmed that not mr)) and

((F has not denied that mr) and (F has not denied that not mr))

(¬conf$_f$ mr ∧ ¬conf$_f$¬ mr) ∧ (¬deny$_f$ mr ∧ ¬deny$_f$¬mr)

Figure 12 illustrates the four descriptions of response events as well as the overlaps of their negations enabled by *confirm* and *deny*.

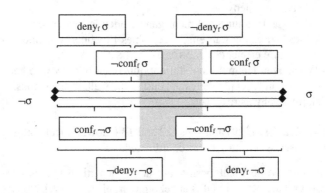

Fig. 12. Epistemic scales: *bestätigen* 'confirm' and *dementieren* 'deny'

Figure 12 illustrates that the following propositions are each equivalent to one another: conf$_f$ σ and deny$_f$ ¬σ, deny$_f$ σ and conf$_f$ ¬σ, ¬deny$_f$ σ and ¬conf$_f$ ¬σ, as well as ¬conf$_f$ σ and ¬deny$_f$σ. There are eight compatible combinations: *i.* ¬conf$_f$ σ ∧ ¬conf$_f$ ¬σ, *ii.* ¬deny$_f$ σ ∧ ¬deny$_f$ ¬σ, *iii.* ¬conf$_f$ σ ∧ ¬deny$_f$ σ, *iv.* ¬conf$_f$ ¬σ ∧ ¬deny$_f$ ¬σ, *v.* ¬conf$_f$ σ ∧ ¬deny$_f$ ¬σ, *vi.* ¬conf$_f$ ¬σ ∧ ¬deny$_f$ σ, and *vii.* ¬conf$_f$ σ ∧ ¬conf$_f$ ¬σ ∧ ¬deny$_f$ σ ∧ ¬deny$_f$ ¬σ (=*i* ∧ *ii*). The last combination corresponds to the Logical Form (29). All combinations are equivalent because conf$_f$ σ and deny$_f$ ¬σ are equivalent. And they are all exemplified in the ZAS database. As to the epistemic square of *deny* in Fig. 13, its original rectangle is rotated 180° horizontally and vertically.

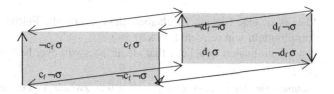

Fig. 13. Epistemic squares: *bestätigen* 'confirm' and *dementieren* 'deny'

What distinguishes the individual combinations? Let us begin with *vii*. Its Logical Form (29) is pleonastic, considering that $conf_f$ σ and $deny_f$ $\neg\sigma$ are equivalent. However, a glance at the syntactic structure (28) shows that it contains two negated different contrary opposites which can be contrasted. Furthermore, the truth value of σ is not fixed. The different contrary opposites distinguish *vii* from *i* and *ii* – cf. (25a, b). The unspecified truth value differentiates it from *iii* and *iv* – cf. (26a, b). The combinations *v* and *vi* are not exemplified because each is tautological in itself.

4.6 Additions to the *Believe*-Case

Apparent Equivalence. Section 4.2 showed that b_f σ and b_f $\neg\sigma$ are complementary intranegative opposites in question-under-discussion contexts because b_f σ and b_f $\neg\sigma$ as well as $\neg b_f$ σ and $\neg b_f$ $\neg\sigma$ are incompatible in these contexts. *Glauben* thus has the compatibility properties $(b_f$ σ $\vee$ b_f $\neg\sigma)$ $\wedge$ $\neg$ $(b_f$ σ $\wedge$ b_f $\neg\sigma)$ from which it follows that $\neg b_f$ σ and b_f $\neg\sigma$ are equivalent. Assuming an Excluded Middle (EM) as a pragmatic presupposition, Bartsch [2] shows that the equivalence of b_f $\neg\sigma$ and $\neg b_f$ σ follows from b_f σ $\vee$ b_f $\neg\sigma$ and $\neg b_f$ σ via disjunctive syllogism. Collins and Postal [3:9] call this an "apparent equivalence". However, the compatibility restrictions of *glauben* 'believe' show that the assumption of this disjunctive syllogism is not necessary. The equivalence of b_f $\neg\sigma$ and $\neg b_f$ σ follows already from the property of *believe* of being a complementary intranegative opposite in a question-under-discussion context.

Neg-Raising or Equivalence. There is a broad discussion as to whether $\neg b_f$ σ is syntactically derived from b_f $\neg\sigma$ via neg-raising or whether $\neg b_f$ σ and b_f $\neg\sigma$ are generated independently of each other and are linked semantically by an equivalence relationship. According to Crowley [4] and Collins and Postal [3], the approaches can be divided into three classes: the pure syntactic accounts (e.g. Fillmore [5] and possibly Horn [12]), the pure semantic-pragmatic accounts (e.g. Bartsch [2], Gajewski [6, 7]), and the mixed accounts (Collins and Postal [3]). Collins and Postal argue that neg-raising only makes sense if there is a syntactic reason for it. As for the cases examined in this paper, there were no syntactic reasons for raising the negation element from the embedded to the matrix clause. Additionally, similar to English, there are many constructions in German that can hardly be explained by neg-raising.

(30) a.   Aber es ist eine Illusion zu glauben, dass jemals alle Flächen frei von
            Kampfmitteln sein werden. DWDS BZ 2001
            'But it's an illusion to believe that all surfaces will ever be free of wea-
            ponry.'

      b.   Glauben Sie daran, dass sich jemals etwas zwischen oben und unten,
            Arm und Reich ändern wird? DWDS Zeit 2008
            'Do you believe that anything will ever change between above and
            below, rich and poor?'

Especially the polar matrix clause in (30b) seems to be an insurmountable hurdle for
the neg-raising approach. This, as well as the fact that there is no syntactic reason to
raise the negation element for the above examples, makes neg-raising obsolete.

5 Summary

The paper focused on the following questions: Why can subjectively veridical predicates like *sicher sein* 'be certain' embed *ob*-clauses in negative contexts, while subjectively veridical predicates like *glauben* 'believe' cannot? And why can't *möglich
sein* 'be possible', which is nonveridical, embed *ob*-questions either?

i. Adapting Adger and Quer's [1] Logical Form of constructions with unselected
   if-clauses, constructions like *Frank isn't certain if Maria is in Rome* are represented as the negated disjunction of two sentences, both sharing the matrix
   predicate but differing with respect to their complementary propositions:

$$\neg c_f \sigma \wedge \neg c_f \neg \sigma.$$

The non-negated disjunctions of all three predicates '$c_f \sigma \vee c_f \neg \sigma$', '$b_f \sigma \vee b_f \neg \sigma$' and '$p_f \sigma \vee p_f \neg \sigma$' cannot be expressed with the help of an *ob*-clause
because *be certain*, *believe* and *be possible* are not objective-veridical.

ii. Predicates like *sicher sein* are regarded as positive contrary intranegative oppo-
    sites because $c_\alpha \sigma$ and $c_\alpha \neg \sigma$ are incompatible and $\neg c_f \sigma$ and $\neg c_f \neg \sigma$ are com-
    patible. This behavior is summarized with the following compatibility restriction:

$$\neg(c_\alpha \sigma \wedge c_\alpha \neg \sigma).$$

This restriction does not exclude the combination $\neg c_\alpha \sigma \wedge \neg c_\alpha \neg \sigma$ which corresponds to the the Logical Form or *Frank is not certain if Maria is in Rome* – cf.
(17). The restriction enables predicates like *be certain* to embed *ob*-clauses in
polarity contexts. These contexts turn them into nonveridical predicates.

iii. *Believe* is defined as a complementary intranegative opposite because $b_\alpha \sigma$ and
     $b_\alpha \neg \sigma$ as well as $\neg b_f \sigma$ and $\neg b_f \neg \sigma$ are incompatible. The negated disjunction
     $\neg(b_\alpha \sigma \vee b_\alpha \neg \sigma)$ would be ruled out by the compatibility restrictions of *believe*:

$$(b_\alpha \sigma \vee b_\alpha \neg \sigma) \wedge \neg(b_\alpha \sigma \wedge b_\alpha \sigma \neg \sigma).$$

The feature '$b_\alpha \sigma \vee b_\alpha \neg\sigma$' is equivalent to '$\neg (\neg b_\alpha \sigma \wedge \neg b_\alpha \neg\sigma)$', which shows that $\neg b_\alpha \sigma$ and $\neg b_\alpha \neg\sigma$ are incompatible. The compatibility restrictions of *believe* imply that $b_\alpha \neg\sigma$ and $\neg b_\alpha \sigma$ are equivalent. Thus neg-raising – a syntactic derivation with the effect that $\neg b_\alpha \sigma$ implies $b_\alpha \neg\sigma$ – is superfluous. However, the assumption of equivalence is a pragmatic presupposition or implication that always takes place in a question-under-discussion context.

iv. As for *be possible*, it is a negative contrary intranegative opposite. Its compatibility restriction prohibits question embedding since $\neg p_\alpha \sigma$ and $\neg p_\alpha \neg\sigma$ are incompatible:

$$(p_\alpha\sigma \vee p_\alpha\neg\sigma).$$

v Due to their implication behavior, predicates like *be certain* and *not be possible* were characterized as strong implication triggers, predicates like *believe* as medium ones and predicates like *not be certain* and *be possible* as weak ones.

vi. Lexically complementary or contrary intranegative opposites like *recall* and *forget* or *confirm* and *deny* were discussed. The latter can appear coordinated on the surface and embed an *ob*-clause in a nonveridical context because they are contrary opposites. The Logical Form of the coordinated structure is pleonastic, but the syntactic structure allows the expression of contrast.

References

1. Adger, D., Quer, J.: The syntax and semantics of unselected embedded questions. Language **77**(1), 107–133 (2001)
2. Bartsch, R.: 'Negative transportation' gibt es nicht. Linguistische Berichte **27**, 1–7 (1973)
3. Collins, Ch., Postal, P.M.: Disentangling Two Distinct Notions of NEG Raising. lingbuzz/003595 (2017)
4. Crowley, P.: Neg-Raising and Neg-movement. Ms, MIT:1–32 (2016). https://semantics archive.net/Archive/TI3NTk3M/neg-raising.pdf
5. Fillmore, C.: The position of embedding transformations in grammar. Word **19**, 208–231 (1963)
6. Gajewski, J.: Neg-raising: polarity and presupposition. Dissertation, MIT (2005)
7. Gajewski, J.: Neg-Raising and polarity. Linguist. Philos. **30**(3), 289–328 (2007)
8. Giannakidou, A.: Inquisitive assertions and nonveridicality. In: Aloni, M., Franke, M., Roelofsen, F. (eds.) The dynamic, inquisitive, and visionary life of phi, ?phi, and possibly phi. A festschrift for Jeroen Groenendijk, Martin Stokhof and Frank Veltman, Amsterdam, pp. 115–126 (2013)
9. Ginzburg, J.: Resolving questions: I/II. Linguist. Philos. 18, 459–527/567–609 (1995)
10. Groenendijk, J., Stokhof, M.: Question. Studies in the semantics of questions and the pragmatics of answers. Dissertation, University of Amsterdam (1984)
11. Hamblin, C.L.: Questions in montague English. Found. Lang. **10**, 41–53 (1973)
12. Horn, L.R.: A natural history of negation, Chicago (1989)
13. Lahiri, U.: Questions and Answers in Embedded Contexts. Oxford Studies in Theoretical Linguistics. Oxford University Press, Oxford (2002)

14. Löbner, S.: Dual oppositions in lexical meaning. In: Portner, P., Maienborn, C., von Heusinger, K. (eds.) Semantics. (=Handbücher zur Sprach- und Kommunikationswissenschaft HSK 33.2), Berlin, pp. 479–506 (2011)
15. Öhl, P.: Veridicality and sets of alternative worlds. On embedded interrogatives and the complementisers that and if. In: Dimroth, C., Sudhoff, S. (eds.) The Grammatical Realization of Polarity, pp. 109–128. Linguistics Today/Linguistik Aktuell 249, Amsterdam (2018)
16. Schwabe, K.: A uniform representation of German embedded polar interrogatives a typology of their embedding predicates, and adaptors. Ms, ZAS Berlin (2018)
17. Schwabe, K., Fittler, R.: Über semantische Konsistenzbedingungen deutscher Matrixprädikate. Sprachtheorie und germanistische Linguistik 24(1–2), 45–75, 123–150 (2014)
18. Spector, B., Égré, P.: A Uniform Semantics for Embedded Interrogatives: an answer, not necessarily the answer. Synthese 192(6), 1729–1784 (2015)
19. Stiebels, B., McFadden, T., Schwabe, K., Solstad, T., Kellner, E., Sommer, L., Stoltmann, K.: ZAS Database of Clause-embedding Predicates. In: Institut für Deutsche Sprache (ed.) Release 1.0 in: OWID[plus], Mannheim. http://www.owid.de/plus/zasembed

A Non-factualist Semantics
for Attributions of Comparative Value

Andrés Soria Ruiz[✉]

Institute Jean Nicod, ENS-PSL-CNRS, Paris, France
andressoriaruiz@gmail.com

Abstract. This paper combines elements from literature on gradability and meta-ethics to offer a non-factualist, hyperplan-based semantics for evaluative adjectives, more specifically for comparative uses of those adjectives. Broadly non-factualist proposals about evaluative language understand value attributions in binary terms, that is, in terms of the expression, on the part of the speaker, of a favorable or unfavorable attitude towards the object under evaluation. But it is not obvious how to extend non-factualism to cover comparative uses of evaluative adjectives, and my purpose is to amend this.

Evaluative adjectives are perplexing for a number of reasons. Among them, they show a particular sensitivity to contextual features (distinct from the context-sensitivity of indexicals and demonstratives, for example); and they are conceptually connected to action, practical goals and motivation in a way that other adjectives are not. Their context-sensitivity has been discussed in philosophy and linguistics, insofar as these adjectives belong to superordinate *geni*, such as gradable [21], multi-dimensional [27,30] or judge-dependent expressions [5,23,35]. By contrast, their connection to action has been long discussed in meta-ethics, but less so in linguistics.

In meta-ethics, the theories that have proved most promising at dealing with the connection between evaluative language and action are those that fall under the *non-factualist* banner[1]. Non-factualists hold that to make a value attribution

---

Funding for this work comes from Obra Social La Caixa and grant numbers ANR-17-EURE-0017 FrontCog and ANR-10-IDEX-0001-02 PSL. Thanks to audiences in Lagodekhi, Berlin, Granada, Lisbon, Paris, and Tokyo. Special thanks to Bianca Cepollaro, Paul Égré, Federico Faroldi, Mora Maldonado, Salvador Mascarenhas, Erich Rast, Isidora Stojanovic, Peter Sutton, Hubert Truckenbrodt, Carla Umbach & Neftalí Villanueva for their valuable comments. Mistakes are mine.

[1] This family of views is often called *expressivism*. I eschew that label here because for some authors 'expressivism' names a sub-family of non-factualist views: those according to which the meaning of the sentences that feature evaluative expressions is characterized or individuated by the types of mental states that they express, which are non belief-like (this includes Gibbard's norm- or plan-expressivism [16,17], Blackburn's quasi-realism [4] and hybrid and neo-expressivist views, such as Bar-On and Chrisman's [1]). Those views are non-factualist because they hold that the main communicative purpose of those sentences is not to communicate facts, but it is a further question whether their meaning is best explained *via* considerations about the mental lives of the speakers who use them.

A. Silva et al. (Eds.): TbiLLC 2018, LNCS 11456, pp. 275–296, 2019.
https://doi.org/10.1007/978-3-662-59565-7_14

is to perform a speech act whose communicative intention is not to describe facts or to represent the world as being a certain way[2]. They hold that, if attributions of value had that communicative intention, then it would be difficult to account for the connection between attribution of value and action, since the recognition of some fact or state of affairs can hardly underwrite by itself a motivation to act. They maintain that, rather, value attributions serve to express the speaker's non-doxastic (non belief-like) attitudes towards the objects under evaluation. More concretely, to ascribe positive [negative] value to something is to express one's support [rejection] of it. The problem of the connection between attribution of value and motivation then vanishes, since attributing value turns out to just be a way of expressing one's practical stance.

The main obstacle for non-factualism is the Frege-Geach problem: very briefly, non-factualists defend that the semantic values of sentences from a fragment of language—in our case, evaluative sentences—are different from the traditional semantic values that semanticists associate with declarative sentences, namely propositions representing worldly states of affairs that can be true or false. The problem is that traditional semantic values play an essential role in explaining (i) the meaning of complex sentences that embed evaluative sentences, and (ii) the cogency of inferences that contain embedded and unembedded evaluative sentences in their premises and conclusions. If traditional semantic values are given up, those explanations are also given up.

We will say more about the Frege-Geach problem and how non-factualists have tried to overcome it, but our purpose here is to focus on a different and more modest problem affecting non-factualism about evaluative adjectives. As we said, these adjectives are gradable, and this is diagnosed *inter alia* by the fact that they admit comparative constructions. But what should non-factualists say about the meaning of such constructions? Non-factualists almost exclusively consider uses of evaluative adjectives in the positive form and pay no attention to comparatives. This is problematic, because non-factualists tend to cash out the meaning of value terms in terms of the expression of *binary* attitudes of support or rejection, approval or disapproval, *PRO-* or *CON*-attitudes, etc; and it is not obvious how to extend this proposal to capture the meaning of sentences like 'Stealing for food is better than murdering for food', which on the face of it do not convey such binary attitudes.

This paper presents comparative constructions as a challenge for non-factualism; and offers a way of solving that challenge. The solution consists in providing a formal framework that captures the intuitive idea that to make comparisons involving evaluative adjectives is to compare the relative degrees of support or rejection of the objects under consideration. Thus, to say that an object of evaluation α is *better* than an object of evaluation β is simply to express *more support* for α than for β.

---

[2] As I will use the term, a *value attribution* is a communicative act by which one attributes value to some object(s). It is the outward, public expression of a value *judgment*, which is a mental act.

The paper has 5 sections. Section 1 characterizes evaluative adjectives and their semantic and conceptual properties, namely the fact that they are liable to figure in "faultless disagreements" both in their positive and comparative form, and their conceptual connection to action and motivation. Section 2 characterizes non-factualism as a theory of the meaning of evaluative adjectives that promises to deliver an account of the properties observed in Sect. 1. Section 3 presents comparative constructions with evaluative adjectives as a challenge for non-factualism, and Sect. 4 presents the proposed solution to that challenge. Section 5 offers some tentative remarks on how this non-factualist framework can deal with the conversational dynamics of value attributions, more specifically with disagreement, and also considers the prospects of extending this proposal to different kinds of evaluative adjectives as well as objects of evaluation of various types. Section 6 concludes.

1 Evaluative Adjectives

Evaluative expressions are expressions of natural language that appear paradigmatically in attributions of value. An attribution of value is a type of communicative act whereby a speaker communicates evaluative information, that is, information to the effect that something has a 'positive or negative standing— merit or demerit, worth or unworth—relative to a certain kind of standard' [39, p. 29]. Evaluative expressions include verbs (*to matter, to justify*), nouns (*idiot, hero*) and most eminently, adjectives[3]. Here's a (partial) list:

> *good, bad, just, justified, credible, beautiful, ugly, virtuous, ethical, important, tasty, advisable, disgusting, courageous, chaste, charming, lazy, stupid, smart, brave, fearful, timid, smart, idiotic, foolish, pretty, elegant, handsome, precious*

These adjectives can be classified along various axes. To mention only two: there are, first, different *flavors* of evaluation. We can speak personal taste (*fun*); of moral (*virtuous*), aesthetic (*elegant*) or epistemic (*credible*) value; there are all-purpose value adjectives as well (*important, good, bad*). Secondly, in the metaethics literature evaluative terms are often divided in *thin* and *thick*. Thin terms are those that appear to have only evaluative meaning. The paradigmatic examples of thin terms are the all-purpose *good/bad*, but we can also consider other adjectives as thin, relative to different flavors of evaluation. So we may talk of, e.g. the pair *beautiful/ugly* as aesthetically thin adjectives. By contrast, thick terms carry both evaluative and descriptive component. Thus, *cruel* is thick,

---

[3] In linguistics, 'evaluativity' means different things. For instance, Rett defines evaluative constructions as those that make reference to a contextually determined standard or threshold ([28, p. 210]; Bierwisch [3] calls that property *normbezug*). Bierwisch [3, p. 87], on the other hand, classifies gradable adjectives as dimensional or evaluative depending on whether their extensions are determined by a single or multiple dimensions, respectively. More recently, Zehr and Égré [12, p. 35] characterize evaluative adjectives as those that can embed under *find*. Neither of these senses of 'evaluativity' are coextensive with ours, although there is considerable overlap.

because in addition to its negative evaluative component, to call something cruel is to describe it in some way [7,39].

Evaluative adjectives are puzzling for a number of reasons. First, they are gradable, as shown by the fact that they are all admissible in comparative form[4]. Gradability gives rise to a few interesting phenomena, most notably vagueness [20] and relatedly, the possibility that speakers disagree faultlessly about whether the positive form applies to an individual. For example, contrast the following dialogues:

(1)   a. Natalia: Darla is smart.

     b. Matheus: No she isn't.

(2)   a. Natalia: Darla is 23.

     b. Matheus: No she isn't.

Whereas in (1) it seems perfectly possible for Natalia and Matheus to both be right, it's clear in (2) that either of them must be mistaken. This is a property that evaluative adjectives share with many other gradable adjectives, like *tall*, *expensive* or *big*: these adjectives in their positive form can give rise to disagreements where both parties seem to be right. But in contrast to those adjectives, evaluative adjectives *also* give rise to this type of disagreement in comparative constructions:

(3)   a. Natalia: Carla is smarter than Dan.

     b. Matheus: No she isn't.

(4)   a. Natalia: Carla is taller than Dan.

     b. Matheus: No she isn't.

Again, it seems that in (4) either Natalia or Matheus has to be wrong, whereas in (3) it isn't so. Natalia and Matheus might be taking different aspects into account in their definition of smartness and thereby ordering individuals differently (for instance, Natalia might be giving more weight to ability with numbers, while Matheus might be giving more weight to social skills). Bylinina [5] calls this property 'scalar variation'. The property of scalar variation does not yet single out evaluative adjectives however, since many non-evaluative adjectives show the same pattern (e.g. multi-dimensional adjectives like *healthy* and *sick* and experiential predicates like *salty* or *soft*). Nonetheless, the possibility of faultless disagreement in positive and comparative constructions suggests that evaluative adjectives have (*i*) variable thresholds and give rise to (*ii*) variable orderings (see [33] for recent experimental evidence about the scope of this phenomenon).

To account for such variability, a natural move is to simply assign different extensions to different uses of these adjectives [32]. This strategy accounts easily for the intuition that such disputes are faultless. However, it does not suffice to account for the intuition that exchanges like (1) and (3) are disagreements: if Natalia and Matheus are assigning different extensions to the word *smart*,

---

[4] With some exceptions, most notably deontic terms like *permissible* or *wrong*.

then they just mean different things by it, and they ought to be perceived to be talking past each other. To accommodate the intuition that dialogues (1) and (3) (or dialogues similar to it) really are disagreements, there has to be something that Natalia and Matheus disagree *about*. This is a thorny issue and we will bracket it for now (although we'll come back to it in Sects. 3 and 6), but let us retain the following idea: that these types of disagreements may arise when these adjectives are used in the comparative form shows that evaluative adjectives induce orderings that can vary cross-contextually (see [38]; [25, ch. 6] for discussion).

Secondly, evaluative adjectives are characterized by their connection to action and motivation. The meta-ethics literature has long expanded on this feature of normative vocabulary (see e.g. [8,18]). The idea is that, when a rational speaker makes a positive [negative] value attribution about an object, they should *eo ipso* be inclined to act in its favor [against it]. In other terms, attributing positive [negative] value invites the inference that one will orient one's action towards [against] the object under evaluation. For instance, if I judge football to be a *great* sport, I am expected to be willing to watch it, play it or follow it somehow. By contrast, if I judge it to be a *popular* sport, there's no expectation about what my practical attitudes towards it will be. This is what Stevenson [36, 16] called the "magnetism" of value vocabulary—the good is attractive; the bad is repulsive.

This property is difficult to capture in linguistic terms, but we may rehearse the following test for its presence. Consider the contrast within each pair of sentences:

(5) a. Matheus thinks volunteering is virtuous although he doesn't have any intention or plan whatsoever of supporting, promoting or doing it.

 b. ? Natalia thinks volunteering is unpaid although she doesn't have any intention or plan whatsoever of supporting, promoting or doing it.

(6) a. Matheus thinks donating is unethical although he is a member of a few charities.

 b. ? Natalia thinks donating is widespread although she is a member of a few charities.

While the *a*-sentences are acceptable, the *b*-sentences sound marked. Why is this so? We venture that the oddness is due to the connective *although*, which suggests a contrast between its arguments. To describe an individual as having a certain evaluative stance—thinking that an activity is virtuous or unethical—while at the same time ascribing to her practical attitudes that are markedly incoherent with that stance constitutes good grounds for such contrast. But regular beliefs about matters of fact—thinking that an activity is unpaid or widespread—do not contrast with practical attitudes in any immediate way.

Thus, the connective *although* marks an inexistent contrast in the *b*-sentences, and the sentences are awkward for that reason. However, the contrast in the *a*-sentences is clearly there, thereby licensing the connective. This feature characterizes evaluative language generally: accepting evaluative claims involves, as a conceptual matter, the adoption of certain practical commitments[5].

Here's another way of making the same point: it is known that cogent practical inferences (that is, inferences whose conclusions are courses of action, or at least attributions of intentions to engage in some course of action) require at least one premise concerning the practical stance of their agent. For example: we often accept advice such as '*it's raining, so take an umbrella*'. But it's clear that, as it stands, that inference is only cogent as an enthymeme whose elided premise is something like '*you don't want to get wet*', or alternatively something like '*it's a bad idea to get wet*'. By contrast, no premise void of any reference *either* to the desires, intentions or plans of the agent, *or* to what the agent takes to be valuable, would do the trick. But note that there is an important contrast between those two types of premises: the former are ascriptions of propositional attitudes (in particular, non-cognitive attitudes), while the latter are not. Thus, we hypothesize that attributions of value are the only type of non-attitudinal claim that can make a chain of practical inference cogent[6]. Accounting for this link between evaluation and action is a *desideratum* that any theory of evaluative language must meet.

2   Non-factualism

Non-factualism names a broad family of theories originally about the meaning of moral terms, but that has been extended to other parcels of natural language, and can of course be extended to other types of evaluative language. In general, a non-factualist about evaluative language would hold that the function of this type of vocabulary is not to describe *ways the world is* [14, p. 471], or states of affairs [6, p. 187]. More specifically, evaluative predicates do not denote properties in the usual sense, and the declarative sentences in which evaluative terms usually figure (which we may call *evaluative sentences*) do not, in general, state that such properties are instantiated. Contrary to appearances, evaluative sentences serve communicative purposes *other* than imparting factual information or representing reality. Those are all negative characterizations; by contrast, the positive stories about the meaning of evaluative terms that non-factualists tell vary widely, but they tend to hold that evaluative vocabulary serves to communicate desires, practical intentions or emotional states.

To take an example, consider sentences (7) and (8):

(7)   Donating money to charity is widespread.

(8)   Donating money to charity is wrong.

---

[5] The significance of this test should not be overstated, however: background or contextual information can very well fill the missing pieces that would make the *b*-sentences acceptable: for example, in a context in which it is common knowledge that Natalia only ever enjoys very exclusive activities, (6b) would be a perfectly natural thing to say. In absence of such contextual clues however, (6b) is odd.

[6] Thanks to Salvador Mascarenhas for this suggestion. See also [18] for related points.

Non-factualists hold that the presence of a value term like *wrong* in (8) marks a profound semantic contrast between the two sentences. In most contexts, an utterance of (7) aims at informing an audience of a fact, namely that the practice of donating has a certain property (*being widespread*). In contrast to this, by uttering a sentence like (8), a speaker does not aim at informing her audience of a similar fact, namely that donating has the property of *being wrong*; rather, an utterance of (8) expresses the speaker's negative attitude towards the practice of donating money to charity. It expresses that the speaker condemns, disapproves or rejects the practice of donating money. Different authors characterize these attitudes differently, but it is common to all of them to stress the opposition between those attitudes and doxastic attitudes like belief or knowledge. Waxing Marxist, one might say that doxastic attitudes aim at *interpreting* the world; while the non-doxastic attitudes associated with evaluative vocabulary aim at *changing* it.

This very brief characterization of non-factualism should already make it clear that this approach is well equipped to account for the features of evaluative adjectives discussed in Sect. 1, namely the intuition that these adjectives give rise to faultless disagreements in positive and comparative form and their action-guidance.

First, the non-doxastic attitudes expressed by using evaluative vocabulary vary from speaker to speaker, thereby allowing for different speakers to express different attitudes in different contexts of use. And it is sensible to think that those different attitudes give rise both to variable thresholds for the application of the positive form of gradable adjectives, as well as to variable orderings of individuals along the relevant value dimensions. This accounts at least for the intuition that speakers in dialogues like (1) and (3) can both be right; in other words, that the fact that either speaker is justified in their assessment does not entail that the other speaker is wrong. To see this, suppose that non-factualism about *smart* is the view that attributions of smartness (calling someone 'smart') express positive attitudes towards the individual under evaluation (perhaps over and above expressing a purely descriptive belief about her cognitive abilities). Thus, when Natalia says that Darla is smart, she expresses a positive attitude about Darla; and when Matheus denies that, he either rejects Natalia's claim or perhaps expresses something stronger, namely a negative attitude towards Darla. Either way, for a non-factualist a disagreement dialogue like (1) need not imply that there is *one right attitude* to have towards Darla, and that this is what is under dispute between the speakers[7].

---

[7] As a reviewer accurately points out, claiming that speakers can have divergent non-doxastic attitudes only shows that disagreements like (1) *can* be faultless. It is a further step to show that a disagreement like this is actually faultless, that is, that both speakers are right. Traditionally however, the "intuition of faultlessness" has been understood in the former, weaker sense. This is because the possibility of faultlessness is enough already to set apart dialogues like (1) from dialogues like (2). Nonetheless, non-factualists can couple the view that speakers can have different non-doxastic attitudes with the independently plausible view that speakers have authoritative access to those attitudes in order to accommodate the stronger sense of faultlessness.

Accommodating the intuition that the exchanges in (1) and (3) really are disagreements is less straightforward, however. The reason for this is that the simplest notion of disagreement is unavailable to the non-factualist. This is the notion according to which two speakers disagree just in case they utter sentences whose conjunction is, or implies, a contradiction. That happens most clearly when speakers utter sentences that express contradictory propositions, such as (2). But recall that non-factualists account for the intuition of faultlessness by appealing to the fact that different speakers use evaluative sentences to express different non-doxastic attitudes. In (1), Natalia and Matheus can both be right because their non-doxastic attitudes can both be *satisfied*: Natalia can continue to enjoy the company of Darla, while Matheus is free to shun her. If that is so, then their sentences simply express different, but compatible contents: in (1) Natalia would be expressing non-doxastic attitude A_N about Darla, while Matheus would be expressing a different non-doxastic attitude about Darla, A_M, in his reply to Natalia. At first sight, there is no contradiction or tension between A_N and A_M, even if the former is a positive attitude about Darla, while the latter is a negative attitude. So how can it be that speakers disagree? What are they disagreeing about? It cannot be a disagreement about how things stand: Natalia has attitude A_N; Matheus has attitude A_M; and this might be transparent to both of them. And if non-factualism is correct, it cannot be either a disagreement about how Natalia and Matheus represent the world, since those attitudes are non-doxastic.

The core strategy followed by most non-factualists has been to construct disagreements like (1) and (3) as a kind of practical disagreement; a disagreement about *what to do*, rather than a disagreement about *how things stand*. The idea is that a disagreement between speakers discussing whether a course of action is 'good', for instance, arises not due to the fact that speakers utter sentences with incompatible truth-conditions, but to the fact that they express conflicting non-doxastic attitudes of support and rejection. By uttering something like 'ϕ-ing is good', a speaker A expresses their support of course of action ϕ; by replying things like 'no', 'I disagree', etc., a speaker B can either express rejection of course of action ϕ or rejection of A's support of it. Speakers in this type of dialogue disagree about what attitude to adopt, not about what is the case [17,29,37]. Recall that the problem of disagreement for non-factualists was that the non-doxastic attitudes that they associate with evaluative sentences were independently satisfiable. The solution consists in noting that, if one assumes that those attitudes are expressed with the communicative purpose of being shared by other speakers, then they no longer are independently satisfiable. Put simply: when Natalia utters (1a), she proposes that both her and Matheus adopt a positive attitude towards Darla; when Matheus utters (1b), he either rejects that they both adopt that attitude or proposes that they both adopt a negative attitude towards Darla. Finally, the impression of faultlessness is preserved since, even though the attitudes cannot be mutually satisfied if there a shared attitude to adopt, there is still no single correct attitude to have towards Darla.

Let us note briefly that writers like Marques [26] and MacFarlane [25, p. 131] have expressed skepticism that the strategy of treating these cases of disagreement as practical or attitudinal disagreements is sufficiently powerful. Marques holds that defenders of this strategy need to say more in order to justify how diverging attitudes can constitute disagreement, and points out that the kind of disagreement that the proposal aims to tackle appears between speakers who share no practical goal or engage in any joint action. MacFarlane points out that these disagreements often involve propositional anaphora (Ann: 'Licorice is tasty'; Bill: 'that's false!') which would suggest that speakers engaging in them really are disputing the truth of a proposition, and not just expressing different non-doxastic attitudes[8].

Secondly, the action-guidance associated with evaluative vocabulary is immediately secured by non-factualists proposals, since evaluative words express attitudes that are, by their very nature, action-guiding. According to non-factualists, to call someone a good person is to express a—perhaps undetermined—positive attitude towards that person; it is to express one's intention of orienting one's actions towards frequenting that person, following her example or promoting her well-being. Thus, for non-factualists, the connection between evaluative language and action-guidance is no mystery; to make a claim of value *just is* to express how one intends to direct one's action. Thus, the *a*-sentences in (5) and (6) are natural because in those sentences we are ascribing someone a relatively irrational state of mind: *e.g.* when we describe Matheus as someone who thinks that volunteering is a virtuous thing to do and at the same time lacks any intention of doing it, it is as though we are saying that he thinks that he has a certain practical stance when he actually does not have it. Marking that contrast with a connective like *although* sounds then perfectly natural. Similarly, for a nonfactualist it is no surprise either that evaluative claims and claims about an agent's desires or intentions play the same role in practical reasoning, given that the former semantically express the attitudes that the latter are *about*.

As we mentioned at the start, the main problem faced by non-factualists is the Frege-Geach problem. If non-factualism about a subset of declarative sentences—let us call it fragment F—is the thesis that elements of F do not describe or represent facts, then it follows that they do not have propositions that can be true or false as their semantic values (at least under standard conceptions of propositions as representations of facts or states of affairs). Truthfunctional operators such as modals or logical connectives take propositions as their arguments. If non-factualism about F is right, then one should expect

---

[8] It is not our purpose to decide this issue here, but simply to point out the lines of research that non-factualists have proposed in order to deal with these phenomena. It is important to note too that authors who criticize the non-factualist strategy of appealing to practical disagreement do not all defend a single alternative account of the semantics of evaluative terms: MacFarlane defends a kind of semantic relativism which he calls *assessor relativism*; see Egan [9–11] as well, who defends a different version of relativism based on Lewis' notion of centered worlds. On the other hand, Marques defend a variety of *contextualism*. For another comprehensive and powerful defense of contextualism, see Silk [32].

that the sentences of F fail to embed under such operators. Now, *all declarative sentences*—including sentences of F—*embed* under such operators. Therefore, non-factualism about F has a problem. More generally, the syntactic and semantic behavior of declarative sentences is by and large uniform. Schroeder puts the matter bluntly, discussing non-factualism about moral terms: '[t]here is no linguistic evidence whatsoever that the meaning of moral terms works differently than that of ordinary descriptive terms. On the contrary, everything that you can do syntactically with a descriptive predicate like 'green', you can do with a moral predicate like 'wrong', and when you do those things, they have *the same semantic effects*' [31, p. 704, his emphasis]. A possible way out is to reject that the semantic values of embedded and unembedded uses of the sentences in F are the same. But if this route is taken, then non-factualists face an even greater hurdle, which is to account for the obvious inferential connections between embedded and unembedded uses of those sentences. For instance: the inference from 'if eating animals is wrong, then I shouldn't be eating this steak' and 'eating animals is wrong' to 'therefore, I shouldn't be eating this steak' relies crucially on the antecedent of the first premise and the second premise having the same meaning. If non-factualists deny that they have the same meaning, then they have to find an alternative explanation for why that inference holds.

Most contemporary non-factualists accounts of evaluative language have found those obstacles too high to overcome, and have adopted some kind of hybrid view, accepting that one needs to make room for traditional semantic values—truth-apt propositions—if one is to avoid the Frege-Geach problem (this move is advocated as early as [15]). Hybrid views thus factor out the meaning of evaluative sentences into a traditional proposition describing a state of affairs that can be evaluated for truth and a non-traditional, non-factual component. The non-factual component is meant to account for cross-contextual disagreement in the way just described (as practical disagreement), as well as the connection of evaluative sentences to action and motivation. On the other hand, the descriptive component is what embeds under truth-functional operators and is therefore productive in explaining the meaning of complex constructions and the inferences they enter into.

Having said that, let us stress that we do not aim to offer any novel solution to the Frege-Geach problem in this paper, and this is not our focus. Nonetheless, since the proposal adopted here builds on Gibbard's *plan-expressivism* [17], it inherits his solution to the Frege-Geach problem. Gibbard's is essentially a hybrid view that associates the meaning of evaluative sentences with the expression of a *plan of action* (as its non-factualist component) and a belief that that very plan of action is adopted by the speaker (as its descriptive component). Consequently, a solution to the Frege-Geach problem along the lines just sketched is the one that we would ultimately endorse. We turn now to the problem that we want to tackle, namely, how non-factualists can and should accommodate comparative uses of evaluative adjectives.

3   Comparatives

As we've seen, non-factualists semantically associate evaluative vocabulary with the expression of positive and negative non-doxastic attitudes. Thus, an utterance of a sentence like (9) might be said to express a positive attitude towards the practice of volunteering for a charity; while an utterance of (10) expresses a negative attitude towards the practice of donating money to charity. The non-factualist approach, however, faces a fundamental shortcoming when faced with comparative constructions like (11):

(9)   Volunteering for a charity is good.

(10)   Donating money to a charity is bad.

(11)   Volunteering for a charity is better than donating money.

When a speaker utters (11), she need not endorse nor reject either *relata*. She is merely comparing the two actions; and her uttering (11) is compatible with adopting almost any combination of positive and negative attitudes towards either of them (with the exception of being in favor of donating money while being against volunteering). This is shown by the fact that (12)–(14) are acceptable, while (15) is not:

(12)   Volunteering for a charity is better than donating money, though both are bad.

(13)   Volunteering for a charity is better than donating money, though both are good.

(14)   Volunteering for a charity is better than donating money; in fact, volunteering for a charity is good whereas donating money is bad.

(15)   # Volunteering for a charity is better than donating money; in fact, volunteering for a charity is bad whereas donating money is good.

That these adjectives pattern in this way actually suggests that *good* and *bad* are relative-standard adjectives in the sense of Kennedy [20], and we venture that many evaluative adjectives show the same pattern [13][9]. The challenge is thus clear: non-factualists defend that evaluative sentences express binary attitudes of support or rejection; but comparative uses of those adjectives express neither. Non-factualists need to enrich their favoured semantics if they want to accommodate comparative uses of evaluative adjectives.

---

[9] Carla Umbach (*p.c.*) pointed out the case of *beautiful* where, in order to cancel the inference to the positive form to either *relatum*, some qualifying particle is needed:

(1)   Anna is more beautiful than Berta, ?? but neither of them is beautiful.

(2)   Anna is more beautiful than Berta, but *in fact* neither of them is beautiful.

However, the fact that the inference can be cancelled shows that this inference cannot be an entailment of *beautiful*.

It bears pointing out, however, that some proposals in the non-factualist camp seem better equipped to meet this challenge than others. Comparatives are a particularly pressing problem for theories that hold that attributions of value express attitudes *towards*, or *about*, the object under evaluation. This is true of most non-factualist proposals, but not all of them (again, see [6] for a recent overview). In particular, Gibbard's *norm-* and subsequent *plan-expressivism* [16,17] is built upon the idea that ascriptions of rationality ('ϕ-ing makes sense'; 'ϕ-ing is rational', etc.) express a non-cognitive attitude of *acceptance* of a set of norms that stands behind such evaluation, or later, of a plan to perform the action under evaluation. For instance, to say that eating vegetables is the rational thing to do is to express one's acceptance of a system of norms that sanctions eating vegetables (according to early Gibbard); or of a plan to eat vegetables (according to later Gibbard). But one does not thereby express any positive or negative attitude towards eating vegetables; one expresses an attitude that has the system of norms or the plan as its object. This feature of Gibbard's proposal makes it immediately better equipped to deal with (11): following his earlier proposal, Gibbard can say that to judge that volunteering is better than donating is to express one's acceptance of a system of norms relative to which volunteering is better than donating; or following his later proposal, he can say that it is to express one's adoption of a plan to volunteer over donating. However, he would still need to say what it is for something to be better than another relative to a system of norms, or what it is for a plan to give preference to one action over another. That is, precisely, the question that we explore here. Gibbard's examples—outright judgments of rationality, sensibleness and the like, were treated in his earlier work as expressing acceptance of systems of norms such that the objects of evaluation either met, failed to meet or were indifferent to them. The problem is that neither of those cases applies in a straightforward manner to comparatives. A similar thing can be said of his later *plan-expressivism*: there, speakers uttering those same sentences express acceptance of plans to do or not do the action under consideration. Nonetheless, his later proposal offers the right formal tools upon which to start building a scalar semantics for evaluative adjectives, so we will rely on that.

In sum, our purpose is to extend the non-factualist insight about *positive* value attributions—i.e. (9) and (10)—to *comparative* ones, like (11). Intuitively, what we need to capture is that, when a speaker utters (11), she is expressing a *higher degree of support* for the action of volunteering than for the action of donating money (without necessarily expressing outright support for either). But what is 'a higher degree of support'? Our answer is that to express more support for volunteering than donating is to express one's *adoption of a plan to volunteer for a charity over donating money in situations that offer a choice between the two actions.*

Importantly, the problem that we raise here for non-factualism is arguably a kind of "incarnation" of the Frege-Geach problem (which has gone seemingly unnoticed). We are concerned with the meaning of comparative uses of evaluative adjectives, and comparatives are morphosyntactically derived from the positive form; i.e. 'smarter' results from combining 'smart' and the comparative suffix '-er'. The semantic counterpart of this is that the meaning of 'smarter' should result from combining the meaning of 'smart' with the meaning of '-er' *via* semantic composition. If non-factualism is presented as a view about the meaning of evaluative adjectives, it must offer a story such that the semantic value assigned to those adjectives can be combined appropriately with comparative morphology in a way to give rise to the expected meaning for the comparative. This problem is essentially the same as the problem of embedding under truth-functional operators. In fact, our concern is arguably even more pressing, because before non-factualists can give a story about how evaluative sentences embed, they need a story about how evaluative sentences are composed (although figuring out how evaluative sentences compose will not by itself offer a solution to how they embed).

Finally, both to preempt a possible (though unlikely) objection to this project as well as to stress the parallelism with the Frege-Geach problem, note that, at this point, it would be open to a non-factualist to claim that the meaning of the positive form of an evaluative adjective is unrelated to the meaning of the comparative—just as it was open to her to claim that embedded and unembedded uses of evaluative sentences did not have the same meaning. And to this we may reply in an exactly parallel way: if that were the case, the obvious inferential relations between positive and comparative uses of these adjectives would be unexplained. To wit: the inference from 'donating is good' and 'volunteering is better than donating' to the conclusion 'volunteering is good' crucially relies on the semantic kinship between 'good' and 'better'. Preserving and predicting that kinship is a *desideratum* of any semantic theory about gradable expressions worth its salt.

4 Ordering Hyperplans

We start out by giving a semantics for the positive form of a thin evaluative adjective taking an action type as its argument, and then for the comparative. As starting point for our semantics, we rely on Gibbard's [17] *plan-expressivism.* In his [17], Gibbard sets out to characterize a basic normative predicate, that of an action *being the thing to do.* To judge that an action is *the thing to do* is to plan to perform that action in the appropriate circumstances; and an utterance of a sentence like 'ϕ-ing is the thing to do' in an appropriate context expresses the speaker's plan to ϕ. Gibbard's view is non-factualist, because he refuses a characterization of normative statements as factual or descriptive. To say that ϕ is the thing to do is not to state that ϕ, or ϕ-ing, possesses a certain property (the property of being the thing to do), but again, it is to express one's plan to ϕ. The action-guiding character of a normative predicate such as *being the thing to do* is thus secured in exactly the way described in Sect. 2.

Gibbard takes plans of action to be perspicuously characterized in terms of what he calls *hyperplans.* Hyperplans are model-theoretical objects that can capture the close connection between normative and evaluative statements and action-guidance discussed in Sect. 1. A hyperplan is a maximally decided planning state: a state that tells you what to do in every conceivable situation that you could find yourself in. A hyperplan will tell what to do if your car breaks, if it doesn't, if there's a fire, if your neighbors fight, if you were Ceasar right before crossing the Rubicon, if you were Catwoman about to push Batman down a flight of stairs after a long fight, etc. (see [17, ch. 3]). More formally, we can think of a hyperplan as a *total* function from the set of conceivable situations S to the set of possible actions A^{10}. A domain H of hyperplans looks very much like the familiar domain W of possible worlds of intensional semantics (i.e. maximally determined states of affairs), and given the usefulness of understanding informational content in terms of set-theoretical operations over W, it is suggestive to understand *evaluative* content in terms of set-theoretical operations over H (see [40,41] for suggestions in this direction).

It bears mentioning that, even though hyperplans are maximally decided, *real* plans of action adopted by agents—contingency plans—are not: as precise as it might be, no contingency plan tells you what to do under absolutely *every* circumstance. We may thus conceive of a contingency plan as a *partial* function from S to A, or alternatively, as a set of hyperplans that agree on what to do in some situations, but not in others.

The notion of a hyperplan allows us to define two functions representing the semantic value of a pair of a positive and negative evaluative adjective in their positive form, VAL^+ and VAL^- ($VAL^{+/-}$ for short)[11]. Four important assumptions to be made about $VAL^{+/-}$ are the following: first, we assume that they take action-types as arguments (we use variable α); that is, they are functions of type $\langle \alpha, t \rangle$ (we consider the question of whether there might be objects of evaluation of other types in the next section). Secondly, we assume that these functions are context-dependent, that is, that their extension varies at different contexts of use. So strictly speaking, they are functions from contexts to functions of type $\langle \alpha, t \rangle$. Thirdly, we assume that, at any context c, these functions have positive and negative extensions, which we may call $pos_{VAL+/-}$ and $neg_{VAL+/-}$ (the set of action-types of which they are true and false, respectively), as well as an extension gap, whose elements fall neither under $pos_{VAL+/-}$ nor $neg_{VAL+/-}$. Thirdly, we assume that the extensions of VAL^+ and VAL^- at any context c mirror each other; that is, $pos_{VAL+} = neg_{VAL-}$ and viceversa.

---

[10] As Gibbard points out, situations should be modeled as centered worlds (in the sense of [24]). After all, if a hyperplan is indeed maximally decided, it ought to tell you what to do even in cases where you've lost track of time, place or who you are. In relation to this, [40] notes that hyperplans can be fully spelled out in terms of *possibilia*, that is, as functions from sets of centered worlds to subsets thereof in which different courses of action are adopted.

[11] The purpose of defining a pair of functions instead of a pair of English expressions is to abstract away from particularities that arise from the semantic features of particular expressions. For example, English *good* and *bad* would be obvious candidates for our definitions, but these adjectives have a beneficiary argument ('good/bad *for* x') that we don't aim to discuss here.

What is the role of hyperplans in relation to these functions? The intuitive idea is the following: given our assumption that VAL^+ and VAL^- are context-dependent, we can say that their extension is determined, at each context, by the hyperplan *of* the context. That is, contexts in this view determine a number of parameters that affect the content of different expressions of language. In particular, we can assume that among those parameters there is a hyperplan parameter representing shared plans of actions between speakers (see [40] for this type of proposal).

What we want, then, is a mapping from the set of *hypothetical instructions* which constitute the content of whatever hyperplan is determined at a context c to the set of actions that fall under the positive, negative and extension gap of VAL^+ and VAL^- at c. There is no obvious way of obtaining this mapping in a precise fashion, but we can say that, relative to a hyperplan h, pos_{VAL+} (and neg_{VAL-}) contains action types that the agent of h *pursuits, prefers* or *supports* to a sufficiently high degree. Conversely, neg_{VAL+} (pos_{VAL-}) relative to h should contain actions that the agent of h tends to *avoid, disprefer* or *reject*. The extension gap of $VAL^{+/-}$ at h contains action types that the agent of h neither pursuits nor avoids to a sufficiently significant degree. We may leave this mapping as informal as this; after all, even though hyperplans are maximally decided, what counts as sufficiently supported or rejected by a hyperplan might be considered rather vague. What is important to keep in mind is that the positive and negative extension of these functions is not determined by properties of the action types that fall in them, but rather by the fact that the relevant hyperplan (dis)prefers them.

We can now give the following lexical entries for VAL^+ and VAL^- at a context c (determining a hyperplan $h \in H$):

(16) $[\![VAL^+]\!]^c = \lambda\alpha.h$ supports α

(17) $[\![VAL^-]\!]^c = \lambda\alpha.h$ rejects α

Recall that VAL^+ and VAL^- are meant to represent a fundamental component of the semantic value of any evaluative adjective in their positive form. As expected, the extension of these functions at a context c is therefore a distribution of action-types into three partitions: supported, rejected and neither supported-nor-rejected.

However, there is intuitively more structure across each cell of the partition; and in fact we require more structure in order to represent comparisons, which is our ultimate objective. Let us assume that, within the extension of VAL^+ and VAL^- relative to a hyperplan, actions can be pairwise ordered in terms of preference or choice (of course, hyperplans are *already* decided, so for a hyperplan to prefer an action over another is not, strictly speaking, a matter of actual choice. Preference here should be understood in relation to features of the situations in which different actions are performed). Let us elaborate on this.

We said before what it was for a hyperplan to support [reject] an action in general terms: it is for it to be a hyperplan to pursue [avoid] that action to a sufficiently high degree. Though vague, it is intuitive: for the action of donating to be supported by a hyperplan is for the hyperplan to choose to perform that action in sufficiently many situations. That is what makes that action type fall in the positive extension of VAL^+ relative to that hyperplan. But that notion of general support relies on a simpler notion of support [rejection] of an action by a hyperplan h *at a given situation*, which is more straightforwardly defined: it is for h to be a hyperplan to [not] *do* that action in that situation. For instance, a hyperplan supports calling the police when your neighbors fight just in case it is a hyperplan to call the police when they fight. Conversely, a hyperplan rejects calling the police when your neighbors fight just in case it is a hyperplan to *not* call the police when they fight.

One and only one action might be performed in each situation, but a single situation might offer many options for action. A situation in which your neighbors fight, for instance, offers a choice between calling the police and approaching them yourself to try to calm them down (again, those possibilities are not choices from the point of view of a hyperplan, but they are choices from a "human" point of view). So preference of an action-type α over another action-type β by a hyperplan can be assessed by considering the situations that offer a choice between α and β. Whatever the hyperplan does, that's what it takes to be "better". Thus, considering situations that offer a choice between at least two actions provides a way of ordering action types by preference relative to a hyperplan.

We arrive at a picture where, as before, the mirror extensions of VAL^+ and VAL^- relative to a hyperplan h is a set of action types partitioned into a positive, negative and extension gap. But in addition to this, considering various sub-hyperplans of h defined for situations that offer choices between different action types induces more structure into the partitions of the hyperplan; in fact, it induces an ordering on them. Formally, sub-hyperplans are equivalent to the real-world contingency plans mentioned at the start of this section—they are hyperplans that are only defined for certain situations; therefore, they might be thought of as partial functions from the set of possible situations to the set of available actions, or equivalently as sets of hyperplans. It seems intuitive to say that this ordering should allow at least for incomparabilities, that is, that it should be a partial order, since judgments of incomparability should hold for pairs of action types that do not share situations where both of them could be performed.

Considering situations where a pair of action types could be performed allows us to define their comparative value based on our previous notions of support and rejection relative to a hyperplan. To do this, we evaluate those actions relative, not to that hyperplan, but relative to some sub-hyperplan thereof that is only defined for those situations that offer a choice between the action types under comparison. So for any pair of action types $\langle \alpha, \beta \rangle$ and context c (determining a hyperplan $h \in H$),

(18) $[\![\alpha$ is VAL^+-er than $\beta]\!]^c = 1$ iff $\exists h'$ such that h' is a sub-hyperplan of h that offers a choice between α & β & h' supports α (and rejects β)

(19) $[\![\alpha$ is VAL^--er than $\beta]\!]^c = 1$ iff $\exists h'$ such that h' is a sub-hyperplan of h that offers a choice between α & β & h' supports β (and rejects α)

That is, in order to obtain an ordering of action types within a hyperplan, we need to evaluate the support and rejection that the relevant action types receive relative to a proper subset of the original hyperplan. If, relative to it, α falls in pos_{VAL+} whereas β falls in neg_{VAL+}, then α has more value than β. If there is no sub-hyperplan of the original hyperplan that offers a choice between the two action types, then α and β are incomparable relative to that hyperplan.

The procedure advocated here is very similar to the one advocated in [22] using the notion of a comparison class. In that work, Klein uses comparison classes to derive a comparative ordering of objects based on applications of the positive form of the relevant adjective. For instance, a pair of objects $\langle x, y \rangle$ falls under the *taller than* relation just in case there is a comparison class relative to which x is tall and y isn't. What we are doing is essentially the same, substituting comparison classes for hyperplans and sub-hyperplans thereof. Moreover, we take there to be an important connection between Klein's approach and ours. Klein is also moved by the observation that, morphosyntactically, comparatives derive from the positive form of adjectives, and basic considerations about compositionality should guide one to derive compositionally the former from the latter. As we saw, a parallel argument can be run in relation to the considerations about non-factualism. Non-factualists cash out the meaning of certain evaluative and normative expressions in terms of simple, binary attitudes of approval, disapproval, etc. If those insights are right, then it is natural to try to analyze judgments of comparison, which do not seem to express those simple attitudes, in terms of some type of operation on the simple attitudes. This is, in a nutshell, what we have just done.

Informally, the intuition is that a hyperplan may be such that any number of actions is supported and rejected by it in different situations, but in order to make a comparative judgment, it doesn't matter whether the actions are actually supported or rejected *tout court*. All that matters is that, given the choice, the hyperplan gives preference to one over the other. This liberality predicts the admissibility of (12)–(14). Nonetheless, predicting the badness of (15) is not so straightforward: in principle, we do not seem to have ruled out the possibility that, *e.g.* relative to a hyperplan h, an action α falls in pos_{VAL+}, another action β is in neg_{VAL+}, and yet given a choice between α and β the hyperplan chooses β. Actually however, it *is* ruled out. Recall that actions in the positive, negative and extension gap of $VAL^{+/-}$ relative to any hyperplan are internally ordered. Actions in the positive extension of $VAL^{+/-}$ at a hyperplan h are chosen over many other actions, and similarly for the negative extension of $VAL^{+/-}$ at h. Given this structure, a principle of *no reversal* holds, such as can be found in similar proposals by Klein [22] and Van Benthem [2]. Informally, the principle says that, for any two pair of actions α and β, if there is a hyperplan (or sub-hyperplan) h such that α falls in pos_{VAL+} and β falls in neg_{VAL+} relative to h, there might be hyperplans or sub-hyperplans h' that superset h such that α and β both fall in pos_{VAL+} or neg_{VAL+} at h', but there can be *no* hyperplan (or sub-hyperplan) h'' that supersets h such that α falls in neg_{VAL+} and β falls in pos_{VAL+} relative to h'. That is, the order of α relative to β cannot be *reversed*. This principle immediately rules out (7), as needed.

To sum up: in this section, we have given a bare-bones semantics for a pair of evaluative adjectives in positive and comparative form. The semantics relies crucially on the possibility of mapping hyperplans—functions from possible situations to courses of action—into sets of action types which are partially ordered by preference and which can be partitioned in cells defining a positive and negative extension, as well as the extension gap for those adjectives.

5 Disagreement, Thick Terms and Value Beyond Action

This section offers some comments on what, in the picture just sketched, is to make an attribution of value and what could constitute a disagreement involving value attributions. We offer some remarks as well about how to extend this proposal to thick terms as well as to objects of evaluation other than action types.

Firstly, we noted at the end of Sect. 1 that accounting for exchanges in which speakers dispute over the extension of an evaluative adjective (cf. (1), (3)) is a difficult task. In Sect. 2, we sketched the way in which non-factualists tend to account for that phenomenon, namely by arguing that disagreements involving attributions of value are practical disagreements about *what to do* (that contrast with factual disagreements, that is, disagreements about *what is the case*). That answer was only partial, but after introducing the hyperplan framework, we can expand on what we take to be the most promising way of enriching that answer. In short, the idea to be defended is that attributions of value have the conversational role of restricting the set of hyperplans accepted by the participants in a conversation. Let us elaborate briefly.

Theories that make use of the notion of common ground (stemming from [19,34]) understand assertion as a speech act that aims at restricting the information available for the speakers participating in a conversation. As we saw, non-factualists claim that attributions of value are not informative, in the same sense that descriptive statements are. But they do convey certain information about their speakers, and they have a conversational purpose.

The proposal here, (essentially that of [40] for deontic modals) is that a parallel operation to the one characterizing informational updates, but this time involving hyperplans, marks the conversational contribution of a value attribution. In a nutshell: a value attribution is a peculiar type of update on the common ground of a conversation, namely an update of the value of a parameter representing a shared plan of action between speakers. The value of this parameter at a context is eminently shifted through the use of evaluative expressions like *beautiful, wrong* or *tasty*. So the result of uttering and accepting a sentence like (9) ('volunteering for a charity is good') at a context c is to rule out from c all hyperplans except those that support volunteering for a charity. In other words, to uptake (9) at a context is for participants in the conversation to adopt a—relatively vague—plan to volunteer for a charity. Similarly, the result of updating c with a comparative value attribution (e.g. (11) 'volunteering for a charity is better than donating money') is to rule out from c all hyperplans whose agents do not choose volunteering over donating in situations that offer a choice between those actions.

The view about disagreement that falls naturally from this proposal is that a value disagreement (a disagreement involving conflicting attributions of value) is an exchange where speakers are offering conflicting ways of updating the hyper-plan parameter in their context. In turn, this means that speakers are pushing different plans of action as the plan to be adopted by participants in the conversation. This is all very sketchy and programatic, but this view about disagreement and the conversational dynamics of value judgment requires far more elaboration than I have space to offer here. Suffice it to say that the hyperplan semantics provides a perspicuous way of capturing the traditional non-factualist *dictum* that a value disagreement is a disagreement about *what to do.*

Secondly, we understood the functions VAL^+ and VAL^- to be capturing an essential semantic component of thin value adjectives, words like *good, bad, beautiful, important,* etc. But most evaluative adjectives are *thick,* that is, they carry descriptive as well as evaluative meaning. Can the view presented here be extended to thick terms? We think it can, based on the following observation: most contemporary and linguistically informed views on thick terms assume that it is possible to factor out the descriptive and evaluative component of these terms into two sentences, one attributing a non-evaluative property or set of properties to whatever the thick term is predicated of and one evaluative sentence containing a thin value adjective like *good* or *bad.* For instance, the adjective *cruel* results from combining a descriptive predicate such as *causes unnecessary suffering* with a general negative evaluation to the effect that whoever causes unnecessary suffering is *bad* for doing so:

(20)   John is cruel.
     (*Descriptive:*) John inflicts unnecessary suffering.
     (*Evaluative:*) Whoever inflicts unnecessary suffering is bad for doing so.

The current debate concerns the relation between the descriptive and evaluative component. For instance, a powerful view is that the descriptive component constitutes the asserted content of a sentence containing a thick term, while the evaluative component is a presupposition [7]. From our part, the fact that the evaluative component features thin evaluation is enough for us to plug in a hyperplan semantics. So when a speaker utters (20), not only is she asserting that John has certain properties, but she is also expressing a negative evaluation of John, based on the fact that he possesses such properties. And regardless of whether this negative evaluation is understood as part of the assertion, pre-suppositions or implicatures of (20), we can say that an utterance of (20) at a context c involves, at some level of its meaning, a proposal to update the hyperplan parameter of c, in such a way that hyperplans that support inflicting unnecessary suffering in others are ruled out from consideration.

Finally, the framework presented here assumes that the primary objects of evaluation are actions. But this framework intends to capture a semantic property that all evaluative adjectives share, and there are many evaluative adjectives that do not seem to evaluate action types. For instance, aesthetic adjectives are most often predicated of individuals. To extend the framework presented here to objects of evaluation that are not action types requires a slight adjustment.

Instead of assuming, as we did, that VAL^+ and VAL^- at a context c are functions from action types, we can assign them arguments of a variable type τ. Then, we can say that for a hyperplan to support [reject] a τ-type object is for it to be a hyperplan to orient the agent's action towards [against] that object. In the limiting case in which the object under evaluation is an action type, to orient one's action towards the object *just is* to perform that action. But in case we are evaluating, say, a song, to orient one's action towards it remains slightly underspecified. Nonetheless, to orient one's action towards a song should intuitively involve actions such as repeatedly listening to the song, learning it, singing it to other people, etc. In this way, a cluster of action types emerges, so that the evaluation of a non-action object can still be understood in terms of action types, and consequently in terms of a hyperplan semantics. We leave considerations in this direction for future work.

6   Conclusion

In this paper, we have characterized evaluative adjectives as gradable adjectives that are subject to faultless disagreement in positive and comparative form and are conceptually connected to action and motivation. We observed that meta-ethical non-factualism offered a promising way of capturing both aspects of their meaning; but we argued that, as far as it has been presented, non-factualists cannot accommodate comparative constructions. We set about amending this, by offering an account in which the non-doxastic attitudes expressed by value attributions that feature comparative uses of evaluative adjectives are understood in terms of the adoption of hyperplans capable of representing the appropriate ordering structure.

References

1. Bar-On, D., Chrisman, M.: Ethical neo-expressivism. Oxford Stud. Metaethics **4**, 133–165 (2009)
2. Benthem, J.V.: Later than late: on the logical origin of the temporal order. Pac. Philos. Q. **63**(2), 193–203 (1982). http://cat.inist.fr/?aModele=afficheN&cpsidt=12379460
3. Bierwisch, M.: The semantics of gradation. In: Bierwisch, M., Lang, E. (eds.) Dimensional Adjectives, pp. 71–261. Springer, Heidelberg (1989). http://www.zas.gwz-berlin.de/fileadmin/mitarbeiter/bierwisch/4_Bierwisch_1989_Gradation.pdf
4. Blackburn, S.: Attitudes and contents. Ethics **98**(3), 501 (1988). https://www.journals.uchicago.edu/doi/10.1086/292968
5. Bylinina, L.: Judge-dependence in degree constructions. J. Semant. **34**(2), 291–331 (2016)
6. Camp, E.: Metaethical expressivism. In: MacPherson, T., Plunkett, D. (eds.) The Routledge Handbook of Metaethics, pp. 87–101. Routledge, Abingdon (2017)
7. Cepollaro, B., Stojanovic, I.: Hybrid evaluatives: in defense of a presuppositional account. Grazer Philosophische Studien **93**(3), 458–488 (2016). http://booksandjournals.brillonline.com/content/journals/10.1163/18756735-09303007

8. Dreier, J.: Internalism and speaker relativism. Ethics **101**(1), 6–26 (1990)
9. Egan, A.: Disputing about taste. In: Warfield, T., Feldman, R. (eds.) Disagreement, pp. 247–286. Oxford University Press, Oxford (2010)
10. Egan, A.: Relativist dispositional theories of value. South. J. Philos. **50**(4), 557–582 (2012)
11. Egan, A.: There's something funny about comedy: a case study in faultless disagreement. Erkenntnis **79**(1), 73–100 (2014)
12. Égré, P., Zehr, J.: Are gaps preferred to gluts? A closer look at borderline contradictions. In: Castroviejo, E., McNally, L., Weidman Sassoon, G. (eds.) The Semantics of Gradability, Vagueness, and Scale Structure. LCM, vol. 4, pp. 25–58. Springer, Cham (2018). https://doi.org/10.1007/978-3-319-77791-7_2
13. Faroldi, F.L.G., Soria Ruiz, A.: The scale structure of moral adjectives. Studia Semiotyczne **31**(2), 161–178 (2017)
14. Frapolli, M.J., Villanueva Fernandez, N.: Minimal expressivism. Dialectica **66**(4), 471–487 (2012)
15. Gibbard, A.: An expressivistic theory of normative discourse. Ethics **96**(3), 472–485 (1986). http://www.journals.uchicago.edu/doi/pdfplus/10.1086/292770, http://www.jstor.org/stable/2381066
16. Gibbard, A.: Wise Choices, Apt Feelings. Clarendon Press, Oxford University Press (1990)
17. Gibbard, A.: Thinking How to Live. Harvard University Press, Cambridge (2003)
18. Hare, R.M.: The Language of Morals. Oxford University Press, Oxford (1952)
19. Heim, I.: On the projection problem for presuppositions. In: Portner, P., Partee, B. (eds.) Formal Semantics: The Essential Readings, pp. 249–260. Blackwell Publishing Ltd/Inc., Hoboken (2002)
20. Kennedy, C.: Vagueness and grammar: the semantics of relative and absolute gradable adjectives. Linguist. Philos. **30**(1), 1–45 (2007). http://link.springer.com/10.1007/s10988-006-9008-0
21. Kennedy, C.: Two sources of subjectivity: qualitative assessment and dimensional uncertainty. Inquiry **56**(2–3), 258–277 (2013). http://www.tandfonline.com/doi/abs/10.1080/0020174X.2013.784483
22. Klein, E.: A semantics for positive and comparative adjectives. Linguist. Philos. **4**(1), 1–45 (1980). http://www.springerlink.com/index/P01U10Q4N62NW277.pdf
23. Lasersohn, P.: Context dependence, disagreement, and predicates of personal taste. Linguist. Philos. **28**(6), 643–686 (2005). http://link.springer.com/10.1007/s10988-005-0596-x
24. Lewis, D.: Attitudes de dicto and de se. Philos. Rev. **88**(4), 513–543 (1979)
25. MacFarlane, J.: Assessment Sensitivity: Relative Truth and Its Applications. Oxford University Press, Oxford (2014)
26. Marques, T.: Disagreeing in context. Front. Psychol. **6**, 257 (2015)
27. McNally, L., Stojanovic, I.: Aesthetic adjectives. In: Young, J. (ed.) The Semantics of Aesthetic Judgment. Oxford University Press, Oxford (2016). http://philpapers.org/rec/MCNAA-3
28. Rett, J.: Antonymy and evaluativity. In: Proceedings of SALT, pp. 210–227 (2007). http://journals.linguisticsociety.org/proceedings/index.php/SALT/article/download/2969/2709, http://elanguage.net/journals/salt/article/download/17.210/1851
29. Ridge, M.: Disagreement. Philos. Phenomenol. Res. **86**(1), 41–63 (2013)
30. Sassoon, G.W.: A typology of multidimensional adjectives. J. Semant. **30**(3), 335–380 (2013). http://jos.oxfordjournals.org/content/early/2012/08/03/jos.ffs012.short

31. Schroeder, M.: What is the frege-geach problem? Philos. Compass **3**(4), 703–720 (2008). https://static1.squarespace.com/static/55505fc8e4b032b4451e4a90/t/55ca3031e4b0274930ad4209/1439313969077/Schroeder_Frege-Geach_Problem.pdf

32. Silk, A.: Discourse Contextualism: A Framework for Contextualist Semantics and Pragmatics. Oxford University Press, Oxford (2016)

33. Solt, S.: Multidimensionality, subjectivity and scales: experimental evidence. In: Castroviejo, E., McNally, L., Weidman Sassoon, G. (eds.) The Semantics of Gradability, Vagueness, and Scale Structure. LCM, vol. 4, pp. 59–91. Springer, Cham (2018). https://doi.org/10.1007/978-3-319-77791-7_3

34. Stalnaker, R.: Assertion. In: Portner, P., Partee, B. (eds.) Formal Semantics: The Essential Readings, pp. 147–161. Blackwell Publishing Ltd/Inc., Hoboken (2002)

35. Stephenson, T.: Judge dependence, epistemic modals, and predicates of personal taste. Linguist. Philos. **30**(4), 487–525 (2007). http://link.springer.com/10.1007/s10988-008-9023-4

36. Stevenson, C.L.: The emotive meaning of ethical terms. Mind **46**(181), 14–31 (1937). http://www.jstor.org/stable/2250027

37. Stevenson, C.L.: Facts and Values: Studies in Ethical Analysis (1963)

38. Stojanovic, I.: Disagreements about Taste vs. Disagreements About Moral Issues (2017)

39. Väyrynen, P.: The Lewd, the Rude, and the Nasty. Oxford University Press, Oxford (2013). https://books.google.fr/books?hl=en&lr=&id=EoJpAgAAQBAJ&oi=fnd&pg=PP1&dq=pekka+vayrynen+the+lewd&ots=_68pSl0E5e&sig=pNosyY2djzH0-Sspc6cRNCZxG54

40. Yalcin, S.: Bayesian expressivism. Proc. Aristotelian Soc. **112**(2), 123–160 (2012). https://academic.oup.com/aristotelian/article-lookup/doi/10.1111/j.1467-9264.2012.00329.x

41. Yalcin, S.: Expressivism by force. In: Fogal, D., Harris, D.W., Moss, M. (eds.) New Work on Speech Acts. Oxford University Press, Oxford (2017)

Spectra of Gödel Algebras

Diego Valota[⊠][iD]

Department of Computer Science,
Università degli Studi di Milano, Via Comelico 39/41, 20135 Milan, Italy
valota@di.unimi.it

Abstract. We exploit the duality between finite Gödel algebras and their homomorphisms, and the category of finite forests and open maps, to compute the duals of non-isomorphic k-element Gödel algebras, for $k \geq 1$. From this construction we obtain a recurrence formula to compute the *fine spectrum* of the variety of Gödel algebras $\mathbb{G}$. Using such a formula we easily compute the set of cardinalities of finite Gödel algebras (the *spectrum* of $\mathbb{G}$), that is equal to the set of positive integers $\mathbb{N}^+$. To complete the picture on spectra of Gödel algebras, we recall the well-known recurrence formula to compute the cardinality of every free k-generated Gödel algebra, that is the *free spectrum* of $\mathbb{G}$.

1 Introduction

Essential knowledge about the finite members of a class of structures $\mathcal{C}$ can be obtained by computing, for every natural number $k \geq 1$:

- *spectrum* of $\mathcal{C}$: $Spec(\mathcal{C}) = \{k \mid k = |C|, C \in \mathcal{C}\}$, that is the set of cardinalities of structures occurring in $\mathcal{C}$;

when $\mathcal{C}$ is a variety of algebras we can also define:

- *fine spectrum* of $\mathcal{C}$: $Fine_{\mathcal{C}}(k)$, that is the function counting non-isomorphic k-element structures in $\mathcal{C}$,
- *free spectrum* of $\mathcal{C}$: $Free_{\mathcal{C}}(k) = |\mathbf{F}_{\mathcal{C}}(k)|$, that is the function computing the sizes of the free k-generated algebra $\mathbf{F}_{\mathcal{C}}(k)$ in $\mathcal{C}$.

Fagin characterizes the sets of integers that are $Spec(\mathcal{C})$ when $\mathcal{C}$ is the class of all models of first-order formulas of a first-order language [24]. The fine spectrum has been introduced by Taylor in [32] when considering $\mathcal{C}$ a variety of algebras, and Quackenbush [31] states that "the fine spectrum problem is usually hopeless" when dealing with ordered structures. Given such a function, the computation of the spectrum of $\mathcal{C}$ is straightforward,

$$Spec(\mathcal{C}) = \{k \in \mathbb{N}^+ \mid Fine_{\mathcal{C}}(k) > 0\}. \tag{1}$$

One of the most famous enumeration problems is the *Dedekind's Problem* [20], that is the free spectrum problem when $\mathcal{C}$ is the class of distributive lattices.

A. Silva et al. (Eds.): TbiLLC 2018, LNCS 11456, pp. 297–311, 2019.
https://doi.org/10.1007/978-3-662-59565-7_15

The variety of Gödel algebras is obtained by the class of Heyting algebras (that are bounded residuated distributive lattices) adding the prelinearity equation. Gödel algebras are the algebraic semantics of Gödel logic, a non-classical logic whose studies date back to Gödel [28] and Dummett [21]. Indeed Gödel logic can be obtained by adding the prelinearity axiom to Intuitionistic logic. Furthermore, Gödel logic is one of the three major (many-valued) logics in Hajek's framework of Basic Logic, that is the logic all continuous t-norms and their residua [25].

Given a finite Gödel algebra $\mathbf{A}$, the set of prime filters of $\mathbf{A}$ ordered by reverse inclusion forms a finite *forest*. Viceversa, given a finite forest F, the collection of all subforests of F, equipped with properly defined operations, is a finite Gödel algebra. This construction is functorial, meaning that it can be extended to obtain a dual equivalence between the category of Gödel algebras and their homomorphisms, and the category of finite forests and open maps (see [2, 16] for details and proofs).

The above duality is a special case of the Birkhoff duality, between the category of finite distributive lattices and complete lattice homomorphisms, and the category of finite posets and open maps. Dropping finiteness restrictions, it is possible to obtain the Priestley duality between the category of bounded distributive lattices and bounded lattices homomorphisms, and the category of Priestley spaces and continuous order-preserving maps (see [18] for both Birkhoff and Priestley dualities). In [19] was observed that open maps dually correspond to Heyting algebras homomorphisms. Hence, the category of finite posets and open maps is dually equivalent to the category of complete and completely join-generated Heyting algebras with complete Heyting algebra homomorphisms [9,10]. Finally, Esakia in [22] establishes a duality between Heyting algebras and so-called Esakia spaces.

In this paper we exploit the category of forests to solve the fine and free spectra problems when $\mathcal{C}$ is the variety of Gödel algebras $\mathbb{G}$. The spectrum problem for $\mathbb{G}$ is easily solved. Indeed $Spec(\mathbb{G}) = \mathbb{N}^+$, because every finite chain carries the structure of a Gödel algebra.

In literature one can find many solutions to the free spectrum problem for $\mathbb{G}$. Indeed, already Horn in 1969 has obtained a recurrence formula to compute the cardinalities of free k-generated Gödel algebras [26]. Another solution to this problem can be achieved by restating the Horn's recurrence in terms of finite forests [16]. Conversely to the best of our knowledge, the fine spectrum problem for $\mathbb{G}$ has never been considered before. In this paper we show how to obtain a set of forests S_k such that for every Gödel algebra $\mathbf{A}$ of cardinality k there exists $F \in S_k$ such that $\mathbf{A}$ is isomorphic to the downsets of F. That is, given a cardinal k we can build the set of finite Gödel algebras with k elements. As a corollary, we obtain a recurrence $Fine_{\mathbb{G}}(k)$ that computes the cardinality of S_k for any $k \geq 1$. Solving in this way the fine spectrum problem for $\mathbb{G}$.

2   Finite Forests and Gödel Algebras

Let P and Q be two disjoint posets, their *vertical sum* $P \oplus Q$ is the poset over $P \cup Q$ obtained by taking the order relation defined in the following way: let x and y be two elements that belong to $P \cup Q$, then $x \leq y$ if the pair (x, y) fall in one of the following three mutually disjoint cases; $x \leq y$ if and only if $x, y \in P$ and $x \leq y$ in P, second $x, y \in Q$ and $x \leq y$ in Q and finally $x \in P$ and $y \in Q$. A special case of vertical sum is the *lifting*, $P_\perp := \{\perp\} \oplus P$, with $\perp \notin P$.

Given a poset $(P, \leq)$ and a subset $Q \subseteq P$, the *downset* of Q is $\downarrow Q = \{x \in P \mid x \leq q,$ for some $q \in Q\}$. By abuse of notation we write $\downarrow q$ for $\downarrow \{q\}$.

Let P and Q be two posets, an *open map* is an order-preserving map from P to Q that sends downsets of P to downsets of Q.

A poset F is a *forest* when for all $q \in F$ the downset $\downarrow q$ is a *chain*, that is $\downarrow q$ is totally ordered. A *tree* is a forest with a bottom element, called the *root* of the tree. A *subforest* of a forest F is the downset of some $Q \subseteq F$.

Finite forests and open maps form a category FF. Let F, F' and G be forests in FF. The *coproduct* of F and F' is the disjoint union $F \sqcup F'$ of F and F'. The product of $F \times F'$ is given by the following isomorphisms,

$$F \times F' \cong F' \text{ when } |F| = 1;$$
$$G \times (F \sqcup F') \cong (G \times F) \sqcup (G \times F');$$
$$F_\perp \times F'_\perp \cong ((F_\perp \times F') \sqcup (F \times F') \sqcup (F \times F'_\perp))_\perp.$$

These characterizations of products and coproducts in FF are fully proved in [5].

A *Heyting algebra* is a structure $\mathbf{A} = \langle A, \wedge, \vee, \rightarrow, \top, \perp \rangle$ such that $\langle A, \wedge, \vee, \top, \perp \rangle$ is a bounded distributive lattice, and $(\rightarrow, \wedge)$ forms a residuated couple, that is $x \wedge y \leq z$ if and only if $y \leq x \rightarrow z$. Negation operation is usually defined as $\neg x = x \rightarrow \perp$. A *Gödel algebra* is a Heyting algebra $\mathbf{A}$ that satisfies the prelinearity condition

$$(x \rightarrow y) \vee (y \rightarrow x) = \top.$$

The variety of Gödel algebras $\mathbb{G}$ is generated by the *standard* algebra $[\mathbf{0}, \mathbf{1}] = \langle [0, 1], \wedge, \vee, \rightarrow, 1, 0 \rangle$ where the operations are defined as follows.

$$x \wedge y = \min\{x, y\} \qquad x \rightarrow y = \begin{cases} 1 & \text{if } x \leq y, \\ y & \text{otherwise} \end{cases} \qquad \neg x = \begin{cases} 1 & \text{if } x = 0, \\ 0 & \text{otherwise.} \end{cases}$$

Since the algebra $[\mathbf{0}, \mathbf{1}]$ singly generates the whole variety $\mathbb{G}$, from universal algebraic facts we have that the free k-generated algebra $\mathbf{F}_\mathbb{G}(k)$ is isomorphic with the subalgebra of the algebra of all functions $f : [0, 1]^k \rightarrow [0, 1]$ generated by the projection functions $x_i : (t_1, \ldots, t_k) \mapsto t_i$, for all $i \in \{1, 2, \ldots, k\}$. Functional and combinatorial representations of $\mathbf{F}_\mathbb{G}(k)$ can be found in [6].

Let $\mathbf{A}$ be a Gödel algebra, then a non-empty subset $\mathfrak{p}$ of A is a *filter* of $\mathbf{A}$ if for all $y \in A$, if there is x in $\mathfrak{p}$ such that $x \leq y$ then $y \in \mathfrak{p}$, and $x \wedge y \in \mathfrak{p}$ for all $x, y \in \mathfrak{p}$. We call *proper* the filters $\mathfrak{p}$ such that $\mathfrak{p} \neq A$.

A filter $\mathfrak{p}$ of $\mathbf{A}$ is *prime* if it is proper and for all $x, y \in A$, either $x \to y \in \mathfrak{p}$ or $y \to x \in \mathfrak{p}$. The set of all prime filters $\mathsf{Prime}(\mathbf{A})^1$ of $\mathbf{A}$ ordered by reverse inclusion is called the *prime spectrum* of $\mathbf{A}$.

When $\mathbf{A}$ is finite, each prime filter $\mathfrak{p}$ of $\mathbf{A}$ is generated by a join-irreducible element a as $\mathfrak{p} = \{b \in \mathbf{A} \mid a \leq b\}$. On the other hand, each join-irreducible element of $\mathbf{A}$ singly generates a prime filter of $\mathbf{A}$. Hence, $\mathsf{Prime}(\mathbf{A})$ is isomorphic with the poset of the join-irreducible elements of $\mathbf{A}$. See Fig. 1 as an example where $\mathbf{A} = \mathbf{F}_{\mathbb{G}}(1)$.

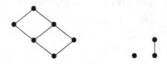

Fig. 1. The free Gödel Algebra on one generator $\mathbf{F}_{\mathbb{G}}(1)$ and its prime spectrum.

In [27], Horn established that the prime spectrum of a finite Gödel algebra forms a forest. Conversely, given $F \in \mathsf{FF}$ we can equip the finite set of subforests $\mathsf{Sub}(F)$ with the structure of a Gödel algebra $\langle \mathbf{Sub}(F), \cap, \cup, \to, \emptyset, F \rangle$, where $F' \to F'' = F \setminus \uparrow (F' \setminus F'')$, for all $F', F'' \in \mathbf{Sub}(F)$. Hence, we obtain the following isomorphisms

$$\mathsf{Prime}(\mathsf{Sub}(F)) \cong F \quad \text{and} \quad \mathsf{Sub}(\mathsf{Prime}(\mathbf{A})) \cong \mathbf{A}.$$

The above equivalence can be extended to a full duality between FF and the category of finite Gödel algebras and their homomorphisms, by making Sub and Prime functors acting also on open maps and Gödel algebra homomorphisms, respectively. These constructions go beyond the scope of the present paper, we refer the interested reader to [2,5,16].

3   Fine Spectrum

To compute the fine spectrum of $\mathbb{G}$ we need to recall some definition on factorizations of natural numbers and to collect some results on subforests.

Let k be a natural number greater than 1. By the fundamental theorem of arithmetic [17], there exists a unique expansion of k into a product of prime numbers $k = p_1 \times \ldots \times p_m$, such that $p_1 \leq \cdots \leq p_m$.

For our purpose we are interested in a slightly different problem. Given a composite natural number k, we want to find all the possible factorizations $n_1 \times \cdots \times n_t$ with $t > 1$, such that $n_1 \times \cdots \times n_t = k$ and $n_1 \leq \cdots \leq n_t$. That is, we need the following set of t-uples,

$$\mathsf{fact}(k) := \{(n_1, \ldots, n_t) \mid k = n_1 \times \cdots \times n_t, n_1 \leq \cdots \leq n_t, t > 1\}. \quad (2)$$

---

[1] In [2,5,16] the set of prime filters of a Gödel algebra $\mathbf{A}$ is usually denoted as $\mathsf{Spec}(\mathbf{A})$, here we adopt $\mathsf{Prime}(\mathbf{A})$ to avoid confusion with $Spec(\mathbb{G})$.

The set $\mathsf{fact}(k) \cup (k)$ is known as the set of *unordered factorizations* (or *multiplicative partitions*) of k. Indeed, since the order of the factors does not matter, elements of $\mathsf{fact}(k)$ are exactly t-uples composed of natural numbers whose product is equal to k. For details and further information on multiplicative partitions see [29]. Finally, we define a function such that

$$pr(k) = \begin{cases} 0 & \text{if } k \text{ is prime;} \\ 1 & \text{otherwise.} \end{cases}$$

From now on, bold faced integers $\mathbf{k}$ will be used to denote the k-element totally ordered tree.

Let F be a finite forest.

Lemma 1. $|\mathsf{Sub}(F)| = k$ *if and only if* $|\mathsf{Sub}(F_\perp)| = k + 1$.

Proof. By definition of lifting, we notice that $F_\perp = F \cup \{\perp\}$, and that $\perp$ has to belong to every subforest of $F_\perp$. Hence, for every subforest F' of F there exists a subforest $F' \cup \{\perp\}$ of $F_\perp$. Now, it is easy to check that the map $F' \mapsto F' \cup \{\perp\}$ is a bijection from $\mathsf{Sub}(F)$ to $\mathsf{Sub}(F_\perp) \setminus \emptyset$. Hence, $|\mathsf{Sub}(F_\perp)| = |\mathsf{Sub}(F)| + |\mathsf{Sub}(\emptyset)|$, that is $|\mathsf{Sub}(F_\perp)| = k + 1$. □

Let F be a forest such that $F = F' \sqcup F''$. Since $\sqcup$ is a disjoint union, then for every forest E' in $\mathsf{Sub}(F')$ there exists a forest E in $\mathsf{Sub}(F' \sqcup F'')$ such that $E = E' \sqcup E''$, for each $E'' \in \mathsf{Sub}(F')$. Then,

$$| \mathsf{Sub}(F') | = n \text{ and } | \mathsf{Sub}(F'') | = m \text{ if and only if } | \mathsf{Sub}(F) | = n \times m. \quad (3)$$

By iterating (3) we derive

Lemma 2. *Let F be a forest such that $F = F_1 \sqcup \cdots \sqcup F_t$, for $t > 1$. Then, $| \mathsf{Sub}(F_i) | = n_i$ with $1 \leq i \leq t$, if and only if $| \mathsf{Sub}(F) | = n_1 \times \cdots \times n_t$.*

Let T be a tree, and denote by r the root of T. Then, the bottom of $\mathsf{Sub}(T)$ is $\emptyset$ and its cover is $\{r\}$. In words, $\mathsf{Sub}(T)$ has a unique atom and it is the root of T. By (3) and the above discussion we conclude,

Proposition 1. *If $|\mathsf{Sub}(F)|$ is prime then F is a tree.*

The converse is not true. For instance, take the tree $\mathbf{1} \oplus \mathbf{1} \oplus (\mathbf{1} \sqcup \mathbf{1})$. It has 6 subforests, and 6 is not a prime number.

Let k be a positive integer. We denote by $A(k)$ the set of non-isomorphic Gödel algebras with k elements. That is $A(k) = \{[\mathbf{A}] \in \mathbb{G} \mid k = |A|\}$, where $[\mathbf{A}]$ is the class of finite Gödel algebras isomorphic with $\mathbf{A}$.

Define the set of of forests corresponding to non-isomorphic k-elements Gödel algebras as:

$$S_k = \{\mathsf{Prime}(\mathbf{A}) \mid [\mathbf{A}] \in A(k)\},$$

then the fine spectrum of $\mathbb{G}$ is

$$Fine_{\mathbb{G}}(k) = |A(k)| = |S_k|. \quad (4)$$

Now define the following set of forests,

$$H_1 = \{\emptyset\} \tag{H_1}$$
$$H_k = P_k \cup Z_k \tag{H_k}$$
$$P_k = \{F_\perp | F \in H_{k-1}\} \tag{P_k}$$
$$Z_k = \{F_1 \sqcup \cdots \sqcup F_t \mid F_1 \in P_{n_1}, \ldots, F_t \in P_{n_t}, (n_1, \ldots, n_t) \in \mathsf{fact}(k)\} \tag{Z_k}$$

Theorem 1. $S_k = H_k$.

Proof. We proceed by strong induction over k. To settle the base case $k = 1$ notice that $S_1 = \emptyset$ because $A(1)$ contains only the trivial Gödel algebra, and hence $H_1 = \{\emptyset\} = S_1$. Notice also that $A(2)$ contains only the two element Gödel chain, whose spectrum is the tree **1**, then $H_2 = S_2$ by definition of H_2.

For the inductive step, we assume that $H_n = S_n$ for every $1 \leq n < k$ and we prove that $H_k = S_k$.

Take a forest $F \in S_k$, either F is a tree, or F can be decomposed in a disjoint union of a family of subforests. Suppose F is a tree $F = (F')_\perp$. By the definition of S_k we have $|\mathsf{Sub}(F)| = k$, and hence by Lemma 1 $|\mathsf{Sub}(F')| = k - 1$, that is $F' \in S_{k-1}$. By induction $F' \in H_{k-1}$, and hence we conclude $F \in H_k$ by (P_k). Now suppose F can be decomposed in $F_1 \sqcup \cdots \sqcup F_n$. Then, $k = n_1 \times \cdots \times n_t$ for $(n_1, \ldots, n_t) \in \mathsf{fact}(k)$, and by Lemma 2 $|\mathsf{Sub}(F_j)| = n_j$, that is $F_j \in S_j$, for $1 \leq j \leq t$. By inductive hypothesis $F_j \in H_j$, for $1 \leq j \leq t$, hence $F \in H_k$ by (Z_k).

We have shown $S_k \subseteq H_k$. Now we prove the opposite relation.

Take a forest $F \in H_k$. By definition, either $F \in P_k$ or $F \in Z_k$. Suppose $F \in P_k$, then by definition $F = (F')_\perp$ with $F' \in H_{k-1}$. By induction $F' \in H_{k-1} = S_{k-1}$, and hence by Lemma 1 $|\mathsf{Sub}(F')| = k - 1$ if and only if $|\mathsf{Sub}(F)| = k$, that is $F \in S_k$. Now suppose that $F \in Z_k$. Then, by definition $F = F_1 \sqcup \cdots \sqcup F_t$ such that $F_j \in P_j \subseteq H_j$, for $1 \leq j \leq t$. By induction $F_j \in S_j$ for every $1 \leq j \leq t$, and by Lemma 2 we have $|\mathsf{Sub}(F)| = \prod_{j=1}^{t} |\mathsf{Sub}(F_j)|$ that is $F \in S_k$.

This concludes the proof. $\qquad\square$

In the Appendix the sets S_k are depicted for $1 \leq k \leq 12$.

Notice that during the recursive building of H_k no forest is created more than once. This fact can be proven by noticing first that to generates two times a forest F in Z_k we need two factorizations $(n_1, \ldots, n_t) \neq (n_{1'}, \ldots, n_{t'})$ in $\mathsf{fact}(k)$, such that $F = F_1 \sqcup \cdots \sqcup F_t$ and $F = F_{1'} \sqcup \cdots \sqcup F_{t'}$. By (Z_k), every F_i for $1 \leq i \leq t$ is a tree belonging to P_{n_i}, and every $F_{i'}$ for $1' \leq i' \leq t'$ is a tree belonging to $P_{n_{i'}}$, we conclude that $t = t'$ and each tree F_i is equal to its corresponding $F_{i'}$. This means that there are at least two $n_i \neq n_{i'}$ such that $P_{n_i} = P_{n_{i'}}$, but this is in contradiction with (P_k). Indeed assuming without loss of generality that $n_i < n_{i'}$, then for each tree T in P_{n_i} there exists a tree $T' \in P_{n_{i'}}$ obtained by lifting T for $n_{i'} - n_i$ times. Therefore, $P_{n_i} \neq P_{n_{i'}}$.

From (4), and the above considerations on the construction of $H_k = S_k$, directly follows a formula to compute the cardinality of H_k, that is the fine spectrum of $\mathbb{G}$.

Corollary 1. $Fine_{\mathbb{G}}(k) = f(k) + pr(k) \times g(k)$ with,

$$f(1) = 1 \qquad\qquad (f_1)$$
$$f(k) = Fine_{\mathbb{G}}(k - 1) \qquad\qquad (f_k)$$
$$g(k) = \sum_{(n_1,\ldots,n_t)\in\mathsf{fact}(k)} f(n_1) \times \cdots \times f(n_t) \qquad\qquad (g_k)$$

Table 1 shows the number of non-isomorphic finite Gödel algebras for cardinalities $1 \leq k \leq 150$ computed using Corollary 1.

Table 1. The number of non-isomorphic k-element Gödel algebras with $1 \leq k \leq 150$.

k	$Fine_{\mathbb{G}}(k)$	k	$Fine_{\mathbb{G}}(k)$	k	$Fine_{\mathbb{G}}(k)$	k	$Fine_{\mathbb{G}}(k)$	k	$Fine_{\mathbb{G}}(k)$
1	1	31	136	61	1484	91	7390	121	25519
2	1	32	162	62	1620	92	7987	122	27003
3	1	33	170	63	1679	93	8123	123	27347
4	2	34	193	64	1868	94	8668	124	29103
5	2	35	199	65	1892	95	8730	125	29249
6	3	36	248	66	2122	96	9627	126	31501
7	3	37	248	67	2122	97	9627	127	31501
8	5	38	279	68	2338	98	10318	128	33559
9	6	39	291	69	2390	99	10528	129	33965
10	8	40	344	70	2631	100	11439	130	36075
11	8	41	344	71	2631	101	11439	131	36075
12	12	42	406	72	2990	102	12418	132	38925
13	12	43	406	73	2990	103	12418	133	39018
14	15	44	466	74	3238	104	13387	134	41140
15	17	45	493	75	3341	105	13713	135	41878
16	23	46	545	76	3651	106	14573	136	44455
17	23	47	545	77	3675	107	14573	137	44455
18	31	48	646	78	4063	108	15947	138	47442
19	31	49	655	79	4063	109	15947	139	47442
20	41	50	740	80	4492	110	17085	140	50619
21	44	51	763	81	4608	111	17333	141	51164
22	52	52	860	82	4952	112	18646	142	53795
23	52	53	860	83	4952	113	18646	143	53891
24	69	54	986	84	5541	114	20119	144	57988
25	73	55	1002	85	5587	115	20223	145	58206
26	85	56	1132	86	5993	116	21604	146	61196
27	91	57	1163	87	6102	117	21955	147	61974
28	109	58	1272	88	6636	118	23227	148	65460
29	109	59	1272	89	6636	119	23296	149	65460
30	136	60	1484	90	7354	120	25455	150	69922

By applying (1), we can use the above recurrence also to compute the spectrum of $\mathbb{G}$,

$$\mathrm{Spec}(\mathbb{G}) = \{k \in \mathbb{N}^+ \mid Fine_{\mathbb{G}}(k) > 0\}.$$

Proposition 2. *Spec*$(\mathbb{G}) = \mathbb{N}^+$.

Proof. The result follows from the fact that every finite chain can be equipped with the structure of a Gödel algebra.

3.1 Sums of Oterms

The sequence of integer numbers reported in Table 1 and produced by $Fine_{\mathbb{G}}$, appears to be sequence A130841 of the *On-Line Encyclopedia of Integer Sequences*. Such a sequence gives the number of ways to express an integer as a sum of so-called *oterms*[2]. In this section, we see how a sum of oterms could be used to syntactically describe a finite forest.

Let w be a word over the language $\{(,), 1, +, \times\}$. We denote $\|w\|$, the number of occurrences of 1 appearing in w. An *oterm* is a word over the language $\{(,), 1, +, \times\}$, inductively defined by the following rules:

(O1) 1 is an oterm;
(O2) $(1 + o_1) \times_1 \cdots \times_{i-1} (1 + o_i)$ is an oterm, when $o_1, \ldots, o_i$ are oterms and $i \geq 1$.

A *sum of oterms* is a string $o_1 + o_2 + \cdots + o_i$, where $o_1, o_2, \ldots, o_i$ are oterms such that $i \geq 1$ and $\|o_1\| \leq \cdots \leq \|o_i\|$. We call SO the class of all sum of oterms.

Example 1. Let $o_1 = (1+1)$ and $o_2 = (1+1+1)$ be two words over the language $\{(,), 1, +, \times\}$. The words o_1, o_2, and $o_1 \times o_2$ are oterms, $1 + (o_1 \times o_2)$ is a sum of oterms, while $(o_1 \times o_2) + 1 \notin$ SO because $\|(o_1 \times o_2)\| \not\leq \|1\|$.

Now we show that sum of oterms are just another way to express finite forests (and viceversa), by defining a bijection between these two classes of objects. Without loss of generality we establish that the disjoint union decomposition of a forest $F \in$ FF into a finite set of trees $T_1 \sqcup T_2 \sqcup \cdots \sqcup T_m$, is such that $|T_1| < |T_2| < \cdots < |T_m|$ and $m > 1$.

Let us define the functions $h : \mathsf{FF} \to \mathsf{SO}$ and $l : \mathsf{SO} \to \mathsf{FF}$ by the following prescriptions,

$$
\begin{aligned}
h(\emptyset) &\mapsto 1 & l(1) &\mapsto \emptyset \\
h(\mathbf{1}) &\mapsto (1+1) & l(1+1) &\mapsto \mathbf{1} \\
h(\mathbf{1} \oplus F) &\mapsto (1 + h(F')) & l(1+o) &\mapsto (\mathbf{1} \oplus l(o)) \\
h(F \sqcup F') &\mapsto h(F) \times h(F') & l(o \times o') &\mapsto l(o) \sqcup l(o')
\end{aligned}
$$

---

[2] The author has not been able to find references to oterms in the scientific literature. The only occurrence of such expressions is in the On-Line Encyclopedia of Integer Sequences, published electronically at http://oeis.org/ [14].

for $F, F' \in \mathsf{FF}$ non-empty forests, and $o, o' \in \mathsf{SO}$ with $o \neq 1 \neq o'$.

Notice that, a decomposition of a forest into a disjoint union of forests is unique up to associativity, and $h(F \sqcup (F' \sqcup F'')) = h((F \sqcup F') \sqcup F'')$. Analogously, $l(o_1 \times (o_2 \times o_3)) = l((o_1 \times o_2) \times o_3)$.

It is easy to see that l is the inverse function of h, and viceversa. Hence, we have established a bijection between FF and SO.

Example 2. Let F be the spectrum of the free Gödel Algebra on one generator $\mathbf{F}_G(1)$ depicted in Fig. 1. Then, $h(F) = (1 + 1) \times (1 + (1 + 1))$. Viceversa, the oterm $o = 1 + ((1 + 1) \times (1 + 1))$ corresponds to the tree $l(o) = \mathbf{1} \oplus (\mathbf{1} \sqcup \mathbf{1})$.

Given integers $k, i \geq 1$, we let $s_k = \{s \in \mathsf{SO} \mid s = o_1 + o_2 + \cdots + o_i$ and $o_1 + o_2 + \cdots + o_i = k\}$. In words, s_k is the set of all possible ways to express k as a sum of oterms (when $i = 1$ we have unary sums). We define $a(k)$ as the cardinality of s_k, that is number of ways to write k as a sum of oterms

Example 3. Let $k = 12$. Then, $a(k) = 12$ because there are 12 ways to write k as a sum of oterms. That is,

$(1+1+1+1+1+1+1+1+1+1+1+1)$, $(1+1+((1+1) \times (1+1+1+1+1)))$,
$(1+1+1+1+1+1+1+1+1+((1+$ $(1+1+((1+1) \times (1+(1+1) \times (1+1))))$,
$1) \times (1+1)))$, $(1+1+((1+1) \times (1+(1+1) \times (1+1)))$,
     $(1+1) \times (1+1+1+1+1+1)$,
$(1+1+1+1+1+1+((1+1) \times (1+1+1)))$, $(1+1) \times (1+1+((1+1) \times (1+1)))$,
$(1+1+1+1+((1+1) \times (1+1+1+1)))$, $(1+1+1) \times (1+1+1+1)$,
$(1+1+1+1+((1+1) \times (1+1) \times (1+1)))$, $(1+1+1) \times (1+1) \times (1+1)$.
$(1+1+1+((1+1+1) \times (1+1+1)))$,

Let S be a set of finite forests. We define the set of sum of oterms generated by the forests in S through the function h as,

$$H(S) = \{h(F) \mid F \in S\}.$$

Lemma 3. $H(S_k) = s_k$.

Proof. We proceed by induction over k. To settle the base case $k = 1$ it is sufficient to notice that $s_1 = 1$ and $S_1 = \{\emptyset\}$.

For the inductive step, we assume that $H(S_n) = s_n$ for every $1 \leq n < k$ and we prove that $H(S_k) = s_k$.

Let F be a forest in S_k. We have two cases. The first $F = \mathbf{1} \oplus F'$ and $F' \in S_{k-1}$. Hence, by induction $h(F') = s \in s_{k-1}$ and by definition of h we have $h(F) = (1 + h(F')) = (1 + s)$. That is, an oterm in s_k.

Second case $F = F' \sqcup F''$ with $F' \in S_n$ and $F'' \in S_m$, for $n \times m = k$. By inductive hypothesis $h(F') = s' \in s_n$ and $h(F'') = s'' \in s_m$. Then, $h(F) = h(F') \sqcup h(F'') = s' \times s''$. Since $n \times m = k$ then $s' \times s'' \in s_k$.

We have shown that $H(S_k) \subseteq s_k$. To prove the opposite relation $s_k \subseteq H(S_k)$, it is now sufficient to notice that, since h is a bijective function, then for every $s \in s_k$ there exists a unique $F \in S_k$ such that $h^{-1}(s) = F$ and $s = h(F) \in H(S_k)$. □

Hence, we conclude that the number of non-isomorphic k-element Gödel algebras is equal to the number of way to express k as a sum of oterms.

Proposition 3. $Fine_{\mathbb{G}}(k) = |S_k| = |s_k| = a(k)$.

Proof. The first equality is given by (4), the second follows from Lemma 3, and the third is the definition of $a(k)$. □

4   Free Spectrum

In this section we show a recurrence formula to compute $Free_{\mathbb{G}}(k)$. Such formula has been obtained for the first time by Horn in [26], and it has been restated in dual terms by D'Antona and Marra in [16]. Here we recall the combinatorial approach given in [2].

In any variety $\mathbb{V}$, the free k-generated algebra is the coproduct of k copies of the free 1-generated algebra $\mathbf{F}_{\mathbb{V}}(1)$. Hence, we can dually describe the prime spectrum of $\mathbf{F}_{\mathbb{G}}(k)$ with

$$\mathsf{Prime}(\mathbf{F}_{\mathbb{G}}(k)) = \prod^n \mathsf{Prime}(\mathbf{F}_{\mathbb{G}}(1)),$$

by knowing that the prime spectrum of $\mathbf{F}_{\mathbb{G}}(1)$ is $\mathbf{1} \sqcup \mathbf{1}_{\perp}$, see Figs. 1 and 2.

Thanks to the nice characterization of products and coproducts in FF recalled in Sect. 2, in the case of Gödel algebras we can do something better. Given $F \in$ FF, we write mF to denote $F \sqcup F \sqcup \cdots \sqcup F$ m-times, that is the disjoint union of m copies of F. Then, defining the following sets of forests,

$$D_0 = \emptyset \qquad D_n = \bigsqcup_{i=0}^{n-1} \binom{n}{i}(D_i)_{\perp},$$

it has been proved that,

Theorem 2 ([2]). $\mathsf{Prime}(\mathbf{F}_{\mathbb{G}}(k)) = D_k + (D_k)_{\perp}$

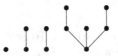

Fig. 2. The prime spectrum $\mathsf{Prime}(\mathbf{F}_{\mathbb{G}}(2))$ of the free Gödel Algebra on two generators, obtained as $D_2 + (D_2)_{\perp}$ where $D_2 = \bigsqcup_{i=0}^{1} \binom{n}{i}(D_i)_{\perp} = \binom{2}{0}(D_0)_{\perp} \sqcup \binom{2}{1}(D_1)_{\perp}$, that is $D_2 = \binom{2}{0}\mathbf{1} \sqcup \binom{2}{1}\mathbf{2} = \mathbf{1} \sqcup \mathbf{2} \sqcup \mathbf{2}$ and $\mathbf{1} = D_1 = (D_0)_{\perp}$.

From the above theorem directly follows a formula to compute $Free_{\mathbb{G}}(k)$ (Table 2),

Corollary 2 ([2]). $Free_{\mathbb{G}}(k) = (d(k))^2 + d(k)$ with,

$$d(0) = 1 \qquad d(n) = \prod_{i=0}^{n-1} (d(i) + 1)^{\binom{n}{i}}.$$

Table 2. The cardinalities of free k-generated Gödel algebras with $1 \leq k \leq 4$. The exact values for $k \geq 4$ can be computed with any software for unbounded length arithmetic.

k	1	2	3	4
$Free_{\mathbb{G}}(k)$	6	342	137186159382	2.05740183252e+64

5   Conclusions

In [7], authors used a brute-force algorithm to compute the number of non-isomorphic k-element residuated structures, including Gödel algebras, for $1 \leq k \leq 12$. In this paper we have tackled spectra problems by looking at duals of finite Gödel algebras, obtaining a recurrence formula to compute the number of non-isomorphic k-element Gödel algebras for any $k \geq 1$.

From this investigation, it appears that spectra problems are easier to handle when restated in dual terms. This seems to be confirmed by the preliminary results on the fine spectrum problem for the class of involutive bisemilattices (the algebraic counterpart of the Weak Kleene Logic) obtained in [12]. Hence, this duality-based approach is worth to be generalized to other varieties related to non-classical logics. A good starting point would be to investigate logics inside Monoidal T-norm Logic hierarchy [23]. Indeed, dual representations for corresponding varieties are already available, and have been used to compute free spectrum for Nilpotent Minimum logic [4,15] and Revised Drastic Product logic (RDP) [13]. The duality of [13] has been used in [33] to obtain subforests representations for varieties corresponding to RDP, DP and EMTL logic [34,1,11]. These logics, together with Gödel logic, are schematic extensions of the Weak Nilpotent Minimum logic [30], whose free spectrum problem has been solved in [3] through subdirect representation. We remark that the varieties corresponding to all the above listed logics, are locally finite.

Free and fine spectra are closely related. Indeed, every k-generated algebra in $\mathbb{V}$ is a quotient of $\mathbf{F}_{\mathbb{V}}(k)$. However, different congruences may generates isomorphic quotients. Hence, studies on spectra can be also used to investigate congruence in varieties. Another interesting topic stemming out from research on spectra is the so-called *generative complexity* [8], that is the function that counts the number of k-generated algebras.

Acknowledgments. The author is indebted with one of the reviewers for reference [19], for a complete clarification of dualities mentioned in the introduction, and for shortening the proof of Lemma 1.

Appendix

We list here the sets S_k for $2 \leq k \leq 12$. Forests in each S_k are separated by dashed lines.

S_{10}

S_{11}

S_{12}

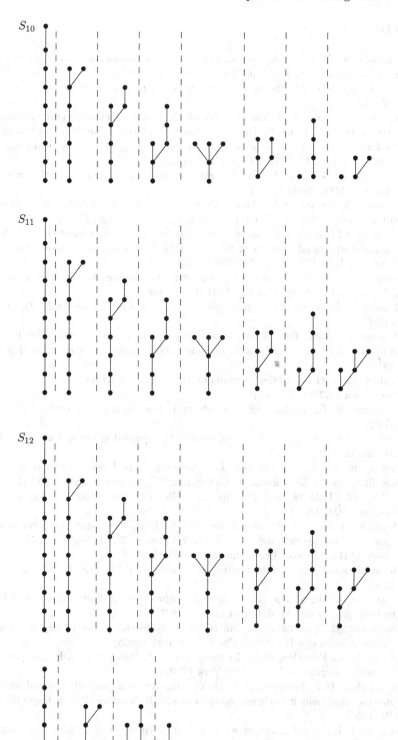

References

1. Aguzzoli, S., Bianchi, M., Valota, D.: A note on drastic product logic. In: Laurent, A., Strauss, O., Bouchon-Meunier, B., Yager, R.R. (eds.) IPMU 2014. CCIS, vol. 443, pp. 365–374. Springer, Cham (2014). https://doi.org/10.1007/978-3-319-08855-6_37
2. Aguzzoli, S., Bova, S., Gerla, B.: Free algebras and functional representation for fuzzy logics. In: Cintula, P., Hájek, P., Noguera, C. (eds.) Handbook of Mathematical Fuzzy Logic. vol. 2, Volume 38 of Studies in Logic. Mathematical Logic and Foundation, pp. 713–791. College Publications (2011)
3. Aguzzoli, S., Bova, S., Valota, D.: Free weak nilpotent minimum algebras. Soft Comput. **21**(1), 79–95 (2017)
4. Aguzzoli, S., Busaniche, M., Marra, V.: Spectral duality for finitely generated nilpotent minimum algebras, with applications. J. Logic Comput. **17**(4), 749–765 (2007)
5. Aguzzoli, S., Codara, P.: Recursive formulas to compute coproducts of finite Gödel algebras and related structures. In: 2016 IEEE International Conference on Fuzzy Systems (FUZZ-IEEE), pp. 201–208 (2016)
6. Aguzzoli, S., Gerla, B.: Normal forms and free algebras for some extensions of MTL. Fuzzy Sets Syst. **159**(10), 1131–1152 (2008)
7. Belohlavek, R., Vychodil, V.: Residuated lattices of size $\leq$ 12. Order **27**(2), 147–161 (2010)
8. Berman, J., Idziak, P.M.: Generative complexity in algebra. Number 828 in Memoirs of the American Mathematical Society 175. American Mathematical Society (2005)
9. Bezhanishvili, G.: Varieties of monadic heyting algebras part II: duality theory. Stud. Logica **62**(1), 21–48 (1999)
10. Bezhanishvili, G., Bezhanishvili, N.: Profinite heyting algebras. Order **25**(3), 211–227 (2008)
11. Bianchi, M.: The logic of the strongest and the weakest t-norms. Fuzzy Sets Syst. **276**, 31–42 (2015)
12. Bonzio, S., Baldi, M.P., Valota, D.: Counting finite linearly ordered involutive bisemilattices. In: Desharnais, J., Guttmann, W., Joosten, S. (eds.) RAMiCS 2018. LNCS, vol. 11194, pp. 166–183. Springer, Cham (2018). https://doi.org/10.1007/978-3-030-02149-8_11
13. Bova, S., Valota, D.: Finitely generated RDP-algebras: spectral duality, finite coproducts and logical properties. J. Logic Comput. **22**, 417–450 (2011)
14. Brown, D.R.L.: Personal communication, March 2018
15. Busaniche, M.: Free nilpotent minimum algebras. Math. Log. Q. **52**(3), 219–236 (2006)
16. D'Antona, O.M., Marra, V.: Computing coproducts of finitely presented Gödel algebras. Ann. Pure Appl. Log. **142**(1–3), 202–211 (2006)
17. Davenport, H.: The Higher Arithmetic: An Introduction to the Theory of Numbers, 8th edn. Cambridge University Press, New York (2008)
18. Davey, B.A., Priestley, H.A.: Introduction to Lattices and Order, Second edn. Cambridge University Press, New York (2002)
19. de Jongh, D.H.J., Troelstra, A.S.: On the connection of partially ordered sets with some pseudo-Boolean algebras. Indagationes Mathematicae (Proceedings) **69**, 317–329 (1966)
20. Dedekind, R.: Über Zerlegungen von Zahlen durch ihre größten gemeinsamen Teiler. Festschrift Hoch. Braunschweig u. ges. Werke **2**, 103–148 (1897)

21. Dummett, M.: A propositional calculus with denumerable matrix. J. Symb. Log. **24**(2), 97–106 (1959)
22. Esakia, L.: Topological Kripke Models. Soviet Mathematics Doklady **15**, 147–151 (1974)
23. Esteva, F., Godo, L.: Monoidal t-norm based logic: towards a logic for left-continuous t-norms. Fuzzy Sets Syst. **124**(3), 271–288 (2001)
24. Fagin, R.: Generalized first-order spectra and polynomial-time recognizable sets. In: Complexity of Computation, SIAM-AMS Proceedings, vol. 7 (1974)
25. Hájek, P.: Metamathematics of Fuzzy Logic. Trends in Logic-Studia Logica Library. Kluwer Academic Publishers, Dordrecht (1998)
26. Horn, A.: Free L-algebras. J. Symb. Log. **34**, 475–480 (1969)
27. Horn, A.: Logic with truth values in a linearly ordered heyting algebra. J. Symb. Log. **34**(3), 395–408 (1969)
28. Gödel, K.: Zum intuitionistischen Aussagenkalkul. Anzeiger Akademie der Wissenschaften Wien **69**, 65–66 (1932)
29. Knopfmacher, A., Mays, M.E.: A survey of factorization counting functions. Int. J. Number Theory **01**(04), 563–581 (2005)
30. Noguera, C.: Algebraic Study of Axiomatic Extensions of Triangular Norm Based Fuzzy Logics. University of Barcelona, Ph.D. Thesis (2006)
31. Quackenbush, R.W.: Enumeration in classes of ordered structures. In: Rival, I. (ed.) Ordered Sets. NATO Advanced Study Institutes Series (Series C – Mathematical and Physical Sciences), vol. 83. Springer, Dordrecht (1982). https://doi.org/10.1007/978-94-009-7798-3_17
32. Taylor, W.: The fine spectrum of a variety. Algebra Universalis **5**(1), 263–303 (1975)
33. Valota, D.: Representations for logics and algebras related to revised drastic product t-norm. Soft Comput. **23**, 2331–2342 (2018)
34. Wang, S.: A Fuzzy logic for the revised drastic product t-Norm. Soft comput. **11**(6), 585–590 (2007)

From Semantic Memory
to Semantic Content

Henk Zeevat[1,2](✉)

[1] SFB991, Düsseldorf, Germany
[2] ILLC, Amsterdam, The Netherlands
henk.zeevat@uva.nl

Abstract. Barsalou (1992), Löbner (2014, 2015) hypothesise that frames form the natural way in which the brain represents concepts and more complicated semantic content built from concepts. An interesting aspect of the hypothesis is that it becomes easy to define stochastic properties of concepts. Particular frames can be seen as a collection of stochastic variables.

This paper develops a simple but powerful notion of semantic memory on the basis of an earlier concept of lexical knowledge under the frame hypothesis. It then tries to answer the question of whether the stochastic information in semantic memory contributes to conceptual content.

1 Introduction

In order to do natural language interpretation one needs an account of lexical meaning. Frame semantics has been developed for precisely the purpose of characterising lexical content and one can therefore look for a theory of lexical content that uses frames. The hardest problem in lexical semantics is the problem of selecting the right meaning for a particular word in context from the large sets of possible and truth-conditionally non-equivalent meanings that one finds for typical words in natural languages. Humans seem to do this perfectly and effortlessly and this ability must be explained somehow. Section 2 recapitulates the formal side of the stochastic account of Zeevat et al. (2015), an account that tries to solve exactly this problem. The resource one needs for this account: a large frame-semantically annotated corpus however also can be used to develop a theory of semantic memory (Sect. 3), a resource that is just as essential for natural language interpretation as the lexicon, as argued in Sect. 4 which shows that semantic memory can solve a number of open problems in natural language interpretation.

This paper can be read as filling two gaps in Zeevat (2014): an account of the lexicon and of the priors needed for the Bayesian interpretation developed there. It is unfortunate that frame-semantically annotated corpora are not currently available. There is good hope though that advances in distributional semantics

ⓒ Springer-Verlag GmbH Germany, part of Springer Nature 2019
A. Silva et al. (Eds.): TbiLLC 2018, LNCS 11456, pp. 312–329, 2019.
https://doi.org/10.1007/978-3-662-59565-7_16

will bring the automatic learning of frames from texts within our reach and with that the required corpora[1].

2   Frames and Lexical Knowledge

2.1   Frame Semantics

Frames are sets of annotated nodes and annotated arrows which form a directed graph. Node annotations (classes) are interpreted as classes of values, arrow annotations (attributes) as functions from values to values. A frame is true under an assignment f to its nodes iff for each node x annotated with label c $f(x)$ falls into the class c and for each arrow a annotated with α and connecting nodes x and y $\alpha(f(x)) = f(y)$.

(0) is an example frame for the noun *father*. n X indicates that node n has annotation X, $n \to L \to m$, that there is an arrow from n to m labeled L and n *central* indicates that n is the central node of the frame.

(0)   father
     n → father → m
     m central

The frame can become more informative in two ways: by acquiring new arrows and by node classes becoming more restricted. In (1) (for Mary's bald father) information gets added in both ways.

(1)   n → father → m
     m central, person
     n person, mary
     m → hair → k
     k none

It is wrong to see frame semantics as the claim that the formal notion of a frame is used to represent concepts and truth-conditions. That would allow arbitrary attributes and arbitrary classes. Arbitrary attributes would include characteristic functions for properties, functions and relations so that any extant format for semantics would be recoverable in frame semantics.

Frame semantics should instead be seen as the claim that meanings are frames constructed from a limited set of universal attributes (the arrow labels) and a set of classes for the range of these attributes (a set that is much less likely to be universal) classifying areas in "conceptual spaces of values" (like sizes, shapes, appearances, smells) as well as classes of non-abstract objects (e.g. children, vegetables, engineers, cars, apples). The attributes map objects to objects and objects to values.

---

[1] This hopeful remark refers to a current project in Düsseldorf by the author and others which tries to abduce frames from binary semantic features whose association with words can be detected by state of the art distributional semantics, cf. Boleda et al. (2013), Baroni and Zamparelli (2010).

The atoms for a frame language are statements $n : c$ and $n\alpha : c$ about the referent x of a node n where α is a natural attribute and c a class in the range of α and saying that x has a value for α in c or that α applied to x falls in class c.

More formally[2], a frame F is a connected directed graph with a central node cn with class labels on the nodes and attribute labels on the arrows. The central node cn allows the interpretation of the frame F as a concept: the concept F applies to an object o if F is true under some assignment f such that $f(cn) = o$.

Frames can be developed as any other logic notation. One specifies a set of types for the special domains possibly with the required extra structure over those types. E.g. if the type collects sizes of objects, it is reasonable to have a linear ordering and a smallest size. In terms of the types, one specifies a signature (the set of attribute labels and the set of class labels with their types) and in terms of the signature the set of frames.

Models for a frame language interpret types a as suitable structures D_a and D_o (o is the type of the objects) as a non-empty set. Class and attribute labels are interpreted by an interpretation function F. $F(c)$ with c of type a is a set of values in D_a, $F(\alpha)$ with attribute α of type $<a,b>$ as a partial function from D_a to D_b.

The conceptual spaces, the relations over them, and attributes mapping one conceptual space to another should be thought of as invariable. D_o, classes over D_o and attributes mapping from and to D_o can change from one model to the other.

Frames F with a central node have the typical ambiguity that Frege criticized in the mental representations of the pre-Fregean logicians. F can be understood as a concept, as a proposition and as a denoting expression.

F can be true or or false (there is an assignment under which F is true or there is no such assignment), it can be interpreted as a concept C such that C applies to o iff $o \in \{f(cn) : F$ is true under $f\}$ or as a definite or indefinite description: F refers to x iff $x = \{f(cn) : F$ is true under $f\}$ or $x \subseteq \{f(cn) : F$ is true under $f\}$.

This ambiguity can be solved by being precise about frame combination, as needed for semantic composition.

The following is one way of making semantic composition explicit (there are other ways). In this approach, numbers will be allowed as extra annotation on nodes (this captures the binding of a frame node by another frame) and frames can be put in a sequence. The idea is that a node numbered n denotes the denotation of a preceding frame F_n in the sequence, a frame that must occur earlier in the sequence.

(2)   Definition
A sequence of frames $< F_1 \ldots F_k >$ is a closed context iff for each $n \leq k$ F_n only has number annotations m such that $m < n$.
A closed context $< F_1 \ldots F_k >$ is true iff there are $o_1, \ldots, o_n$ such that each F_i is true of o_i if n occurring in F_i is interpreted as o_n.

---

[2] See Petersen (2007) for the motivation for central nodes.

A negated context is a context C prefixed with $\neg$. Negated contexts can be added at the end of a context D iff the concatenation of D and C forms a closed context DC.

$\neg C$ is true relative to its precontext D and an assignment f of values to D iff there is no extension g of f to the frames of C such that DC is true.

The negation is enough for capturing the expressive power of FOL, by well known results on DRT[3] Kamp and Reyle (1993). Next to negation one can add definite contexts $def C$, using a definiteness operator def that checks that in its precontext the frames in C have a unique reference.

The following are two examples. Frames are given between square brackets. (Counting is carried on under the definiteness operator.)

(3)   John likes a donkey. He feeds it the carrots in the garden.
$< john_1, donkey_2, [like_3, experiencer1, theme2], 1_4, 2_5,$
$def < [garden_6, owner1], [carrots_7, loc6] >,$
$[feed_8, agent5, goal2, theme7] >$
If a man likes a donkey, he feeds it .
$< \neg < man_1, donkey_2, [like_3, experiencer1, theme2],$
$\neg < 1_4, 2_5, [feed_6, agent5, goal6] >>>$

Frame semantics traditionally attempts to find generalisations in the meaning of lexical items. It is a linguistic semantics, trying to find natural structure in lexical meaning. As such, it can also be interpreted as a theory about natural concepts: natural concepts are the concepts that tend to get lexicalised.

3   Lexical Knowledge

Dictionaries typically give many readings of words and it is easy to see from the informal descriptions of these readings that these readings entail different truth-conditions. Language users seem to converge effortlessly on the right reading in a given context and an account of lexical knowledge needs to be able not just to account for the fact that there are so many readings but also for the way in which language users will find the right reading in the context.

The approach developed in Zeevat et al. (2015) for this problem starts from the notion of a semantically annotated corpus. This is a model of a person's linguistic experience. The utterances produced and understood are stored together with their meaning, indexing which part of the utterance is responsible for which part of the semantics.

A semantically annotated corpus is a set of items:

$$(n, u, F, f)$$

---

[3] Frame semantics as presented here can be seen as a more structured version of DRT in which frames are both discourse markers (their referential function) and conditions (seeing them as formulas).

where n is a unique natural number, u a sentence, F its frame representation and f a function from the components of u to the relevant subframes of F[4].

The semantically annotated corpus induces a lexical database of items

$$(n, w, G)$$

where n is the number of the utterance in the semantic corpus, w a word in u and $G = f(w) \subseteq F$

This can be turned into a generation dictionary of items (G, w, k) where G is the frame, w the word used to express it and k the number of occurrences of w as an expression of G. This is a generation lexicon since it allows finding the right word or words for a given frame.

The generation dictionary can be used to define both $p(G|w)$ and $p(w|G)$ and also to define $p(A|w)$, where A is a component of G. All three play a role both in lexical choice (given G what w is to be used in context C) and in lexical interpretation.

From $p(G|w)$ and $p(A|w)$, one can derive p_w : a joint distribution over components of G expressed by w. $\leq_w$ is defined from p_w by (4).

(4) $A_1 \ldots A_n \leq_w B_1 \ldots B_k$ iff $p_w(A_1 \ldots A_n) \leq p_w(B_1 \ldots B_k)$.

$\leq_w$ —unlike p_w— allows an abstraction over differences between different speakers. Linguistic experience will be different, but can lead to an identical set of inequalities in $\leq_w$. There is a $\leq_w$ determined by all the linguistic experience of all speakers, which one can take as the norm for an ideal speaker. But even real speakers can agree on all inequalities even if they both fall short of the ideal norm.

Moreover $\leq_w$ allows a symbolic representation of the integrated semantic content, a symbolic description of ambiguous words. (5) gives an overview of 4 useful operators that are available, in each case followed by an informal explanation and a formal definition.

---

[4] Such corpora do not currently exist and their construction would require considerable annotation effort. The construction of f is also non-trivial since frames —at this point not an improvement from logical formulas— do not have a unique decomposition. One would need —for the attribution of specific subframes to words— to see what generalises best over different uses of the same word. It should be possible in this way to obtain a full semantic lexicon from just a pairing of utterances and their frame-semantic meaning representation.

(5) $A; B$
      w has A by default, else B
      $X, B < X, A$

      $select(A_1, \ldots, A_n)$
      pick one of the A_i s
      $X, A_i \& A_j \leq 0$ and $X, \bigvee_i A_i = 1$

      $\neg A$
      A is excluded
      $X, A \leq 0$

      $A \rightarrow B$
      if A holds, so does B
      $X, A \& B = X, A$

      (A)
      A is optional
      $X, A < 1$

The symbolic representation can be exploited for efficient representation of different possibilities of interpretation and is an effective of way of representing equivalence classes of p_w in a human readable way.

Given resolution to the context and consistency with the context, the over-specified representation determines a "meaning in use": a proper frame, without any occurrence of the new operators. The next example from Zeevat et al. (2015) is an integrated representation of 84 readings of the verb *fall*. For the details of the formalism and the underlying data, see the paper.

(6)   THEME $\in$ {concrete,light,precipitation,task,date,judgment,proposal}
       ABSOLUTE
       nocause(THEME)
       nocontrol(THEME) ABSOLUTE
       SOURCE: location(DIMENSION)
       POSITION: location(DIMENSION)
       DIMENSION $\in$ { space:DEFAULT, posture, life, health, moral, quantity,
       level, outcome(PROCESS)} ABSOLUTE
       SOURCE= PART1 If split ABSOLUTE
       split(PART1, PART2, THEME)
       at(POSITION, down(SOURCE,DIMENSION)) ABSOLUTE

       ===
       at(THEME,POSITION) DEFAULT, NEW
       at(PART2, POSITION) If split ABSOLUTE, NEW
       movement(THEME,SOURCE) If at(THEME, SOURCE) ABSOLUTE
       NEW

The representation gives all attested possibilities of meanings in use for a range of languages. It can be regarded as the meaning of the ambiguous verb *fall*.

Disambiguation needs the context, both the syntactic context (the words surrounding w), the wider linguistic context and the common ground of the speakers involved. The context is not given by the generation lexicon and it is not clear how it could be.

The role of the context in disambiguation is to rule out the more probable meanings within p_w. Without the context, those would be the best bet, the context changes the probabilities and thereby rules out certain readings, making other readings the most probable ones. Defaults should be adopted, because they increase the likelihood of the hypothesis: they contribute to the interpretation as an explanation of the word. This is counterbalanced by the prior defined by the context which determines how well the interpretation fits in. Part of the prior is consistency with the context, another important part is formed by the requirements the interpretation puts on the context. (In the *fall* example above, these requirements are everything above the line, below the line there is the new information from the concept).

An approximation is to go for the interpretation that adopts as many defaults as possible and that fully fits the context without coming into conflict with the common ground. It is an approximation, since defaults can make the interpretation less probable in specific contexts to an extent that overrides the gain in likelihood attendant on their adoption.

Finally, the interpretation must be checked for semantic blocking. A natural example is the meaning of *run/rennen* in English and Dutch. English engines run, while dutch engines walk. The concept of running seems however largely the same in both languages for the two verbs which also clearly have a common origin. That means one would expect an interpretation for dutch: de machine rent, that corresponds to english: the engine runs. And it is obvious why that does not happen: for that interpretation, a dutch speaker would use: de machine loopt. Semantic blocking is the case where the competing expression gets all the probability putting the likelihood of the expression under interpretation to zero for this particular reading.

The generation corpus supports both the checking of semantic blocking and disambiguation as a model of likelihood. The disambiguation process needs however extra resources for determining the prior. The prior should be determined by semantic memory and the context. This paper makes the point that the resources for the lexicon are the same as the resources one needs for a model of semantic memory and makes an additional case for semantic memory in semantics and pragmatics.

While the proposal can hardly be the last word on the matter, it seems that the joint distribution over semantic features will have to be part of any future proposal that wants to predict meanings in use from the context and information about lexical items.

4   Semantic Memory

If annotated semantic corpora can deliver on the problem of deriving meanings in use for highly polysemous words by allowing the estimation of relevant probabilities, it is also possible to use the same kind of estimations to give an approximation of the general stochastic knowledge that goes under the label of semantic memory. This term refers to the aspects of memory that are not episodic memory: memories of concrete events, states and objects that the subject has been witnessing in the past. Semantic memory does not depend on personal experience, it is the general memory that one shares with others. This paper is concerned only with the content of semantic memory that can be called conceptual knowledge.

Semantic memory is directly necessary for the Bayesian interpretation of lexical items and larger utterances, since it defines the priors needed to determine the most probable information in the context.

A semantically annotated corpus already embodies a lot of experience because the sentences on which it is based are taken from natural communication. Ideally, it should be replaced by a full representation of all past experience. On our assumptions, the linguistic items in the corpus are annotated with frames. Frames come with a subframe relation where a subframe has a subset of the same nodes including the central node, a subset of the arrows and possibly more general classes on nodes. A subframe has less information than the full frame. $F \subset G$ will be used for proper subframes. The frame of *father* is a subframe of the frame for *Mary's bald father* given above.

Frames as types can occur many times in the same corpus. For this, we identify two frames if they are identical but for the identity of the nodes.

Armed with these two notions, one can count the number of occurrences of a subframe F in the corpus that are extended to G and similarly, the number of occurrences of F that are extended into an H that is not compatible with G. In terms of the example, it is possible to count how often a father is a bald father (slightly more often than men are bald, since fathers are older) and how often a father is the father of Mary (highly dependent on personal experience).

There is one complication: the number of occurrences of F may be too small to give a reliable estimate of $p(G|F)$. Abstractly this is not a problem. One can start with the set of attested subframes and consider all probability distributions over this set. The corpus can now be considered as learning data for Bayesian learning over all the elements of this set. Learning will reduce the set to the probability assignments that are plausible given the corpus and will assign intervals to $p(G|F)$ that are small when F occurs often and approach to the full interval $[0, 1]$ when F occurs only rarely.

One way to deal with the indeterminacy of many F is to consider a partial function $E(F, G)$ that is undefined in case the remaining distributions span $[0, 1]$ (which is the same as saying that there is no information about these probabilities) and else give the resulting interval. Other functions can however be considered as well.

A restriction that may be adopted or not is to limit F and G to frames expressible by linguistic expressions. Looking at Barsalou (2009), that would be too restrictive since many of the concepts needed for that account are not linguistically expressible.

This frame-based take on semantic memory is related to other models of semantic memory in psychology[5] and can be directly used to explain lexical priming and activation of semantic fields. In the following, the focus will be on linguistic applications.

Attribute Distributions

Let α be an attribute not defined in F and n the central node of F. Let $h(c) = E(F \cup \{n\alpha c\}|F)$. $h(.)$ collects the probabilities for the classes that may come to annotate the node $n\alpha$ if further information comes in. One can introduce the notation $F^\alpha = h$ for this expectation.

h is the expectation of the values for α in F. If one uses h on a set of length classes (e.g. intervals between n and $n+1$ cm), it differentiates the probability of the different classes and makes them more and less expected and can be displayed as a distribution over sizes. If α is *size* applied to the concept stick, it predicts how long sticks will be. If α is *profession* and is applied to *man* it gives a distribution over professions as in (7).

(7) man^{size}
      large man^{size}
      $man^{profession}$

5   Semantic Memory Is Important for Linguistics

First of all, through the notion of extension, semantic memory models subcategorisation: it indicates whether attributes like agent, theme, experiencer, and others apply to verbs and other categories. They come moreover with stochastic expectations for the values of these attributes and so support parsing and identity inferences when the value is given by the context and not by syntactic integration.

But there are also new applications. If one makes the reasonable assumption that in the frame-theoretic interpretation of integrated text and in experience, causal relations are marked by an attribute *cause*, semantic memory provides the basis for making causal inferences as in the examples in (8).

(8)   John fell. Bill pushed him.
       (Why did John fall? Because Bill pushed him.)
       John ran towards the bus stop. He just caught the bus.
       (Why did he run? He wanted to catch the bus)

Affordances (couches are for sitting in, the value of *purpose* for chairs) would likewise support identity inferences (John sat down on the couch).

(9)   In the back there was a couch. John sat down.

Putting John in an ownership relation with respect to the coat solves the lexical presupposition of give that one can only give what one owns.

---

[5] Jones et al. (2015) is a good overview.

(10)   John wore a brown coat. He gave it to Mary when she was feeling cold.

Zeevat (2016) makes causal inferences a special case of identity inferences and relates the latter to a forced choice for any new object introduced: will the new object be understood as different from all given objects or be identified with one of them. Given that identity information is extra information, the prior of an interpretation with an identity inference is always at least as high as that of the interpretation without it. But this is not the case with respect to the interpretation with the additional information that the new object is distinct from all given ones.

Given that causal and identity inferences are the bread and butter of building integrated representations of experiences and text and dialogue, it is hard to see how they would be possible without the stochastic information supplied by semantic memory.

But the same applies for the less ambitious tasks in interpretation, where syntactic and morphological marking may play an important supporting role. Many syntactic ambiguities are decided by having good priors for the different choices.

Semantic memory supplies the prior for any scheme of Bayesian interpretation. Frame semantics can model the implicit binding of attributes that are not syntactically bound using attribute distributions and identity inferences.

5.1   Defaults for Unbound Attributes

Any node in a given frame F can be expanded by our function E. Its class give rise to a distribution over its subclasses (if they have any), all attributes with a probability for their being defined for the node given F (in addition to the ones already given in the frame) can be added with the relevant probability. Also incoming arrows can be similarly assumed, with a probability for their number.

This operation can be applied to all nodes of a given frame and iterated.

One thus obtains different notions of a stochastic frame. Expanded in some particular nodes, in all nodes, with n-times iteration, completely iterated. The last notion will nearly always add the whole of E to a given frame and does not seem useful anymore as an additional data structure.

As a variant, one can form the normal extension of a node (or normal frames). This is a dominant extension G of F at node n which replace the class of n with its dominant subclass -if any- and adds the dominant incoming and outgoing (with a number) attribute arrows for the revised node, given F. The simplest definition is just in terms of $E(F, G)$ where one replaces F by the largest G such that $E(F, G) > 0.5$.

Ellipsis examples like (11) (Saeboe 1996) can be treated by dominant classes for argument nodes.

(11)   a. John gave to the Red Cross.
        b. John gave 5 dollars.
        c. John ate.

Argument ellipsis comes in two different kinds, In one case (b), the omitted argument is an anaphor and needs to be bound from the context. In the other case (a), the omitted argument is bound by a normal value for the attribute. In (11a) some sum of money that is customary (not one million euro, not 1 cent). Crucially also not a button, one's life, heart or body, even though these are perfectly acceptable objects of giving. What is standard moreover seems to depend on the other arguments.

Similarly, in (11c) what is eaten must be some normal food, not fire, nails, or one's heart, even if those are fine fillers of the object of eating.

5.2   Semantic Memory and Semantic Content

F^α and its generalisations $F^{\alpha_1,\dots,\alpha_n}$ however also contain information about conceptual content that is not captured by the frame itself.

Some of this can be expressed in principles like (12).

(12)   On anybody's experience, no man is 3 meters or 0.4 meter long, a fact
      directly recoverable from semantic memory.

The probabilities in question can be adapted for new experience and are thus not certain. They are however certain enough for everybody to know that John is not 3 m tall if one knows he is a man and has no information about his size. This knowledge comes from semantic memory and is good enough for making assertions like (13).

(13)   John isn't 3 meters tall.

As conceptual knowledge (13) is analytic, but it is also based on experience and therefore a posteriori. It is revisable under new admittedly unexpected experience. For all we know, it is non-necessary knowledge.

There is a lot of such analytic a posteriori knowledge.

From a frame perspective, cats and dogs are rather similar. The concepts are defined for a range of attributes like (14) and take similar values in these.

(14)   size
      weight
      color
      shinyness
      furryness
      legs
      claws
      teeth
      head
      shape
      behaviour
      sound
      smell

The problem is that is considerable overlap in the attribute values.

(15) Small dogs are the size of big cats.
      Most cat colors are dog colors.
      Not every cat makes good meows,
      Not every dog is a good barker.

The one exception is perhaps that cats have whiskers and they make in certain circumstances purring noises. But not all cats have whiskers and not all of them purr.

But humans, cats and dogs are rarely wrong in distinguishing cats from dogs. Apparently there is an accurate poly-attribute discriminator that also functions without whiskers, barking and purring. This discriminator must be regarded an essential ingredient of both the naive cat and the naive dog concept. The discriminator is based on data in semantic memory, and clearly not directly accessible to consciousness. It is not itself analysable as a frame or as a logical formula.

Binder et al. (2016) treat this discrimination problem by a range of typical features: having characteristic sound, shape, smell, biomotion, skin etc. The natural counterpart in frame theory is to have an attribute for each of the features that maps the different kinds to characteristic classes: cat walks, cat skin, cat smell etc. while dogs have dog walks for biomotion, dog smell for smell etc. There is then a number of ways of recognising cats and dogs with other characteristic attributes giving independent evidence, to make recognition of the kind by one attribute much more reliable and to generate further predictions about the relevant kind.

What holds for cats and dogs holds by extension for other natural kinds and for derived natural kinds. (16) lists some of such derived natural kinds: activities produced by different animals or parts of the bodies of such animals.

(16) walking
      running
      trotting, stepping and galopping
      yawning
      laughing
      squeezing
      flirting
      barking
      meowing
      cat claw
      cat tail
      duck egg

The various behaviours are part of the biological repertoire of biological species.

The resulting concepts cannot be regarded as definitions of the species in any scientific sense. They do not give an analysis of the natural kind concepts: the setting of values and indeed the creation of such values is due to parameter setting by learning and not further penetrable to analysis. But they come a long way in explaining recognitional power for the natural kinds and for inferences on these concepts.

5.3 Individual Semantic Memory as the Basis for Non-referential Uses Of quantifiers

Quantifiers can range over contextually given sets, identified by the nomen in which case they are referential. Their treatment is adequately captured by the standard treatments of generalised quantifiers. Often though no sets are involved and the use should be regarded as a non-referential one. The examples in (17) could be such uses.

(17)   All academics are rich.
        Academics are rich.
        Many academics are rich.
        Few academics are rich.
        Some academics are rich.
        No academics are rich.

Acquaintance with small sets allows counting and forms the basis for cardinal determiners.

5.4 Non-referential Quantifiers

Non-referential uses of quantifiers are the uses where on the one hand the nomen is not used to refer to a given set of objects (a restrictor set) and on the other hand the set of objects given by the scope within the class given by the nouns is not identified. The quantifier then describes the proportion of the A & Bs among the As. The proportion correlates with $p(B|A)$ and since there is no way of checking the truth of the quantifier by inspection in these case, the basis of assertion containing them must be an estimate of that probability, i.e. semantic memory.

Notice that *all, some, many, few, no, most* are very naturally defined by probabilities as in (18).

(18)   all: $p(x) = 1$,
        some: $p(x) \neq 0$,
        many: $p(x) > d$,
        few: $p(x) < d$,
        no: $p(x) = 0$
        most: $p(x) > 0.5$.

These quantifiers are useful in arguing about proportions.

It is unsurprising that Moxey and Sanford (1986) find that these quantifiers often have the goal of resetting the proportions in the semantic memory of the interlocutor, even at the expense of their literal meaning.

(19)   Not just some, but many doctors are rich.
        All doctors are rich.

The lexemes present in nearly all languages are natural classifiers for proportions, but much less useful for reporting about given sets. There counting and cardinality and definite descriptions are much more directly relevant.

(20)   John and Bill met.
        Two boys met.

The difference with analytic a posteriori is that the proportion classifiers are
not based on shared semantic memory, but on the private semantic memory of
the interlocutors, that can differ substantially based on differences in experience.
One can judge certain proportions to hold on the basis of semantic memory. It
is a feature of modern culture that there are procedures that even for large sets
one can give reliable estimates of such proportions based on sampling, but it
does not belong to human culture before modern times.

5.5   Generics

Most of the things that can be said about non-referential quantification also holds
for generics. Indeed one can try to reduce generic sentences like the ones in (21) to
non-referential quantification, making them unmarked for universal/existential
import and in this sense analogous to attested modal operators in Norwegian and
Stat'imc'ets. This misses an important generalisation about generics however.
Generics differ from their quantificational counterparts in having a built-in extra
restriction on quantification based on the preconditions of the predicate Kasper
(1992).

(21)   Dogs eat grass.
        All dogs eat grass.
        Dogs often/rarely eat grass.
        Many/few dogs eat grass.

The stochastic connections are as before.

6   Stochastic Frames: A Formal Concept of Natural Concepts

In this section, an attempt is made to be more precise about the stochastic
conceptual knowledge that codetermines natural concepts. A choice is made
here to consider only the distribution of subclasses of given node classes and to
remain silent about the possibility of adding extra attributes. This is one choice
among a large number of possibilities.

Stochastic frames seem a good candidate for natural concepts, if the arguments given above are on the right track. Under that view content is partly
defined by stochastic parameters. Stochastic frames could be defined as (F, d)
where F is a frame and d is a set of joint distributions over the node denotations
of F.

The case where all classes at the nodes are minimal and d just holds one
single distribution (that assigns those values to the nodes with probability 1)
is the exception. In other cases, the distributions assign joint probabilities to
combinations of subclasses of the classes of the nodes.

The most important argument for stochastic frames is given above. Traditional frame analyses of concepts need extra primitives referring to proper and derived natural kinds as values. The distributions for the typical values for attributes like smell, shape, sound and others can be used instead of these extra primitives for characterising the common sense version of such natural kinds.

A second argument is that joint distributions also give a handle on relational constraints between nodes. Such a constraint is of the form $C(n_1 \ldots n_m)$, where $n_1 \ldots n_m$ are nodes of F. If $u_1 \ldots u_m$ are values for $n_1 \ldots n_m$ but do not meet $C(n_1 \ldots n_m)$, (F, d) can exclude $u_1 \ldots u_m$ because for all $p \in dp(n_1 = u_1, \ldots, n_m = u_m) = 0$.

In the other cases, elements of d allow the combination of values. Elimination of elements from d thereby may increase the set of constraints that hold on (F, d) and one can update the set d by absolute or stochastic constraints.

As an example of a stochastic constraint, weight and length for humans are positively correlated in a imprecise way: sometimes heavier people are smaller. But the tendency that weight and length are positively correlated can be modeled by requiring that $p(length = x|weight = y) < p(length = v|weight = z)$ if $v < x$ and $y > z$.

It is assumed that the frames are normal classical frames with a central node. This is needed for making them concepts of objects. Further one would like distributions to be limited. The inverse of the distribution should also omit areas in the various continous domains. E.g. the size of adult human should be around 1.80 m with a high probability, around 1 m with a much lower probability and 0 for such values as 0.1 m or 4 m.

Moreover the area where the inverse is defined should be convex in the sense of Gärdenfors (1988). This is a demand on individual distributions. But one would like a similar convexity for the set d of distribution. If $<u_1, \ldots, u_n>$ obtains a non-zero value for some $p \in d$, its image under d should be an open interval on $[0, 1]$. (For this, in a continous domain, the values must be areas, not points).

6.1   Unification

The unification of two concepts (F, d) and (G, e) can be defined in a relatively straightforward way. For F and G take the standard unification $F \bigwedge G$ (classes are intersected and fail to unify if their intersection is empty, the central nodes must be the same) where H is the DAG shared by F, G and $F \bigwedge G$, with classes as in $F \bigwedge G$. For the distributions, consider d' and e' obtained by renormalisation after omitting the values for nodes in H that are not in the classes that $F \bigwedge G$ assigns to these nodes and renormalisation. The distributions for $F \bigwedge G$ are now obtained as the renormalised union of any two distributions in d' and e' that coincide on H.

(22) Unification

$(F,d) \bigwedge (G,e) = (F \bigwedge G, h)$

Let $H \subseteq F \bigwedge G$ maximal such that all nodes of H are nodes of F and G.

Let $d' = \{norm(restr_H(p) : p \in d\}$ and $e' = \{norm(restr_H(p) : p \in e\}$.

$restr_H(p) = \{< x,y >:< x,y >\in p,$ x restricted to H is a vector of values in the node classes of $H\}$

$norm(p)(x) = \frac{p(x)}{\sigma_{x \in dom(p)} p(x)}$

Then $h = \{p \cup q : p \in d', q \in e', p$ restricted to $H = q$ restricted to $H\}$

6.2 Application of a Concept to Objects

Assume for every attribute α of F of (F,d) there is a natural operation O_α that can be applied to objects and values. That allows us to define an isomorphic concept (really a set of isomorphic concepts, due to fact tat the central node is not the top of the graph) for a given object a. Assign a to the central node of F. For outgoing arrows labeled with α use O_α to determine the value of α for a. For incoming arrows labeled α, find all objects b such that $O_\alpha(b) = a$ (each of these gives their own copy). Repeat this until an isomorphic copy of F is reached. For application, one of the isomorphic copies should unify with (F,d).

It is difficult to know an isomorphic copy for a given object a and for the natural attributes normally assumed it is to some extent indeterminate what the natural operation is that corresponds to the attribute. Length can be measured fairly precisely, but does it include the hair and the shoe soles? Are conception and giving birth essential conditions on motherhood? Clearly, the legal procedures in these cases are stricter and less natural than the common sense procedures.

It is however the case that the application notion defined here continues to work with fairly sloppy natural procedures that would assign rather big intervals for size and would find the "best" mother when the different criteria do not give the same result.

If in some domain certain values are properly excluded for the concept, also the notion of exclusion (the concept excludes the object from its extension) can be defined. But marginal cases are quite conceivable.

6.3 Judgment

More theoretically interesting is probably the concept of judgment, where a is only partly known and a degree of uncertainty is part of the notion. This requires an extra notion of diagnosticity.

If a is partly known, this partial knowledge can be represented by a classical frame A_a where wide classes cover the values of the partly known attributes. If (F,d) is a natural concept, e.g. of a mountain, a cow or a tribe, it will have parts (concepts (G,e) such that $(G,e) \bigwedge (F,d) = (F,d)$) whose presence makes the full (F,d) probable. This is core semantic memory: $E(G,F)$ is high. Assume that A_a entails some such parts (G,e) $(A_a \bigwedge (G,e) = A_a)$ Then (F,d) can be inferred with a high probability, especially if the different (G, e) independently predict (F, d). (F, d) can then also be used to replace A_a by $(F, d) \bigwedge A_a$: a is known by bringing it under a concept. Feedback from the new concept can be used to further confirm that a is an (F, d).

Natural concepts of this kind seem indispensable for arriving at an account of knowledge by experience. Such concepts arise within the larger set of concepts arising naturally by evolution: better diagnosticity and better prediction define a fitness function[6]. Learning a language with many words for good natural concepts is the human way of acquiring many natural concepts beyond the ones obtained from experience.

7   Conclusion

A properly annotated semantic corpus gives joint probabilities. These are needed for lexical disambiguation and for defining conceptual content.

Both the view of the lexicon as a list of readings per word and the view that lexical content is a frame with classes labeling nodes are inherited from the traditional view that describing meaning is a question of giving necessary and sufficient conditions. Such views have problems with accounting for effortless disambiguation on the one hand and with accounting for the lexical content of words for natural phenomena.

Semantic memory is not more complicated than the lexicon of a natural language and as unmysterious. Both are based on large amounts of data and have the task of dealing with everything there is.

Semantic memory is essential for natural language interpretation in the following three ways.

1. as a resource that codetermines prior probabilities
2. as the basis for causal and identity inferences and therefore for treatments of discourse coherence and bridging reference
3. as what defines the classification of natural objects by natural kinds and the classification of human and animal behaviour
4. as part of a full account of natural kinds which brings in diagnosticity

Truth for natural language utterances is often indeterminate. But abstracting from that indeterminacy, concepts as fixed by the semantic memory of the competent speaker determine their extensions as properly as any other notion, under the right assumptions.

Finally, lexical knowledge itself should be seen as part of semantic memory. Nodes in the corpus of experiences can have an attribute *lexeme* (they are then the central node for a frame) and their extensions with the lexeme attribute as given are the concepts expressible by the value of lexeme. It is possible that all talk of lexical concepts is no more than a convenient abstraction for a particular kind of generic knowledge.

---

[6] The importance of diagnosticity for natural concepts is stressed by Annika Schuster in ongoing work.

References

Baroni, M., Zamparelli, R.: Nouns are vectors, adjectives are matrices: Representing adjective-noun constructions in semantic space. In: Proceedings of the 2010 Conference on Empirical Methods in Natural Language Processing, EMNLP 2010, pp. 1183–1193. Association for Computational Linguistics, Stroudsburg (2010)

Barsalou, L.: Frames, concepts and conceptual fields. In: Lehrer, A., Kittay, E.F. (eds.) Frames, Fields and Contrasts: New Essays in Semantic and Lexical Organisation, pp. 21–74. Lawrence Erlbaum, Hillsdale (1992)

Barsalou, L.: Simulation, situated conceptualization, and prediction. Philos. Trans. R. Soc. Lond. Ser. B Biol. Sci. **364**(1521), 1281–9 (2009)

Binder, J.R., et al.: Toward a brain-based componential semantic representation. Cogn. Neuropsychol. **33**(3–4), 130–174 (2016). PMID: 27310469

Boleda, G., Baroni, M., Pham, T.N., McNally, L.: Intensionality was only alleged: on adjective-noun composition in distributional semantics. In: Proceedings of the 10th International Conference on Computational Semantics (IWCS 2013) - Long Papers, Potsdam, Germany, pp. 35–46. Association for Computational Linguistics (2013)

Gärdenfors, P.: Knowledge in Flux: Modeling the Dynamics of Epistemic States. MIT Press, Cambridge (1988)

Kamp, H., Reyle, U.: From Discourse to Logic: Introduction to Modeltheoretic Semantics of Natural Language, Formal Logic and Discourse Representation Theory. Kluwer Academic Publishers, Dordrecht (1993)

Kasper, W.: Presuppositions, composition, and simple subjunctives. J. Semant. **9**(4), 307–331 (1992)

Löbner, S.: Evidence for frames from human language. In: Gamerschlag, T., Gerland, D., Osswald, R., Petersen, W. (eds.) Frames and Concept Types. Studies in Linguistics and Philosophy, vol. 94, pp. 23–67. Springer, Cham (2014). https://doi.org/10. 1007/978-3-319-01541-5_2

Löbner, S.: Functional concepts and frames. In: Gamerschlag, T., Gerland, D., Osswald, R., Petersen, W. (eds.) Meaning, Frames, and Conceptual Representation. Studies in Language and Cognition, vol. 2, pp. 13–42. Düsseldorf University Press, Düsseldorf (2015)

Moxey, L.M., Sanford, A.J.: Quantifiers and focus. J. Semant. **5**(3), 189–206 (1986)

Jones, N., Jon Willits, M., Dennis, S.: Models of semantic memory. In: Busemeyer, J.R., Townsend, J.T. (eds.) Oxford Handbook of Mathematical and Computational Psychology, pp. 232–254. Oxford University Press, Oxford (2015)

Petersen, W.: Representation of concepts as frames. In: Skilters. J., et al. (eds.) Complex Cognition and Qualitative Science. The Baltic International Handbook of Cognition, Logic and Communication, vol. 2, pp. 151–170. University of Latvia (2007)

Saeboe, K.J.: Anaphoric presuppositions and zero anaphora. Linguist. Philos. **19**(2), 187–209 (1996)

Zeevat, H.: Language Production and Interpretation. Linguistics Meets Cognition. Jacob Brill, Leiden (2014)

Zeevat, H.: Local satisfaction explained away. In: Moroney, M., Little, C.-R., Collard, J., Burgdorf, D. (eds.) Proceedings of the 26th Semantics and Linguistic Theory Conference, vol. 26 (2016)

Zeevat, H., Grimm, S., Hogeweg, L., Lestrade, S., Smith, E.A.: Representing the lexicon: Identifying meaning in use via overspecification. In: Balogh, K., Petersen, W. (eds.) Proceedings of Workshop Bridging Formal and Conceptual Semantics (BRIDGE 2014). Düsseldorf University Press (2015)

Explaining Meaning: The Interplay of Syntax, Semantics, and Pragmatics

Yulia Zinova[(⊠)]

Heinrich Heine University of Düsseldorf, Düsseldorf, Germany
zinova@phil.hhu.de

Abstract. Russian verbal prefixes are traditionally analysed as polysemous and their contribution often appears to be unpredictable. Exploring in detail two of them, *po-* and *na-*, I show how most of the variation in their interpretations can be predicted by combining the Frame Semantic analysis offered in Zinova (2017) with pragmatic computation within Rational Speech Act theory. The set of possible competing expressions is determined on the basis of morphological information and context-dependent syntactic restrictions on the verb. Such a system where morphology, syntax, semantics, and pragmatics work together allows to explain (most of) the apparent polysemy and non-transparency of verbal prefixes.

Keywords: Russian · Frame semantics · Lexical semantics · Pragmatic competition · RSA · Verbal prefixation

1 Introduction

Consider the following sentence:

(1) Pomjal   eë, ponjuxal   xvojnyj zapax, pogrel   v rukax.
    po.crush.PST.SG.M she po.smell.PST.SG.M pine   smell po.warm.PST.SG.M in hand

   He squeezed it in his palm for some time, smelled it's pine smell, warmed it in his hands.

Mixail Gigolašvili. Čertovo koleso (2007)

In this sentence there are three verbs with the prefix *po-*:

- *pomjal* 'squeezed for some time' referring to a certain time span of the action denoted by the base verb (squeezing);
- *ponjuxal* 'smelled' referring to an action denoted by the base verb without further notable meaning component apart from the perfective aspect;
- *pogrel* 'warmed up' referring to a certain degree of temperature change of the object associated with the action denoted by the base verb (warming).

© Springer-Verlag GmbH Germany, part of Springer Nature 2019
A. Silva et al. (Eds.): TbiLLC 2018, LNCS 11456, pp. 330–352, 2019.
https://doi.org/10.1007/978-3-662-59565-7_17

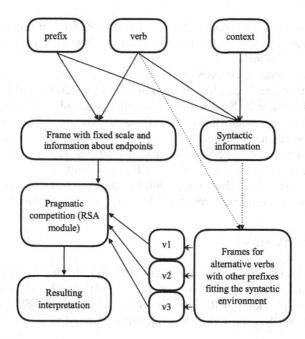

Fig. 1. General architecture of the proposed analysis

Apart from various scales that come into play when the same prefix is attached to different verbs, Russian speakers typically also infer that the time spent for squeezing was not too long and the temperature reached after warming the object in the hands was not too high (can be higher if the warming activity is continued).

In this paper I address the question of how and why such inferences emerge. I show that the existing analyses do not account for the whole range of data and argue that the perceived 'low degree' component is not a part of the semantic representation of the prefix, but a result of a pragmatic competition.

I discuss in detail two prefixes (*po-* and *na-*) and propose an integral account that allows us to derive both the semantics of the prefixed verb and its interpretation after pragmatic computation. In order to achieve this, I provide justification of particular (highly underspecified) semantic analysis based on the relevant corpus data (see Sects. 2 and 3) that is then implemented using Frame Semantics (Kallmeyer and Osswald 2013, see Sect. 4). At this step, a part of the interpretation variation of the prefix gets explained away by the interaction between the semantic representation of the verb and that of the prefix. The relevant attributes of the obtained representations then enter the pragmatic computation implemented within Rational Speech Act Theory (Frank and Goodman 2012, see Sect. 5).

The architecture of the proposed system is presented on Fig. 1, whereby the dotted lines stand for the parts that are not yet implemented. In this paper,

I provide motivation and results of the computation for the upper part of the model[1] and show a possible implementation for the lower left part of the scheme (pragmatic computation).

The analysis shown here relies on two ideas. First is the scalar nature of prefix contributions that has been already proposed by Filip (2000) and Kagan (2015) and is pushed further in Zinova (2017), where the number of usages obtained through a single representation of the prefix is extended. The second idea is rooted in game theory and optimality theory: whenever the semantics of two or more lexical items (prefixed verbs formed from the same stem in our case) overlaps, their usage gets restricted in such a way that the uncertainty of the listener is minimized. This line of reasoning follows the recent research on vague language usage, see, e.g., van Deemte (2009) and references therein.

2   The Prefix *po-*

In this section I motivate the decision of not including the 'low degree' component into the semantics of the prefix *po-*. Let us start with the list of possible usages of the prefix *po-* provided in Russian grammar by Švedova (1982, pp. 364–365). It consists of five situation types that the verbs prefixed with *po-* can refer to:

1. to do the action that is denoted by the derivational base with low intensity, sometimes also gradually: *poprivyknut'* 'to get somehow used', *poiznosit'sja* 'to get somewhat worn out', *pomaslit'* 'to put some butter on something' (productive, especially in spoken language, often called *attenuative*);
2. to do the action that is denoted by the derivational base repeatedly, with many or all of the objects or by many or all of the subjects: *povyvezti* 'to take out many/all of something' (productive, especially in spoken language, *distributive*);
3. to do the action that is denoted by the derivational base for some (often short) time: *pobesedovat'* 'to spend some time talking' (productive, also called *delimitative*);
4. to start the action that is denoted by the derivational base: *pobežat'* 'to start running' (productive, *inceptive*);
5. to complete the action denoted by the derivational base: *poblagodarit'* 'to thank' (productive, *resultative*).

As is evident from the list above as well as from the ex. (1), the variation in the semantic contribution of the prefix *po-* is huge. In studies that aim to unify usages one would find the first and the third usages put together by extracting the 'below the standard' semantic component (Filip 2000 and Kagan 2015). As for the rest, there is no obvious connection between distributive, inceptive, and

---

[1] Semantic representations are paired with syntactic analysis within Tree Adjoining Grammars (Joshi 1985, 1987; Joshi and Schabes 1997) as formalized in Kallmeyer and Osswald (2013). The syntactic part of the modelling is implemented for the prefixes discussed here, but is not shown in this paper. For more details in this respect, see Zinova (2017).

resultative usages, but as it shown in Zinova (2017), all of them can be unified. The semantics of the prefixed verb in this case emerges from the interaction between the prefix and the scale that is determined by the contribution of the verbal base or the context.[2]

For the purpose of the current paper, the cases of interest with respect to the prefix *po-* are usages that are classified as delimitative or attenuative (the first and the third classes of the *po*-prefixed verbs in the list above)[3] Such usages of *po-* are traditionally associated with some characteristic of an event being lower than the expected value: for example, an event lasting for a short period of time, a small quantity of the theme consumed, etc. According to Filip (2000, pp. 47–48), who compares it with accumulative *na-*, "[t]he prefix *po-* contributes to the verb the opposite meaning of a small quantity or a low degree relative to some expectation value, which is comparable to vague quantifiers like *a little, a few* and vague measure expressions like *a (relatively) small quantity/piece/extent of.*"

Examples of the delimitative usage of the prefix *po-* include such sentences as (2), whereby the sentence (2-a) is taken to mean that the walk around the city was relatively short, and (2-b) – that the quantity of the apples eaten was relatively small.

(2) a. Ivan poguljal   po   gorodu.
        Ivan po.walk.PST.SG.M around town
        'Ivan took a (short) walk around the town.'

                                        = example (9c) in Filip 2000

    b. Ivan poel   jablok.
        Ivan po.eat.PST.SG.M apple.PL.GEN
        'Ivan ate some (not many) apples.'

                                        = example (3) in Kagan 2015 (p. 46)

---

[2] It is worth noting that researchers that adopt the idea of distinguishing two classes of prefix usages in Russian, lexical and superlexical, never classify the two latter usages of *po-* in the list above as superlexical (Ramchand 2004; Svenonius 2004a; Romanova 2006; Tatevosov 2007). According to Svenonius (2004b, p. 229), superlexical prefixes are distinguished in that they: (i) do not allow the formation of secondary imperfectives; (ii) can occasionally stack outside lexical prefixes, never inside; (iii) select for imperfective stems; (iv) attach to the non-directed form of a motion verb; (v) have systematic, temporal or quantizing meanings, rather than spatial or resultative ones.

Although predominant in the literature, this distinction is problematic, as discussed rather briefly in Kagan (2015) and extensively in Zinova (2017). One of the main reasons for the criticism is that there is no pair of criteria that would apply to the same set of prefixes, which leads to different classification in every individual paper on the topic. This strongly indicates that if there is a distinction, it is not categorical.

[3] Other usages, such as distributive or inceptive, emerge from underspecified representations as soon as the relevant scale type is selected, so no further refinement is needed in these cases as long as the proposed computation is performed. In Zinova (2017) this is done within the Frame Semantics, so we will be using the same framework here in order to move towards an integrated system.

Indeed, the standard (contextually specified or implied) degree is normally not reached if the event is referred to by a *po*-prefixed verb. For example, Braginsky (2008, p. 183) applies the following test in order to show the difference between the verbs prefixed with the resultative prefix *za*- and the verbs prefixed with *po*-: one has to continue the given sentence with 'but it is hard to call it X' where X is the result state corresponding to the derivational base. Such a continuation is only possible if there is no implication that the maximum/contextually specified as sufficient degree is reached on the relevant scale by the end of the event. Braginsky (2008, p. 183) provides two examples repeated under (3) and (4) here. They show that when sentences are headed by the *po*-prefixed verb, the result state does not have to be reached when the event terminates, which is not the case with the *za*-prefixed resultative verbs.

(3) a. Varen'je pogustelo^{PF}, no ego ešče trudno nazvat' gustym.

     Jam   PO-thickened but it yet hard to call thick

     'The jam thickened a bit, but it is hard to define it as thick yet.'

   b. *Varen'je zagustelo^{PF}, no ego ešče trudno nazvat' gustym.

     Jam   ZA-thickened but it yet hard to call thick

                               = example (49) in Braginsky 2008 (p. 183)

(4) a. Gvozd' poržavel^{PF},   no ego ešče trudno nazvat' ržavym.

     Nail   PO-became rusty but it yet hard to call rusty

     'The nail became a bit rusty, but it is hard to define it as rusty yet.'

   b. *Gvozd' zaržavel^{PF},   no ego ešče trudno nazvat' ržavym.

     Nail   ZA-became rusty but it yet hard to call rusty

                               = example (50) in Braginsky 2008 (p. 183)

The observations about the low degree on some scale, associated with the discussed usage of the prefix *po*-, are commonly accepted and seem to be well established. However, if one assumes that this is always the case, sentences that contain *po*-prefixed verbs in combination with a degree adverbial pose a problem. Consider the following examples from the corpora, where the verbs *pobrodit'* 'to wander for some time' (semantically close to *poguljat'* 'to take a short walk' from (2-a)) and *poest'* 'to eat' (same verb as in (2-b)) are modified with *mnogo* 'a lot' and *očen' plotno* 'very tightly' (in context of eating meaning 'to eat until becoming very full'), respectively.

In (5-a) the verb *pobrodil* 'wandered', that presumably contains the delimitative prefix *po*-, refers to a lot of wandering, and in (5-b) the verb *poel* 'ate' refers to a situation of eating a lot. If the semantics of the delimitative prefix *po*- would include the semantic component 'the degree is lower than the expected value', such sentences would be unacceptable or would trigger an additional pragmatic inference, i.e., be interpreted sarcastically. This is not the case: both (5-a) and (5-b) are unmarked. What is also important is that the same verbs can be also used in combination with the adverbials denoting small quantity (such as *nemnogo* 'a bit'), as in (6).

One possible solution is to say that we are dealing with two different usages of the prefix *po-*: a delimitative usage in the examples (2-a) and (2-b) and some other usage in the examples (5-a) and (5-b) (probably corresponding to the last, resultative, usage of *po-* in the list provided above). This solution does not seem right to me: the verb *poel* 'ate' in (2-b) and the verb *poel* 'ate' in (5-b) seem to have the same meaning. In a dictionary, there is just one interpretation for the verb *poest'* 'to eat' that reflects the usages of the verbs *poel* 'ate' in both (2-b) and (5-b). This can be either 'to eat not much' (Ušakov 1940) or 'to eat' (Efremova 2000) meaning. Another evidence in favor of the single meaning is that the verbal phrase in the example (2-b) can also be modified with an adverbial denoting sufficient quantity, as evidenced by the example (7), that is taken from the corpora.

(5) a. Znat', mnogo po svetu pobrodil,   vsjakogo raznogo uspel
         know, a lot on world po.wander.PST.SG.M, all   different have time
         naslušat'sja- nasmotret'sja.
         na.hear.INF.refl, na.look.INF.refl

         'You know, he wandered a lot around the world, he had time to see and hear all kinds of different things.'
                                    Marija Semenova. *Volkodav: Znamenie puti* (2003)

     b. Kogda do   stolicy ostavalos'   tridcat' kilometrov, našël stolovuju i
         when before capital stay.PST.SG.N.refl thirty kilometers, found canteen and
         očen' plotno poel,   poskol'ku do   sledujuščego priëma pišči
         very tight po.eat.PST.SG.M, because before next   reception food
         neizvestno skol'ko   vremeni.
         unknown   how much time

         'When I was about 30 km away from the capital, I found a canteen and had a very square meal, as I didn't know how long it would take until my next chance to eat something.'
                                            Anatolij Azol'skij. *Lopušok* (1998)

(6) a. On pobrodit   nemnogo i sejčas že   ujdet.
         he po.wander.PRES.SG.3 a bit   and now same u.go.PRES.SG.3
         'He will wonder around a little bit and immediately leave.'
                                    Anna Berseneva. *Vozrast tret'ej ljubvi* (2005)

     b. My kupim   ptičkam kormu i sami   poedim   nemnogo.
         we buy.PRES.PL.1 birds   food and ourselves po.eat.PRES.PL.1 a bit
         'We will buy food for the birds and eat something small ourselves.'
                                    V. P. Kataev. *Bezdel'nik Èduard* (1920)

(7) Togda on poel   jablok   vdovol'.
     then he po.eat.PST.SG.M apple.PL.GEN enough
     'Then he ate apples to his heart's content.'
                                    Aleksandr Iličevskij. *Matiss* (2007)

This naturally leads to the idea of defining the semantics of the delimitative usage[4] of *po-* in such a way that the verb prefixed with it can remain neutral with respect to the state achieved by event termination. In this case the meaning component 'quantity/degree is lower than some expectation value' does not belong to the semantics of the prefix and has to be derived separately. I claim that it arises due to pragmatic competition of the prefixed verb with alternative expressions.[5] I show how exactly this functions in Sect. 5, after discussing the general approach to the prefix *na-* in the next section and providing frame representations for both prefixes in Sect. 4.

3 The Prefix *na-*

In this section we consider the prefix *na-* and the 'above the standard' inference often associated with *na*-prefixed verbs. As well as in the case of the prefix *po-*. I argue that this inference is not s part of the semantic contribution of the prefix. According to the grammar by Švedova (1982, p. 360), the following six categories of derived verbs emerge when attaching this prefix to a verb:

1. to direct the action denoted by the derivational base on some surface, to place on or come across something: *nakleit'* 'to paste' (productive);
2. to accumulate something by performing the action denoted by the derivational base: *navarit'* 'to cook a lot of' (productive, *cumulative*);
3. to perform the action denoted by the derivational base intensively: *nagladit'* 'to iron thoroughly' (productive, colloquial);
4. to perform the action denoted by the derivational base weakly, lightly, on the go: *naigrat'* 'to strum' (non-productive, colloquial);
5. to learn something or acquire some skill by performing the action denoted by the derivational base: *natrenirovat'* 'to train until some level', *nabegat'* 'to train to run' (productive, used in professional slang);
6. to perform the action denoted by the derivational base until the result: *nagret'* 'to heat up', *namočit'* 'to make wet', *napoit'* 'to give something to drink' (productive, *resultative*).

In case of the prefix *na-*, the range of prefix contributions varies from low intensity (fourth, although non productive usage) to completion (fifth and sixth usages) and high intensity (second and third usages). Here as well there is a common division between the second usage that is often unified with the third one (Filip 2000 and Kagan 2015) and considered superlexical, while the other usages are classified as lexical ones.

The cumulative usage we are going to start with in this section appears under (2) in the above list by Švedova (1982) and is usually unified with the third usage

---

[4] I will use the term *delimitative* to refer to the discussed usage in order to differentiate it from the distributive (second in the list) and inchoative (fourth in the list) usages, but I will not imply attenuativity.

[5] Such an analysis naturally leads to the unification of the resultative (last in the list by Švedova 1982) usage of the prefix with the delimitative usage.

(Filip 2000; Kagan 2015). It shares some properties with the delimitative usage of the prefix *po-*: both prefixes are claimed to denote a vague measure function (Filip 2000; Součková 2004). Součková (2004) formulates two differences between these prefixes: the direction of the relation and the dimensions of the scales they select for.

There are two main usages of the cumulative prefix *na-* in Russian: transitive and reflexive. Transitive usage is exemplified by (8-a), where the prefix measures the quantity of the undergoer (potatoes) that has been peeled. Reflexive usage is exemplified by (8-b); here, the prefix *na-* measures the degree to which the subject (Katja) is full after eating potatoes.[6]

(8)   a.   Katja načistila   kartoški.
          Katja na.clean.PST.SG.F potato.GEN
          'Katja peeled a lot of potatoes.'

      b.   Katja naelas'   kartoški.
          Katja na.eat.PST.SG.F.refl potato.GEN
          'Katja became full by eating potatoes.'

Another usage of *na-*, the sixth in the list by Švedova (1982), is usually analysed as related to the cumulative usage exemplified by (8-a). Consider the verb *namočil* 'made wet', as in (9). It refers to an event of making something wet that is non-cumulative in every respect: a single actor made a single object wet with a single move. Note also the difference with respect to the source of the scale: in (8-a) the event is measured along the quantity scale provided by the direct object, in (8-b) the scale comes from the subject due to the postfix contribution, and in case of (9) the relevant wetness scale is provided by the verb.

(9)   Petja namočil   kistočku   v stakane   vody.
     Petja na.wet.PST.SG.M brush.SG.ACC in glass.SG.PRP water.SG.GEN
     'Petja made the brush wet by putting it into a glass with water.'

To account for this, one can either accept the polysemy among the productive usages of the prefix *na-* or try to unify them. An observation that can be made if one considers the list of *na-*prefixed verbs that are associated with cumulative semantics is that for verbs in this list there is always another way to express the completion of the event denoted by the derivational base. For example, instead of (8-a) the speaker could have uttered (10) which is neutral with respect to the quantity of the potatoes peeled or (11) that means that Katja peeled all of the potatoes.

---

[6] A full analysis of this case requires providing a formal representation of the postfix *-sja* but for this paper I will limit myself to the claim that the contribution of the prefix remains the same and the role of the postfix is to select one of the subject-related scales as a measure dimension of the event. For more information and formal analyses of the reflexive usage of the prefix *na-*, see Kagan and Pereltsvaig (2011a), Kagan and Pereltsvaig (2011b), Součková (2004), Filip (2000), and Filip (2005).

(10) Katja počistila   kartoški.
      Katja po.clean.PST.SG.F potato.GEN
      'Katja peeled some potatoes.'

(11) Katja počistila   kartošku.
      Katja po.clean.PST.SG.F potato.ACC
      'Katja peeled the potatoes.'

(12) Liza navarila   supa.
      Liza na.cook.PST.SG.F soup.GEN
      'Liza cooked a lot of soup.'

(13) Liza svarila   sup.
      Liza s.cook.PST.SG.F soup.ACC
      'Liza cooked soup.'

The same happens in the pair of sentences (12) and (13). The *na*-prefixed verb refers to an event of cooking that results in producing some quantity of soup that exceeds the standard amount. The *s*-prefixed verb is neutral with respect to the quantity of soup produced while it also refers to a completed cooking event, similarly to the *na*-prefixed verb. (These sentences differ with respect to case marking of the object, but I consider it to be the consequence and not the source of the difference: *na*- requires an open scale as a measure dimension of the event and this is realized in genitive case marking if this scale is the amount of the object.)

At the same time, similar considerations as with the prefix *po*- also apply here: there are speakers that produce utterances where a *na*-prefixed verb is modified by an adverbial denoting a low degree, such as *nemnogo* 'a few' in (14). Note that this is, however, not as good as (5-a) or (5-b): it is not accepted by all the speakers and such utterance patterns are only found on the web but not in the corpora.

(14) Nemnogo pribrala   doma,   prigotovila   syrnyj sup s   krabovym
      a bit   tidy.PST.SG.F at home prepare.PST.SG.F cheese soup with crabb
      mjasom, napekla   nemnogo ovsjanyx   blinčikov.
      meat   na.bake.PST.SG.F a few   oatmeal.PL.GEN pancake.PL.GEN
      'I tidied up the house a bit, cooked a cheese soup with crab meat, baked some oatmeal pancakes.'

                                                              www.diary.ru

On the basis of these observations, I propose that the core contribution of the prefix *na*- is:

1. selecting a contextually specified open scale, whereby the preference is given to a noun/context-related scale;
2. identifying the final stage of the event as being associated with some degree on the measure dimension that is at or above the (contextually specified) threshold.

This approach allows us to unify the second, the third, the fifth, and the sixth usages in the list by Švedova (1982), so that the only other productive usage not covered here is one associated with the spatial[7] scale.

The description provided above is very close to that of Kagan (2015), who offers the semantic representation of the prefix *na*- that is shown in (15). Kagan

---

[7] This should be possible as well, but the detail are not worked out.

(2015, p. 55) proposes that *"na-* looks for a verbal predicate that takes a degree, an individual and an event argument and imposes the '$\geqslant$' relation between the degree argument and the contextually provided expectation value d_c. As a result, the degree of change is entailed to be no lower than the standard."

(15) $[\![na-]\!] = \lambda P \lambda d \lambda x \lambda e.[P(d)(x)(e) \wedge d \geqslant d_c]$
where d = degree of change (Kennedy and Levin, 2002)

$$= (17) \text{ in Kagan 2015, p. 55}$$

The semantic representation proposed by Kagan (2015) also allows us to capture the semantics of both the cumulative and the resultative usages of the prefix *na-*. What is not addressed by the author is when exactly the cumulative/exceeding the threshold interpretation is obtained. In other words, the question that remains is what is the difference between such verbs as *čistit'* 'to peel' and *gret'* 'to warm' that leads to the verb *načistit'* 'to peel a lot of' acquiring a cumulative interpretation while the verb *nagret'* 'to warm up' is interpreted as referring to any event of heating that terminates when the threshold temperature is reached. I show how this difference can be derived in Sect. 5 after introducing frame representations for both *po-* and *na-* in the next section.

4   Frame Semantic Representations

The idea of using frame representations in linguistic semantics and cognitive psychology has been put forward by Fillmore (1982) and Barsalou (1992), among others. The main ideas that motivate the use of frames as a general semantic and conceptual representation format can be summarized as follows (cf. Löbner 2014):

- conceptual-semantic entities can be described by types and attributes;
- attributes are functional relations, i.e., each attribute assigns a unique value to its carrier;
- attribute values can be also characterized by types and attributes (recursion);
- attribute values may be connected by additional relational constraints (Barsalou 1992) such as spatial configurations or ordering relations.

A number of recent studies offer further formalization of the frame theory (Petersen 2007; Petersen and Osswald 2009; Kallmeyer and Osswald 2012 2013; Kallmeyer et al. 2015; Löbner 2014, among others). This paper is based on the formalization provided in Kallmeyer and Osswald (2013). Frames in the sense of Kallmeyer and Osswald (2013) are finite relational structures in which attributes correspond to functional relations. The members of the underlying set are referred to as the *nodes* of the frame.

An important restriction of the formalism is that any frame must have a *functional backbone*. This means that every node has to be accessible via attributes from at least one of the *base nodes*: nodes that carry *base labels*. Importantly, feature structures may have multiple base nodes. In such a case often some nodes that are accessible from different base nodes are connected by a relation. Base

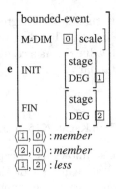
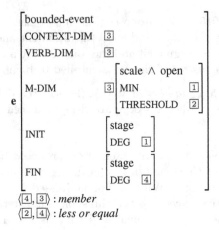

Fig. 2. Frame representations of the prefixes *po-* (left) and *na-* (right) following Zinova (2017)

labels serve as unique identifiers, that is, a given base label cannot be assigned to more than one node. Due to the functional backbone requirement, every node of the frame can be addressed by a base label plus a (possibly empty) finite sequence of attributes.

The frame on the left side of Fig. 2 represents the proposed analysis of the semantic contribution of the prefix *po-*. It consists of the following components:

1. the type of the base node **e** is *bounded event*;
2. the measure dimension (M-DIM) of the event is of type *scale*;
3. there are an initial (INIT) and a final (FIN) stages of the event that are associated with some degrees on the measure dimension scale (first two relations below the attribute value matrix);
4. the degree associated with the initial stage is less than the degree of the final stage of the event (third relation).

In other words, the attachment of the prefix leads to the introduction of an initial and a final stages of the event, but they remain underspecified. This semantics for the prefix *po-* allows to derive all the usages listed in the grammar, using additional constraints associated with particular scale types[8].

The right side of Fig. 2 shows the representation of the prefix *na-*. In addition to what is required by the prefix *po-*, *na-* has the following contribution:

---

[8] The most tricky usage in this respect is inceptive. It is derived from an underspecified representation by using a constraint that the movement can only start at the point that is the minimum of the path scale (in case *path* is the measure dimension of the event). It is a formulation of an idea that movement can only start from the location where the undergoer of the movement is located at the start of the translocation event, expressing the asymmetry between the start and end points of the path description. More details are provided in Zinova (2017).

1. contextually determined dimension (CONTEXT-DIM) is the measure dimension (M-DIM);
2. the dimension of the verb (VERB-DIM) has to be the same as the contextually determined dimension[9]
3. the measure dimension is of type *open scale*;
4. degree of the initial stage is the minimum of the measure dimension;[10]
5. degree of the final stage is at or above the threshold value on the relevant scale.

Such semantic representation allows to derive four out of six usages of the prefix *na-*: all of the productive verb classes except the first one (that is related to a path scale).

Apart from the above-mentioned benefits of frame representations it is important to note that the goal of this study is to contribute to an integrated system that allows to predict how complex verb are interpreted. Of course the contributions of the prefixes *po-* and *na-* can be provided in any other semantic framework, but crucially frame representations paired with a TAG allow us to obtain representations of prefixed verbs on the basis of underspecified prefix representations. And although the usages of *po-* and *na-* that are of most interest for this paper can be derived straightforwardly, it is not so in other cases. And as for the pragmatic competition the system requires the relevant semantic facts not only for the verb of interest, but also for alternative verbs, semantic representations for them must be at hand when needed.

5 Pragmatic Competition

5.1 RSA: A Very Brief Introduction

To model pragmatic competition, we are going to use the Rational Speech Act (RSA, Frank and Goodman 2012, Goodman and Frank 2016, Goodman and Tenenbaum 2016, Scontras and Tessler 2017) framework, that models communication as recursive reasoning. The RSA model is an implementation of a social cognition approach to the understanding of utterances. It is based on the Gricean idea (Grice 1975) that speakers are cooperative and aim to produce utterances that provide a balance between saving effort and being informative. Pragmatic listeners then interpret the utterance by inferring what a speaker must have meant, given the expression they uttered.

---

[9] This instantiates the idea that there are two possible sources of relevant scales (verb and context, often in form of the direct object) and various prefixes have different preferences. At the same time, if a verb lexicalizes a scale, as in the case of *gret'* 'to warm up' (temperature scale), it cannot be overwritten and the contextual scale must be of the same type. For more details, see Zinova (2017).

[10] Note that the measure dimension only stores a relevant segment of the scale, not the whole segment. For instance, in case of the temperature scale the minimum of the measure dimension is the temperature of the object before the event start.

In this paper, I limit myself to a basic version of RSA introduced in Frank and Goodman (2012) for modelling referential choice. The underlying idea is that the (pragmatic) listener L_1 interprets an utterance by reasoning about a speaker. L_1 uses Bayesian inference in order to find out which state of the world is likely given the speaker's choice of utterance. The pragmatic speaker S_1, in turn, performs this choice by reasoning about how a literal listener L_0 is most likely to interpret available utterances.

Let us briefly discuss the functions that allow the computation sketched above. In the first step, the literal listener reasons about the probability of various states given the utterance. As this is a naive (non-pragmatic) listener, the assumption is that interpretations in this case correspond to the provided semantic meaning of the utterances. This is formalized as shown in 16, where u refers to one of the utterances from the given set, s is a state of the world from the given set, $[\![u]\!] : S \rightarrow \{0,1\}$ is a denotation function that maps states to Boolean values, and $P(s)$ is an a priori belief about how likely the speaker is to refer to a given state.

(16) $P_{L_0}(s|u) \propto [\![u]\!](s) \cdot P(s)$

On the next level the pragmatic speaker chooses an utterance based on its utility, approximating the rational choice with a *softmax* function with a parameter α that controls speaker's level of rationality. The computation is done according to the formula (17), where $C(u)$ stands for utterance cost.

(17) $P_{S_1}(u|s) \propto \exp(\alpha(\log L_0(s|u) - C(u)))$

The last level that is used in a basic model is the pragmatic listener that reasons about the process of selecting the utterance by the pragmatic speaker, bearing in mind prior probabilities of different states. The probability of state s given the utterance u is then computed according to the formula (18).

(18) $P_{L_1}(s|u) \propto P_{S_1}(u|s) \cdot P(s)$

5.2   Russian Data from the Pragmatic Perspective

Recall the examples (2-b) and (5-b). I claim that the crucial difference that leads to varying interpretations of the verb *poest'* 'to eat' is that in the first case there is a pragmatic competition between the verb *poest'* 'to eat' and alternative expressions whereas in the second case such competition is blocked for syntactic reasons.

Consider the sentence (2-b). Crucially, there are alternative ways of referring to an event of eating apples, such as (19) and (20). If the speaker wants to describe an event of eating all of the apples, they can utter (19). The most appropriate description of the situation of eating the apples until becoming full is (20). Due to such competition when the sentence (2-b) that literally means that some apples were eaten is uttered, it gets enriched with an additional inference that the quantity of the apples eaten is lower than the number of apples available or the amount of apples necessary for the actor to become full. At the same time, in the context of (5-b) none of the verbs *s"est'* 'to eat all of' and *naest'sja* 'to eat until becoming full' can substitute *poest'* 'to eat', as both alternatives require an object (either in accusative or in genitive case), but the left context (considered linearly up to the point where the verb appears) only allows for an intransitive verb.

(19) Ivan s"el   jabloki.   (20) On naelsja   jablok.
     Ivan s.eat.PST.SG.M apple.PL.ACC   he na.eat.PST.SG.M.refl apple.PL.GEN
     'Ivan ate the apples.'   'He became full by eating apples.'

As a further step, I propose to compute these inferences using the Rational Speech Act model, implementing the described intuitions.

5.3 Implementation Using WebPPL

The implementation is done using a probabilistic programming language (WebPPL[11]) with a basic three-layered RSA model outlined above. This model includes (i) a literal listener that interprets the utterance according to the provided literal semantics; (ii) a pragmatic speaker that selects an utterance from the available options based on the probability of the literal listener inferring the desired state of the world; (iii) a pragmatic speaker that interprets the utterance by reasoning about the pragmatic speaker. Six things need to be provided as an input to the model:

1. the world model (set of states s);
2. probability distribution over possible world states ($P(s)$);
3. set of alternative utterances (u);
4. their cost ($C(u)$);
5. a meaning function from utterances to states ($[\![u]\!](s)$);
6. a value of the optimality parameter α.

Let us go through the list and set a model that computes the interpretation of the verb *poest'* 'to eat' in a context where there are two alternative verbs – *s"est'* 'to eat all of' and *naest'sja* 'to eat until becoming full'.[12] First is the world model that in our case it contains four states based on two scales that are relevant in this case (quantity of the theme consumed and saturation of the actor after the consumption):[13]

---

[11] https://WebPPL.org/.

[12] RSA code is provided in Appendix A.

[13] More research has to be done to find out the best world model to represent the relevant states. Alternative solutions include distinguishing between 'not all of the theme is consumed' and 'the quantity of the theme consumed is not relevant' case as well as adding 'excess' state of the world. I stay with the proposed version as a first option that has to be further adjusted and tested.

1. the actor is not full and all of the theme is consumed;
2. the actor is full and not all of the theme is consumed;
3. the actor is full and all of the theme is consumed;
4. the actor is not full and not all of the theme is consumed.

Next is the probability distribution over different states. In the implementation provided here I have assumed a flat prior over four world states which means that they are supposed to be equally likely. In order to later assess the predictions of the model against speakers' intuitions the prior has to be either estimated from the data or the experimental design should allow for an explicit prior (e.g., providing provabilities of certain kinds of events).

I assume that the set of alternative utterances in case of a context-free setup is the set of all prefixed verbs formed from the same stem.[14] Such a set, however, can be very large, so an additional assumption I adopt here is to limit the set of alternatives to the verbs that have the same or smaller degree of morphological complexity (in terms of the number of derivational morphemes) with respect to the target verb. If more complex verbs are to be added, they would probably be associated with higher cost. For the purposes of this paper, I assume the same cost for the verbs of the same complexity, which is encoded as a flat probability distribution over utterances. A better estimation can be done on the basis of corpora data and is planned as one of the next steps of the project.

The next piece of input is the meaning function that maps utterances to world states. For *s"est'* 'to eat all of' I assume that the world state is 'all of the theme is consumed'. For *naest'sja* 'to eat until becoming full' it is 'the actor is full' (or overfull if this would be included in the model[15]). The last verb, *poest'* 'to eat', can refer to all the states of the world. The remaining parameter that has to be set is alpha, the optimality parameter. In the current implementation, the value of alpha is 1[16].

The output of computing the pragmatic listener's probability distribution for inferring any of the four states after the speaker utters the verb *poest'* 'to eat' in a context where it can be substituted with either *s"est'* 'to eat all of' or *naest'sja* 'to eat until becoming full' is shown on the left side of Fig. 3. As desired, the most likely interpretation is that of the actor not being full and the theme being not consumed completely.

The graph on the right side of Fig. 3 represents the listener's probability distributions over the four possible situations when no competing utterances are available. As the *po*-prefixed verb can refer to any of the situations, this

---

[14] The question of competition between verbs that have different stems but are semantically close is left for future work.

[15] In this case, however, one would also have to include *obest'sja* 'to eat too much' that is true only if the actor is overfull, so I have omitted this option for simplicity and illustration purposes with respect to the behaviour of the prefix *po-*.

[16] This is an arbitrary selected value. By varying this parameter one can model different behaviour: more or less dependent on the rational considerations. If alpha equals zero, pragmatic listener's behaviour will not differ from that of a literal listener.

distribution corresponds to the world prior distribution (in this case a uniform one). This explains the absence of the low degree inference in (5-b).

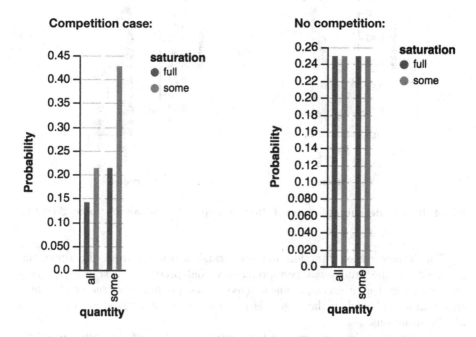

Fig. 3. RSA model output: *poest'* 'to eat'

Now let us consider one more example mentioned above: the verb *gret'* 'to heat'. Three prefixed verbs – *pogret'* 'to heat somehow', *nagret'* 'to warm up', *peregret'* 'to overheat' can be formed from it. The fact that I want to explain is why there is no inference of 'more than the contextually specified degree' in this case (in other words, why this usage is traditionally classified as a resultative and not as a cumulative one). In this case, the account I propose naturally follows from the competition between the three verbs mentioned above. Let us again go through the parameters of the model and see the resulting probabilities of various possible interpretations.

The world model in this case consists of three possible situations.[17] Due to the presence of only one special point on the relevant scale fragment (standard temperature for the theme to be considered warm), situations of being below, at, and above this standard are worth distinguishing. This is reflected in the three states of the world in the code: *degree* can be *normal*, *low*, or *excess*. I again assume a flat distribution over these situations as well as over the different utterances.

---

[17] RSA code (the part of it that differs from the code for *poest'* 'to eat') is provided in Appendix A.

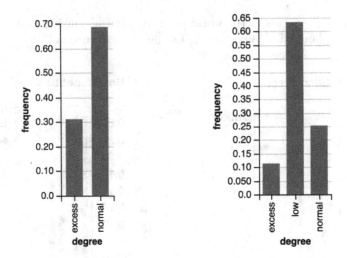

Fig. 4. RSA model output for *nagret'* 'to warm up' (left) and *pogret'* 'to heat' (right)

The interpretation function follows straightforwardly from the semantics assumed for the prefixes: a *pere*-prefixed verbal predicate is true of a situation with the degree *excess*, a *na*-prefixed verbal predicate is true of the same situation and also when the degree is *normal*, and a *po*-prefixed verb is true in all the situations.

The output of the computation for the interpretation of the verb *nagret'* 'to warm up' is shown on the left side of Fig. 4. A pragmatic listener, according to the model, will most likely (almost 0.7 probability) understand this verb, if uttered by a pragmatic speaker, as referring to a situation of warming up to the contextually specified degree. With a lower probability (around 0.3) the listener will arrive at a world where the heating stopped at a higher degree (excess interpretation). The *po*-prefixed verb *pogret'* 'to heat', while being completely neutral in its semantic representation, is interpreted as referring to a lower than the standard degree of heating with probability greater then 0.6 (right side of Fig. 4).

The last illustration I would like to provide involves the verb *žarit'* 'to fry'. As in the previous case, three prefixed verbs can be derived from it: *perežarit'* 'to fry too much', *nažarit'* 'to fry enough/a lot of', and *požarit'* 'to fry'. What is different in this case with respect to the warming up case is that in the situation of frying there are two scales available as potential measure dimensions of the event: a scale for the degree of being fried (let us call it *degree* scale) and a quantity scale of the fried object produced as a result of the frying activity. In this situation, the prefix *pere*- selects the degree scale, the prefix *na*- selects the quantity scale, and the prefix *po*- can apply to both. This is due to the different preferences of prefixes, encoded in their semantic representations: *pere*- prefers the verbal scale (degree of being fried), while *na*- is looking for a scale supplied by the context, which is the quantity of the fried food. As for the prefix *po*-, the

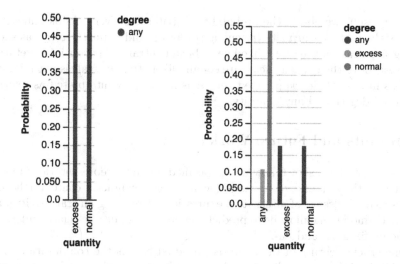

Fig. 5. RSA model output for *nažarit'* 'to fry enough/a lot of' (left) and *požarit'* 'to fry'

source of the scale is not specified (see the frame representation), which means that the event can be quantified over any available scale.

As a result, no competition arises between either *pere-* and *na*-prefixed verbs, so the excess interpretation stays available for both of them, but with respect to different scales. At the same time, the interpretation of the *po*-prefixed verb gets shifted to referring to the standard degree of being fried, so the degree scale interpretation turns out to be more likely than the quantity scale interpretation. In this case we observe how the competition among various available alternatives influences both the scalar selection and the degree assignment process[18]). Let us have a look at the implementation in this case.

The most problematic part is modelling the possible world states, as with the presence of two scales in each case only one of them is relevant. In the code provided below (Appendix C) I have decided to use three values on each scale: *normal* (event ends at the contextually specified standard), *excess* (the value on the scale exceeds the contextually specified standard), and *any* (this scale is not relevant for the event description). Further research is needed here to provide more elegant solutions. A model that takes into account the question under discussion (QUD) may serve as an appropriate extension in this case.

The rest of the input to the model stays similar to the previous case: distributions are assumed to be flat and meaning function only changes with respect to the world parameter, as *pere-* and *na-* apply to different scales.

---

[18] Note that in case of the verb *est'* 'to eat' considered above the presence of two different scales does not lead to *na-* acquiring the cumulative interpretation. This is not reflected by the code provided in the appendix, but is due to the competition with the verb *ob"est'sja* 'to eat until becoming overfull' that selects the same scale.

As a result, we obtain the probability distributions over possible interpretations of the verbs *nažarit'* 'to fry enough/a lot of' and *požarit'* 'to fry' as shown on Fig. 5. In case of the *na*-prefixed verb, both neutral and excess interpretations with respect to the quantity scale are equally likely (degree scale is not relevant), whereas in case of the *po*-prefixed verb it is most likely interpreted as referring to normal degree and any quantity.

6   Results and Future Work

In this paper I have shown how underspecified semantic representations in combination with semantic restrictions (scale types), syntactic context (the case of presence or absence of alternative expressions depending on transitivity), and pragmatic modelling allow us to predict the presence of degree-related inferences for the prefixes *po-* and *na-*.

One crucial feature of the analysis proposed here is the computation of the probabilities instead of a direct prediction of how the verb must be interpreted. On one hand, this kind of approach creates difficulties such as setting world priors, as they influence the resulting probabilities to a great extent. On the other hand, it creates the space for testing the predictions of the theory in a more precise way, taking into account not only qualitative, but also quantitative results of the computation.

In future work I plan continue implementation within the RSA framework parallel to the experimental work that would allow to verify the predictions of the proposed approach.

A   RSA Code: *poest'* 'to eat'

```
// possible states of the world //
var worldPrior = function() {
  return categorical({ps: [1, 1, 1, 1],
  vs: [{saturation: "full", quantity: "all"},
       {saturation: "some", quantity: "all"},
       {saturation: "some", quantity: "some"},
       {saturation: "full", quantity: "some"},]})}

// possible one-word utterances //
var utterances = ["sjest","naestsja","poest"]
// possible preferences of utterances//
var utterancePrior = function() {
  return categorical({ps: [1, 1, 1, 1], vs: utterances})}
// meaning function to interpret the utterances//
var meaning = function(utterance, world){
  return utterance == "sjest" ? "all"==world.quantity :
  utterance == "naestsja" ? "full"==world.saturation :
  true}
```

```
// literal listener //
var literalListener = function(utterance){
  Infer({method:"enumerate"}, function(){
    var world = worldPrior();
    var uttTruthVal = meaning(utterance, world);
    condition(uttTruthVal == true)
    return world})}

// define speaker optimality //
var alpha = 1
// pragmatic speaker //
var speaker = function(world){
  Infer({method:"enumerate"}, function(){
    var utterance = utterancePrior();
    factor(alpha * literalListener(utterance).score(world))
    return utterance})}

// pragmatic listener //
var pragmaticListener = function(utterance){
  Infer({method:"enumerate"}, function(){
    var world = worldPrior();
    observe(speaker(world), utterance)
    return world})}
```

B RSA Code: *nagret'* 'to warm up'

```
// possible states of the world //
var worldPrior = function() {
  return categorical({ps: [1, 1, 1],
        vs: [{degree: "normal"},
        {degree: "low"},
        {degree: "excess"}]})}

// possible one-word utterances
var utterances = ["peregret","pogret","nagret"]

// possible preferences of utterances
var utterancePrior = function() {
  return categorical({ps: [1, 1, 1], vs: utterances})}

// meaning function to interpret the utterances
var meaning = function(utterance, world){
  return utterance == "peregret" ? "excess"==world.degree :
  utterance == "nagret" ? "excess"==world.degree
  || "normal"==world.degree :
  true}
```

C   RSA Code: *nazharit'* 'to fry a lot of'

```
// possible states of the world
var worldPrior = function() {
  return categorical({ps: [1, 1, 1, 1], vs:
                [{degree: "normal", quantity: "any"},
                 {degree: "any", quantity: "excess"},
                 {degree: "excess", quantity: "any"},
                 {degree: "any", quantity: "normal"}]})}


// possible one-word utterances
var utterances = ["perezharit","pozharit","nazharit"]

// possible preferences of utterances
var utterancePrior = function() {
  return categorical({ps: [1, 1, 1], vs: utterances})}

// meaning function to interpret the utterances
var meaning = function(utterance, world){
  return utterance == "perezharit" ? "excess"==world.degree :
  utterance == "nazharit" ? "excess"==world.quantity
  || "normal"==world.quantity : true}
```

References

Barsalou, L.W.: Frames, concepts, and conceptual fields. In: Lehrer, A., Kittay, E.F. (eds.) Frames, Fields, and Contrasts, New Essays in Semantic and Lexical Organization, Chap. 1, pp. 21–74. Lawrence Erlbaum Associates, Hillsdale (1992)

Braginsky, P.: The semantics of the Prefix ZA - in Russian. Ph.D. thesis, Bar-Ilan University, Department of English (2008)

Efremova, T.F.: Novyj slovar russkogo jazyka. Tolkovo-slovoobrazovatel'nyj [New dictionary of Russian. Explanatory and interpretational]. Russkij jazyk, Moscow (2000)

Filip, H.: The quantization puzzle. In: Pustejovsky, J., Tenny, C. (eds.) Events as Grammatical Objects, pp. 3–60. CSLI Press, Stanford (2000)

Filip, H.: Measures and indefinites. In: Carlson, G.N., Pelletier, F.J. (eds.) References and Quantification: The Partee Effect, pp. 229–288. CSLI Publications, Stanford (2005)

Fillmore, C.J.: Frame semantics. In: Linguistic Society of Korea (ed.) Linguistics in the Morning Calm, pp. 111–137. Hanshin Publishing Co., Seoul (1982)

Frank, M.C., Goodman, N.D.: Predicting pragmatic reasoning in language games. Science **336**(6084), 998 (2012)

Goodman, N.D., Frank, M.C.: Pragmatic language interpretation as probabilistic inference. Trends Cogn. Sci. **20**, 818–829 (2016)

Goodman, N.D., Tenenbaum, J.B.: Probabilistic Models of Cognition (2016). http://probmods.org/v2. Accessed 2 Mar 2018

Grice, H.P.: Logic and conversation. In: Cole, P., Morgan, J.L. (eds.) Syntax and Semantics 3: Speech Acts, pp. 41–58. Academic Press, New York (1975)

Joshi, A.K.: Tree adjoining grammars: how much contextsensitivity is required to provide reasonable structural descriptions? In: Dowty, D., Karttunen, L., Zwicky, A. (eds.) Natural Language Parsing, pp. 206–250. Cambridge University Press, Cambridge (1985)

Joshi, A.K.: An introduction to Tree Adjoining Grammars. In: Manaster-Ramer, A. (ed.) Mathematics of Language, pp. 87–114. John Benjamins, Amsterdam (1987)

Joshi, A.K., Schabes, Y.: Tree-adjoining grammars. In: Rozenberg, G., Salomaa, A. (eds.) Handbook of Formal Languages, pp. 69–123. Springer, Heidelberg (1997). https://doi.org/10.1007/978-3-642-59126-6_2

Kagan, O.: Scalarity in the Verbal Domain: The Case of Verbal Prefixation in Russian. Cambridge University Press, Cambridge (2015)

Kagan, O., Pereltsvaig, A.: Bare NPs and semantic incorporation: objects of intensive reflexives at the syntax-semantics interface. In: Browne, W., Cooper, A., Fisher, A., Kesici, E., Predolac, N., Zec, D. (eds.) Formal Approaches to Slavic Linguistics 18: The Cornell Meeting, pp. 226–240. Michigan Slavic Publications, Ann Arbor (2011a)

Kagan, O., Pereltsvaig, A.: Syntax and semantics of bare nps: Objects of intensive reflexive verbs in russian. In O. Bonami and P. C. Hofherr, editors, Empirical Issues in Syntax and Semantics 8, 221–238 (2011b)

Kallmeyer, L., Osswald, R.: A frame-based semantics of the dative alternation in lexicalized tree adjoining grammars. Submitted to Empirical Issues in Syntax and Semantics, vol. 9 (2012)

Kallmeyer, L., Osswald, R.: Syntax-driven semantic frame composition in lexicalized tree adjoining grammars. J. Lang. Model. 1(2), 267–330 (2013)

Kallmeyer, L., Osswald, R., Pogodalla, S.: Progression and iteration in event semantics - an LTAG analysis using hybrid logic and frame semantics. In: Colloque de Syntaxe et Sémantique à Paris (CSSP 2015) (2015)

Kennedy, C., Levin, B.: Telicity corresponds to degree of change. Unpublished MS, Northwestern University and Stanford University (2002)

Löbner, S.: Evidence for frames from human language. In: Gamerschlag, T., Gerland, D., Osswald, R., Petersen, W. (eds.) Frames and Concept Types, Studies in Linguistics and Philosophy, pp. 23–67. Springer, Cham (2014). https://doi.org/10.1007/978-3-319-01541-5_2

Petersen, W.: Representation of concepts as frames. Balt. Int. Yearb. Cogn. Log. Commun. 2, 151–170 (2007)

Petersen, W., Osswald, T.: A formal interpretation of frame composition. In: Proceedings of the Second Conference on Concept Types and Frames, Düsseldorf (2009, to appear)

Ramchand, G.: Time and the event: the semantics of Russian prefixes. Nordlyd 32(2) (2004)

Romanova, E.: Constructing Perfectivity in Russian. Ph.D. thesis, University of Tromsø (2006)

Scontras, G., Tessler, M.H.: Probabilistic language understanding: an introduction to the Rational Speech Act framework (2017). https://michael-franke.github.io/probLang/. Accessed 12 Oct 2018

Součková, K.: Measure prefixes in Czech: Cumulative na- and delimitative po-. Master's thesis, University of Tromsø (2004)

Svenonius, P.: Slavic prefixes and morphology. An introduction to the Nordlyd volume. Nordlyd 32(2), 177–204 (2004a)

Svenonius, P.: Slavic prefixes inside and outside VP. Nordlyd 32(2), 205–253 (2004b)

Tatevosov, S.: Intermediate prefixes in Russian. In: Proceedings of the Annual Workshop on Formal Approaches to Slavic Linguistics, vol. 16 (2007)

Ušakov, D.N. (ed.): Tolkovyj slovar' russkogo jazyka. [Explanatory Dictionary of the Russian Language]. Izdatel'stvo Akademii Nauk SSSR, Moscow (1935–1940)

van Deemter, K.: Utility and language generation: the case of vagueness. J. Philos. Log. **38**(6), 607 (2009)

Švedova, N.J.: Russkaja Grammatika, vol. 1. Nauka, Moscow (1982)

Zinova, Y.: Russian verbal prefixation. Ph.D. thesis, Heinrich-Heine University, Düsseldorf (2017)

Correction to: Finite Identification with Positive and with Complete Data

Dick de Jongh and Ana Lucia Vargas-Sandoval

Correction to:
Chapter "Finite Identification with Positive
and with Complete Data" in: A. Silva et al. (Eds.):
Language, Logic, and Computation, **LNCS 11456,**
https://doi.org/10.1007/978-3-662-59565-7_3

By mistake the originally published version of this chapter did not include the acknowledgement text. This has been corrected so that the updated version of the chapter now contains the following acknowledgement: We want to thank Sebastiaan A. Terwijn for his assistance in the methods of the proof of Theorem 11. We thank the two anonymous referees that helped us to clarify a number of issues and improve the paper.

The updated version of this chapter can be found at
https://doi.org/10.1007/978-3-662-59565-7_3

Author Index

Printed in the United States
By Bookmasters